Constructing Science

Constructing Science

Connecting Causal Reasoning to Scientific Thinking in Young Children

Deena Skolnick Weisberg and David M. Sobel

The MIT Press
Cambridge, Massachusetts
London, England

The MIT Press would like to thank the anonymous peer reviewers who provided comments on drafts of this book. The generous work of academic experts is essential for establishing the authority and quality of our publications. We acknowledge with gratitude the contributions of these otherwise uncredited readers.

This book was set in Stone Serif and Stone Sans by Westchester Publishing Services. Printed and bound in the United States of America.

Library of Congress Cataloging-in-Publication Data

Names: Weisberg, Deena Skolnick, author. | Sobel, David M., author.
Title: Constructing science : connecting causal reasoning to scientific thinking in young children / Deena Skolnick Weisberg and David M. Sobel.
Description: Cambridge, Massachusetts : The MIT Press, [2022] | Includes bibliographical references and index.
Identifiers: LCCN 2021045987 | ISBN 9780262044684 (paperback)
Subjects: LCSH: Science—Methodology. | Reasoning in children. | Scientific ability. | Science—Study and teaching—Psychological aspects. | Constructivism (Education)
Classification: LCC Q175.32.R45 W45 2022 | DDC 501—dc23/eng/20211214
LC record available at https://lccn.loc.gov/2021045987

10 9 8 7 6 5 4 3 2 1

For Michael—D.S.W.
For Lisa, Paulina, and Nate—D.M.S.

Contents

Acknowledgments

Dave and Deena met at the 2007 meeting of the Society for Research in Child Development. We immediately skipped a bunch of talks in favor of arguing for three hours about Superman. This led to a collaboration on a set of projects on children's understanding of fiction and fantasy worlds. When the fantasy papers became reality, we started a new collaboration on children's causal reasoning and scientific thinking. This led to a funded collaborative grant, which led to collecting a lot of data, which in turn led to writing this book.

We would first like to acknowledge that much of the empirical work described in this book was supported by that collaborative grant from the National Science Foundation (*Young children's beliefs about causal systems: Learning about belief revision in the lab and in museums*, grants 1661068 to Dave and 1929935 to Deena). Any opinions, findings, and conclusions or recommendations expressed in this material are those of Dave and Deena and do not necessarily reflect the views of the National Science Foundation. Some of the work described in this book was the result of an earlier grant that Dave received from the National Science Foundation (*Explaining, exploring, and scientific reasoning in museum settings*, grant 1420548). Dave was also supported by NSF grants 1917639 and 2033368 during the process of writing this book, and Deena received additional support for this book's completion from Villanova University's Subvention of Publication Program.

Both Deena and Dave have collaborators, students, and staff who were instrumental in making a lot of this research happen. At the top of the list are a set of postdoctoral researchers (many of whom are now in faculty positions), graduate students (many of whom are now postdoctoral researchers, faculty, or successful members of the private sector), and staff who collaborated on many of the projects we describe in detail throughout the book. They include Deon Benton, Emily Blumenthal, David Buchanan, Nadia Chernyak, Natasha Chlebuch, Elysia Choi, Chris Erb, Philip Fernbach, Zoe Finiasz, Amanda Haber, Emily Hopkins, Garrett Jaeger, John Knutsen, Susan Letourneau, Elena Luchkina, Deanna Macris, April Moeller, Sarah Munro, Jane Reznik, Laura Stricker, Tiffany Tassin, Jordan Taylor, Yuyi Taylor, Kristen Tummeltshammer, Lea Ventura, Aiyana Willard, and Dahe Jessica Yang. Marta Biarnes and Jayd Blankenship also contributed to our joint research environment during the last few years and provided us with helpful discussion and insight on these and related projects.

We are also grateful to have collaborated with a fantastic set of advisers, who mentored us during our own graduate and postdoctoral careers: Paul Bloom, the late Lisa Capps, Alison Gopnik, Kathy Hirsh-Pasek, Frank Keil, and Angeline Lillard. Many of the projects we describe here started under their tutelage and with their collaborative effort. We also want to acknowledge our collaborators on many of the projects we described: Sophie Bridgers, Daphna Buchsbaum, Emily Bushnell, Maureen Callanan, Kathleen Corriveau, Douglas Frye, Roberta Golinkoff, Tom Griffiths, Hande Ilgaz, Jennifer Jipson, Melissa Kibbe, Natasha Kirkham, Tamar Kushnir, Asheley Landrum, Cristine Legare, Jin Li, David Rakison, Laura Schulz, Beate Sodian, Jessica Sommerville, and Josh Tenenbaum. Working with each of these individuals has been a privilege and a pleasure. We are grateful for the time you have spent with us and for what we have learned from you. In addition to the people we've directly collaborated with, we have also had productive conversations about various topics in the book with Melissa Koenig, Alan Leslie, Tania Lombrozo, Nora Newcombe, Rebecca Richert, and Mark Sabbagh, and we thank them for their insights.

We are both also lucky to work with many undergraduate and graduate research assistants who helped to recruit participants, collect data, and code responses for this work. Among these dedicated students are Jaclin Boorse, Michal Clayton, Calais Cronin, Camra Davis, Kelsey Decker, Brianna Doherty, Matthew Dong, Julia Donovan, Sirmina Dremsizova, Iman Fathali, Estee Feldman, Sarah Fracasso-Francis, Annie Freeman, Janis Ripoll Garriga, Meghan Gonzales, Josh Goodstein, Katie Green, Nicie Jenell Grier-Spratley, Isobel Heck, Naomi Heilweil, Yubin Huh, Antonija Kolobaric, Colton Lacy, Maya Lennon, Candace Mariso, Emily Marucci, Carol Medina, Miranda Meketon, Keren Mir-Almaguer, Anne Moody, Sydney Morris, Christine Odenath, Joyce

Pan, Marcella Parisi, Rebecca Patey, Elizabeth Rawson, Katie Ridge, Tess Rossi, Elena Schiavone, Aileen Scullin-Turcich, May Stern, Nate Stauffer, Emily Stoddard, DJ Williams, Emily Yang, Helen Yu, and Martha Yumiseva. Many of these students have gone to (or completed) graduate programs, or are applying now to graduate schools, and we wish them all the best.

Our colleagues and collaborators at Providence Children's Museum (Robin Meisner, Janella Watson, and Kristen Read) and at the Academy of Natural Sciences of Philadelphia (Jacquie Genovesi) were also instrumental in making a lot of this work happen, and we thank them for welcoming us and our projects to their museums. We would also like to acknowledge the late John Andrews-Labenski of the University of Pennsylvania Psychology Department Shop and Ben Dugan at Current Designs in West Philadelphia, who built the fancy new blicket detectors that were used to conduct many of the studies described in this book. At the Springfield School District, Anthony Barber, Cynthia Mattei, and Lori Schmidt collaborated with us to implement the studies described in chapter 7; we are grateful for their support.

We also want to thank the members of the advisory board for our collaborative NSF grant: Lisa Brahms, Maureen Callanan, David Danks, Joshua Gutwill, Hyowon Gweon, and Jennifer Jipson. These advisers took time out of their busy schedules to travel to both Providence and Philadelphia to give us excellent advice on all of our projects.

We also could not have done this work without the help of school and museum staff, teachers, parents, and especially the children who participated in our experiments. Thank you all for generously volunteering your time to help us learn more about child development.

Special thanks to Phil Laughlin, our editor at MIT Press, who was incredibly patient with all our (many) questions and our (many) requests for extensions. Thanks, Phil, for shepherding this book through the process and helping it come to fruition.

Finally, to our spouses, Michael Weisberg and Lisa Faille, and to our children, Brandon Weisberg (9 years old), Melora Weisberg (5 years old), Paulina Sobel (12 years old), and Nate Sobel (10 years old): Thank you for your patience through our long hours, your insightful, silly, and thought-provoking questions, and your unwavering support and love. You help us do the work that we do and inspire us to do it better. We love you.

I   The Foundations of Scientific Thinking

1   How Do We Develop the Capacity to Think Scientifically?

Science is more than a body of knowledge. It is a way of thinking; a way of skeptically interrogating the universe with a fine understanding of human fallibility.
—Carl Sagan in an interview with Charlie Rose, 1996

Science is a remarkable human achievement. It has allowed us to send people into outer space, to explore the depths of the ocean, and to probe the complexities of our own thoughts. We have been able to accomplish these feats not only because we have learned vast quantities of scientific content information, but also because we have developed the capacity to think scientifically. As former *Science* editor-in-chief Bruce Alberts wrote, "We must teach our science students to *do something* in science class, not to memorize facts" (Next Generation Science Standards, 2013). This quote places a strong emphasis on teaching children the capacity for scientific thinking, perhaps even over and above their knowledge of science content. But what is scientific thinking? And how do young children become capable of engaging in such thinking? These are the two questions we consider in this book.

Demonstrating that children are capable of scientific thinking does not mean we are going to start asking them to do research on viruses, black holes, or neurons. But knowing how scientific thinking develops and how it is connected to other early-emerging reasoning capacities can help us improve science education, public engagement with science, and perhaps even the practice of science itself.

To illustrate, consider the many public debates (especially in the United States) about scientific issues: the teaching of evolution (e.g., Miller et al., 2006; National Academy of Sciences, 2008), the administration of vaccines (e.g., Offit, 2011), the existence of climate change (e.g., Leiserowitz et al., 2010; Zehr, 2000), and whether masks are effective protection against COVID-19 (e.g., Stutt et al., 2020). Although all these topics concern particular scientific content, our everyday discussions of them would benefit from greater knowledge about how science is practiced and better scientific thinking skills (see Weisberg et al., 2021). Yet adults possess robust misunderstandings about a variety of scientific concepts (e.g., McCloskey, 1983; Shtulman, 2017) and often fail to apply certain principles of scientific thinking in their problem-solving (e.g., Kuhn, 2007a; Schauble, 1996). Discovering whether and how children can reason scientifically can give us insight into how we can become a more scientifically literate society.

Our primary goal, then, is to explore the nature of children's scientific thinking. We examine the development of various types of scientific thinking skills, particularly focusing on their origins in children's causal reasoning abilities and how these early abilities combine with other developmental achievements into a systematic framework for scientific thought. Before we begin, we want to briefly explain what we mean by scientific thinking and why causal reasoning is a good place to begin our investigation of this ability.

What Is Scientific Thinking?

We conceptualize scientific thinking as the suite of skills that allows people to generate hypotheses, solve problems, and explain aspects of the world. These skills match what scientists do in their work, as they identify problems or ask questions about the world. Doing science involves constructing methods to investigate questions and to generate evidence. It involves coming up with answers, often by making informed guesses about possible causes, reasons or mechanisms, as well as drawing conclusions by logical and inferential processes for evaluating that evidence. Doing science also involves communicating results and analyzing related investigations done by others (Fischer et al., 2014; Koslowski, 1996).

We want to note at the outset that our focus on the *process* of doing science—on the domain-general capacities that individuals use to engage in the tasks outlined above—differs from other definitions of science (e.g., Klahr et al., 2011). Other definitions tend to emphasize the process of scientific thinking, as we do, but also the idea that science is about certain topics.

We disagree with this latter emphasis, mostly because we believe that it is too difficult to draw a clear line between which content is scientific and which content is not. Historically, the term "science" has been used to refer to problems in the physical and natural worlds, but similar kinds of reasoning processes are used in many other domains. For example, the social sciences emphasize the systematic study of topics beyond the ones that are usually considered to be standard scientific topics. Political science, linguistics, and history focus on causes and reasons, just as much as do fields like chemistry or biology. There may be folk conceptions that certain topics are more scientific than others (Keil et al., 2010), but even nonscientists see that these topics exist on a continuum and are not defined by a clear categorical boundary.[1]

So here is our definition: Scientific thinking involves understanding how to frame a problem, how to judge the objectivity of a test, how to examine the integrity of data, how to integrate past knowledge with present observations, and how to evaluate the validity of the process used to reach a conclusion. Importantly, these skills are necessary tools for reasoning and processing information in everyday settings, not just lab-based ones. And they apply in many domains, not just ones that are designated as sciences by universities.

But do young children possess any of these abilities? The answer to this question depends on what you count as evidence.

Scientific Thinking in Childhood I: Content Knowledge

Although we believe that knowledge about particular scientific content should not be part of the definition of scientific thinking, studying what children know about scientific content (sometimes called *domain-specific knowledge*) has become part of the research program that examines the skills that we do consider to be part of scientific thinking. For example, Dündar-Coecke and Tolmie (2020) suggest that "one aspect of scientific thinking is acquiring knowledge about causal mechanisms" (p. 130). Similarly, in their excellent review on scientific thinking, Dunbar and Fugelsang (2005) initially define scientific thinking as "the mental processes used when reasoning about the content of science (e.g., force in physics)" (p. 705) as well as more domain-general reasoning about that content.

Documenting the domain-specific knowledge that children possess does two things. The first is obvious—it tells researchers, educators, and parents what children usually know at particular ages. For example, preschool-age children have a good base of scientific content knowledge about biology (e.g., Carey, 1985; Inagaki & Hatano, 2006; Slaughter & Lyons, 2003; Wellman & Gelman, 1998) and psychology (e.g., Astington et al., 1988; Wellman, 2014). Even infants seem to know many things about the domain of physics, like the fact that physical objects are solid and cannot pass through one another, and the fact that they generally retain their forms (e.g., Baillargeon, 1994; Spelke et al., 1992).

Incidentally, there is a long-standing debate in cognitive development over whether infants possess knowledge about the nature of objects' properties and relations at birth or learn it from early exposure to the environment (consider, for example, the contrast between Spelke et al., 1992, and Haith et al., 1993). One can argue that infants are born with the capacity to divide objects into their component parts based on certain static perceptual features, and what limits the emergence of understanding object segmentation is certain neurological developments.[2] For example, Needham and Baillargeon (1997) showed that, when two different objects are touching each other, 8-month-olds look longer at a display in which the objects move as if they were stuck together than at one in which the first object moves independently from the second. Younger infants generally do not do this, unless they are given particular kinds of environmental exposure. One example of such environmental exposure is the ability to manipulate their environment through "sticky mittens," which are infant-sized mittens made with Velcro that allow infants to experience manipulating objects that have Velcro strips on them (Needham et al., 2002; see van den Berg & Gredebäck, 2021, for an overview of this paradigm). These studies suggest that infants may understand object segmentation all along, but are limited by their perceptual and motor capacities (see Carey, 2009, for a broad and compelling version of this argument).

Others, in contrast, argue that object segmentation is a learned skill, based on statistical regularities in the environment (e.g., Wu et al., 2011). For example, we see a lot of heads connected to glasses, and a lot of heads connected to necks. It is not that surprising when we see someone take their glasses off their head, but it would be really surprising if someone took their head off their neck (without any negative consequences). Even though heads and glasses and heads and necks are both correlated, one correlation is stronger than the other, and this can be learned from observing the environment.

This debate is not the topic of this book (although we do take a position on it later in this chapter, which we expand on in chapter 3). For our purpose, the point of bringing up the development of science content knowledge is to illustrate that a great deal of this knowledge is in place early in development. Frankly, we could write an entire book about the scientific content knowledge that children possess. But what is important about this content knowledge is that it is independent of what children know about it. Regardless of how infants are able to segment objects in the world, they do not know what the word "segment" means, nor do they possess the metacognitive understanding that they are using object segmentation capacities when they make responses in laboratory settings or act on objects in everyday settings. It is that kind of higher-order understanding that we take to be crucial to scientific thinking.

Documenting the content of children's science knowledge allows us to learn about the first-order development of this knowledge. But it also does another important thing: It sets boundary conditions on what and how children can learn at any particular point in time—what Metz (1995) calls the "complex interaction between domain-specific knowledge and scientific inquiry" (p. 115). For example, a set of intervention studies has used storybooks to introduce young children to the concept of adaptation by natural selection (Emmons et al., 2017; Kelemen et al., 2014; Ronfard, Brown, et al., 2021). Although this topic is often not taught until late in elementary school or even until high school, prior work on young children's understanding of the domain of biology indicates that 5-year-olds have the requisite background knowledge to understand natural selection (see Kelemen et al., 2014). Indeed, children in these studies can successfully learn principles of natural selection after hearing storybooks about fictional animal populations that adapt to changing environments. Perhaps unsurprisingly, the older children in these studies (7- and 8-year-olds) learn and transfer their new knowledge to novel animal populations somewhat better than the younger children, arguably because they enter the task in possession of more well-developed frameworks for understanding animal traits and are able to build on that knowledge.

To take another example, during the preschool years, children develop the capacity to appreciate that others can have false beliefs about the world (Perner et al., 1987; Wimmer & Perner, 1983).[3] Understanding that others can act based on false beliefs potentially allows children to understand when they should discount verbal information in favor of the actual state of the world. There are many examples of this, but some of our own work illustrates this process. In one study (Sobel, 2015), we presented 3- and 4-year-olds with stories about characters who claimed to have or have not learned how to do something (for example, solve a puzzle). Those characters then proceeded to be able or not to be able to do what they claimed to have learned (that is, they could or could not solve the puzzle). When the character's claims and demonstrative actions agreed, children in the study had no trouble answering a question about whether the character had really learned how to solve the puzzle. When the character's claims and demonstrative actions were in conflict, however, children were more confused about how to answer that question—unless they understood that others can have false beliefs about the world. In that case, even the youngest children in the sample correctly said that the character who demonstrated the action (solved the puzzle) had learned and that the character who could not demonstrate the action had not learned, regardless of what the character claimed. If children understood that others' claims could be false, they discounted the claim in favor of the demonstrable action; if they did not, they tended to weigh the claim and demonstration equally. This is just one of a multitude of examples of how children's understanding of beliefs affords them insight into other aspects of human behavior (see e.g., Wellman, 2014, for a review).

There is also now a large literature on the extent to which children can learn from testimony—information that other people communicate directly (see, e.g., Harris et al., 2018; Mills, 2013; Sobel & Kushnir, 2013, for reviews). One big question in that field is the extent to which children should forgive informants when they make errors. Typically, if you make an error, you are considered a less reliable source of future knowledge. For instance, if I ask you to label a common object and you fail to generate a correct label (e.g., you call a pen a "shirt"), I'm going to be less likely to trust you when you call a novel object a "blicket." And if a more reliable source of knowledge calls that object a "wug," I might trust that latter source's label for the object (Koenig & Harris, 2005).

But what is interesting about this situation is that it requires an appreciation of others' mental states; when the speaker labels an object, the speaker believes that the object has that label and is trying to communicate that reference. Indeed, some research has shown that children's performance on a battery of theory of mind measures, including measures of understanding others' false belief (Wellman & Liu, 2004), predicts the extent to which children generalize the labels of previously accurate informants over previously inaccurate ones (Brosseau-Liard et al., 2015). There is definitely debate on this issue, but there does seem to be some relation between mental state understanding and social learning.

Another classic debate in cognitive development is illustrated by this discussion—the extent to which development involves the acquisition of individual pieces of knowledge (or, more technically, *domains* or bodies of knowledge) or involves the growth in a set of more general cognitive capacities (such as attention, memory, and cognitive control). On one extreme, all of children's knowledge in one domain may be independent from knowledge in every other domain (*encapsulated,* as some philosophers might say; see Fodor, 1983). On the other side of the continuum, knowledge may be irrelevant; what matters may be only the domain-general representational structures and operations that instantiate that knowledge.

Here, we are going to take a position up front, which we illustrate with a brief example: If the light in my office does not go on when I flip the switch, I can reasonably intuit that the bulb is burned out, and I might run to the supply room to get another. I only do that, however, if I can determine that the building has electricity. If everyone's lights are off and my office computer is dead, then getting a new bulb is probably not necessary. That is, to solve this problem, I'm using specific mechanistic knowledge about light bulbs and electricity. But I'm also using more general causal reasoning capacities, particularly about mechanisms and enabling conditions (i.e., knowledge of how certain things work). As this case illustrates, we think that there is a broad, domain-general reasoning system that allows for making causal inferences and engaging in scientific thinking, rather than sets of reasoning capacities that are each specialized for individual domains. That domain-general reasoning system can take domain-specific knowledge as input, and it can be constrained by the nature of this domain-specific knowledge, but the reasoning itself is based on general processes that do not care (to anthropomorphize for a moment) what they are reasoning about. Some researchers suggest that a majority of scientific thinking is domain-specific (e.g., what Sinatra & Chinn, 2012, refer to as the development of "epistemic cognition"), but we believe that it is more reasonable to conclude that scientific thinking involves an interaction of domain-specific knowledge with domain-general processes.

This means that we can distinguish between two types of development when we talk about scientific thinking. One type is just about the development of content knowledge. For example, as described above, developing an understanding of others' false beliefs allows children insight into cases when others' claims can be false, and thus discounted. One of the central hypotheses of our approach is that, in order to make causal inferences, children must recruit their existing knowledge, even if it is impoverished.

But, as noted above, domain-specific content is not the same as the domain-general processes that underlie causal inference. Those processes develop separately. Further, the mechanisms that children have for reasoning about causality undergo change. So here is our take on the major debate in cognitive development about whether children have knowledge at the start. The answer is yes: Infants seem to start with a small set of simple concepts, such as temporal priority (i.e., causes precede their effects in time), and possibly some impoverished representations of content knowledge as well, particularly in domains like physics. The content knowledge itself changes over the course of development. But what also changes is how that knowledge is represented and how new knowledge in these domains is inferred from the environment (see Sandoval et al., 2014). We describe these processes in more detail in chapters 2 and 3.

Importantly, we wish to stress the second point—that the way that knowledge is inferred from data in the environment changes over the course of development. This claim is fundamental to the bridge we want to build in this book between children's causal reasoning abilities and their scientific thinking abilities. The difference between causal reasoning and scientific thinking does not involve content knowledge. The difference involves the development of thinking processes. Scientific thinking involves some inferential abilities that are part of causal reasoning and that can be seen early in development. But these abilities continue to develop long past infancy and undergo important changes along the way.

Scientific Thinking in Childhood II: Doing Science

As we argued above, it is important to distinguish scientific content knowledge from the *process* of doing science. Thinking scientifically includes skills and habits of mind that cut across domains (Zimmerman, 2007), coupled with the idea that scientific thinking is fundamentally a form of reasoning and problem-solving (Dunbar & Fugelsang, 2005; Klahr & Dunbar, 1988). Scientific thinking requires numerous cognitive abilities, including forming hypotheses, designing interventions to test those hypotheses, observing data based on those interventions, evaluating those data and the possibility of error, and integrating data and theory (e.g., Schauble et al., 1995).

Under this definition, the field of cognitive development has historically viewed whether children have the cognitive abilities necessary for scientific thinking with a great deal of pessimism. One of the most influential developmental theorists of the twentieth century, Jean Piaget, argued that young children were "precausal"—incapable of making even simple causal inferences, that is, of understanding the relation between an event and its effects. Given this, he argued that young children were also incapable of constructing sophisticated explanations of causal structures and of making counterfactual inferences (Piaget, 1929, 1930; Piaget & Inhelder, 1951).

Some literature on children's scientific thinking has tended to support these pessimistic conclusions, finding that preschoolers and even elementary-school-age children struggle greatly with designing unconfounded experiments and engaging in other basic scientific thinking tasks. According to this work, one major locus of this difficulty is young children's inability to understand the difference between data and theory. This confusion renders them incapable of understanding that these two elements are in dialogue and that data can be used to confirm or disconfirm a theory (e.g., Klahr, 2000; Kuhn, 1989; Zimmerman, 2000).

Within the field of cognitive development, however, evidence has accumulated over the past twenty years to demonstrate (contra Piaget) that children can engage in causal reasoning and problem-solving. Beyond registering simple principles, like spatial and temporal contiguity (e.g., Oakes & Cohen, 1990), children and even infants can consider the importance of mechanisms (how causal relations work), interventions (doing things to change the environment), agency (the ability to effect change volitionally), and differences between covariation that indicates causality and spurious associations (e.g., Buchanan & Sobel, 2011; Gopnik et al., 2001; Madole & Cohen, 1995; Saxe et al., 2005; Saxe et al., 2007). Importantly, the researchers cited here have different underlying views about how this development occurs, running the gamut from strong nativist to strong empiricist conceptions of children's interaction with the world. This illustrates the importance of causal reasoning in cognitive development, and the extent to which researchers across the theoretical spectrum agree that Piaget fundamentally underestimated children's understanding of causality.

In addition to this foundation of causal reasoning, scientific thinking also involves curiosity (e.g., Jirout & Klahr, 2012; Morris et al., 2012). Curiosity is also reflected across theories of cognitive development. Piaget, for example, advocated for the idea of the child as an active learner, which could be used to translate classrooms into dynamic environments (e.g., Elkind, 1976). This idea is shared among various theoretical positions in cognitive development, including nativism (Carey, 2009; Spelke et al., 1992) and the constructivist ideas we will describe and endorse later in the book: Children are not passively absorbing information, but instead are actively interpreting the data they observe and seeking out novel experiences from which to learn. This kind of active learning also involves selecting one's own interventions, which has clear benefits for learning in adults (Gureckis & Markant, 2012; Lagnado & Sloman, 2004; Sobel & Kushnir, 2006). All of these views suggest that children act on their curiosity to gain information about the world. A fundamental theme in both causal reasoning and scientific thinking, and across a variety of theories of cognitive development, is that young children are active learners.

Taken together with the research we described in the previous section, which finds that children have a good deal of science content knowledge, many researchers have used children's abilities in causal reasoning and active learning to draw an analogy between children and scientists. Children have representations of theories (just like scientists), which change with evidence (just like scientists); their everyday understanding of the world develops via processes similar to that of scientific thinking. This approach was originally labeled *theory theory* or the *child-as-scientist* view (Carey, 1985; Gopnik & Wellman, 1994; Gopnik & Meltzoff, 1997; Gopnik et al., 1999), though more recently it has been described as *rational constructivism* (Gopnik & Wellman, 2012; Xu & Kushnir, 2012), reflecting the active work that children do to construct an understanding of the world around them.

In this view, children are nascent scientists because there are many similarities between how they learn about the world and scientific thinking. Children represent causal knowledge in an abstract and coherent manner. Those representations support predictions about what will happen given the present state of the world, explanations of what has happened, and inferences about different types of counterfactual possibilities—what might have happened had alternative events been in place, or what would have been necessary for a particular outcome to have occurred, given that a different one did. Most importantly, as we concluded in the last section, those representations change over the course of development (Gopnik & Meltzoff, 1997).

Although we are generally in agreement with this view, which we describe in more detail in chapter 2, the work on rational constructivism does not show that children have the capacity for scientific thinking as it has been traditionally construed. As Morris et al. (2012) wrote, "Young children are sometime deemed 'little scientists' because they appear to have abilities that are used in formal scientific reasoning. . . . At the same time, many studies show that older children (and sometimes adults) have difficulties with scientific reasoning. For example, children have difficulty in systematically designing controlled experiments, in drawing appropriate conclusions based on evidence, and in interpreting evidence" (p. 61). That is, although children have some early abilities to reason causally and to construct explanations from sets of data, they are not yet capable of engaging in other cognitive activities that are necessary for mature science, such as systematically testing hypotheses or reasoning about uncertainty (see also Shtulman & Walker, 2020).

Even more importantly, although young children demonstrate capacities for succeeding at some aspects of scientific thinking, they lack metacognitive awareness of these capacities. That is, children might explore the world in order to learn, and their exploration might even be rationally related to causal learning (Cook et al., 2011; Schulz & Bonawitz, 2007), but it is not clear that young children know they are engaging in rational inference when they do it. For example, there are situations in which toddlers will systematically ask for help as opposed to trying to work things out themselves (Gweon & Schulz, 2011), cases in which preschoolers will learn better from their own actions as opposed to from similar data generated by another person (Sobel & Sommerville, 2010), and cases in which children will use behavioral cues to reason about how they and others gather evidence and gain new information (Leckey et al., 2020). While these studies may make it seem like young children are behaving like scientists, none of these cases necessarily show that children themselves are aware that they are learning better or differently through their own actions or choices. More explicit and strategic information-seeking behaviors emerge later in development (e.g., Ruggeri et al., 2016), and it is these more explicit strategies that are the hallmark of more mature scientific thinking.

Put another way, scientific thinking is intentional: While scientists sometimes stumble on answers randomly (and then go about confirming these hypotheses in a systematic way), scientific thinking usually occurs in the other direction. Scientific thinking also involves reflection: Making explicit changes in one's beliefs should be based both on evidence and on the awareness that one is doing so. As noted above, it is not clear that young children understand the distinction between theory and evidence (Kuhn, 2002), making them incapable of reflecting on the process of testing causal claims. Elementary-school-age children do not always metacognitively reflect on their actions (as any parent will tell you). But more critically, they do not necessarily test hypotheses in a systematic way (e.g., Kuhn et al., 1988). They do, however, like to make things happen (see e.g., McCormack et al., 2016; Schauble et al., 1991). This tendency has often been connected to their failure to design experiments that change only one variable at a time (Tschirgi, 1980). Rather than systematically testing hypotheses, children (and sometimes adults) favor a strategy that tests multiple ideas at once or that leads to noticeable changes. While children can be instructed to use a more systematic strategy for controlling variables in elementary school (e.g., Chen & Klahr, 1999), and even preschoolers may be able to use this strategy given significant support (e.g., van der Graaf et al., 2015), it might not be a natural way that children interact with the world. Similarly, while children can determine whether an experiment is going to yield confounded or unconfounded data by age 6 (Sodian et al., 1991), it is not clear that children are aware that they are using such a strategy (Kuhn & Pearsall, 2000). In general, then, children might implicitly explore the world in a rational manner and hence appear to be acting like little scientists. But scientific thinking additionally involves explicitly knowing the "whys" behind your behavior—and this skill might have a prolonged developmental trajectory.[4]

We revisit this argument in more detail throughout the book, but for now, we emphasize that a variety of cognitive capacities are under consideration when we talk about scientific thinking. These include (but are not limited to) children's ability to evaluate evidence to make a conclusion, to revise articulated beliefs in light of counterevidence, to recognize when evidence is confounded, to generate evidence that might allow them to draw appropriate conclusions, and to be aware that they are doing all of these things with the explicit purpose of learning or making better inferences. Contrary to arguments from rational constructivism, we will argue that some of these capacities do not develop early in life, but rather continue to develop substantially past the preschool years.

We will also argue that there are ways in which young children might be able to engage in aspects of scientific thinking, such as using the control of variables strategy, which would allow them to design systematic interventions to learn about causal structure. That is, we do not want to underestimate the causal reasoning capacities of young children that serve as the backbone for scientific thinking. Causal reasoning is the ability to understand whether and how events are related to one another, as well as the nature of that relation; we describe our view about the nature of this ability in more detail in chapter 2. Right now, what is important is that children's abilities to relate events together causally are in place early in development, although these abilities do undergo change throughout the preschool and early elementary school years, as children develop more sophisticated reasoning abilities. Previous work may have underestimated these abilities because the methods used to test children's causal reasoning abilities differ in complexity and context, and in both conceptual and metacognitive requirements, from measures used to test children's scientific thinking abilities. These latter tasks require developmental milestones beyond just causal reasoning abilities, but there are ways to bridge the gap. We argue this point more fully in chapter 4.

Scientific Thinking in Childhood III: Defining "Science"

In the previous two sections, we have argued that there are two different aspects to science: content knowledge and thinking skills. We also noted that a metacognitive awareness that one is using these skills is a fundamental part of scientific thought, and one that might be out of children's reach potentially until the elementary school years. Along these lines, we wish to propose that there is a particular aspect of children's understanding that is important to scientific thinking: the ability to understand what science is, particularly to define it as a set of thinking skills, as we have. That is, just as there should be a connection between children's abilities to think scientifically and their understanding of how and why they are doing so, there might also be a connection between children's abilities to think scientifically and their explicit understanding of what science is.

This may seem like an odd claim at first. Why should the ability to define "science" in a particular way matter at all to one's ability to solve a scientific thinking task? After all, this is not the case for many (perhaps most) activities that human beings engage in: One does not need to know that one is conjugating a verb or using the present perfect tense in order to do so (as an example). Babies learn language without any awareness of what they are doing, or possibly of what language even is. Given this, it may not seem relevant that children explicitly understand science as a process of knowledge creation in order to control variables in an experiment.

But scientific thinking might not be like other mental capacities, and understanding what science is and that one is doing science is important, perhaps even necessary, to reasoning scientifically. Why might this be so? One possibility is that the activity of science itself may be facilitated by one's knowledge that one must approach a problem as a scientific one—through the lens of a particular kind of reasoning process. Specifically, if children think of science as a process of learning about the world, they may approach a problem described as "science" ready to carefully and objectively examine their beliefs and to look for patterns in sets of data. They may expect a science problem to be of a certain type and so may approach it with a particular mindset that can help them to succeed. Conversely, without this explicit knowledge of what science is, a child who is told that something is scientific might not bring the same mental resources to engage in the task in the same way. There may be some kind of connection between children's definitions and their scientific thinking abilities because having a particular definition of science may help activate the kinds of thought processes that enable scientific thinking. This means that, if children lack an explicit understanding of how the practice of science creates knowledge, or even an understanding of the fact *that* the practice of science creates knowledge, children should not be considered fully capable scientific thinkers (Kuhn, 2007a; Kuhn & Pearsall, 2000; Sandoval et al., 2014).

Some prior work has examined related effects of children's mindsets about science. Rhodes and colleagues presented young children with tasks that could be easy or challenging by asking them to either "be scientists" or to "do science" (Rhodes et al., 2019). The fundamental difference between these two phrases is the kind of inferences they support. When a child is told to "be a scientist," they might assume that, because an adult is telling them to be a scientist, they are not currently a scientist. Rather, they must become one in order to do the task. The problem with this directive is that children *essentialize* social categories: They think that being a scientist involves having some special nonobvious property that makes a scientist a scientist (for similar findings, see Archer et al., 2010; Foster-Hanson et al., 2020; Lei et al., 2019; Rhodes et al., 2020; see also Gelman, 2003, for a review of essentialism). Because of this, when children encounter challenges in this context, they might simply give up, because they infer that they do not have that nonobvious "scientist" property. *Doing* science, in contrast, is just an activity. There is no special, essentialized property that one must have in order to do the activity of science. On this view, encountering a difficult problem or performing incorrectly should be seen as just another challenge, and should not necessarily affect whether one perseveres at a task. These arguments are consistent with how children (particularly girls) behaved on Rhodes's measures. When children were told to "be scientists" they persevered less on a challenging task than when they were told to "do science." Telling children to "do science" rather than to "be a scientist" potentially helped them feel more confident and competent in their abilities.

These findings strongly suggest that the framing of a science problem is important to how children approach it. However, this work always began with a contrast between "doing science" or "being a scientist." As we report in chapter 8, we took the prior step of asking about how children define science in the first place, and we contrasted children who do and do not possess a definition of science that includes the aspects of scientific thinking that we have highlighted here: learning or engaging in other active processes. Using this open-ended query for children's own definitions of science allows us to determine whether children's beliefs about what science is relates to the causal reasoning or exploratory capacities that we think form the foundation for scientific thinking. More broadly, this line of investigation can provide more insight into the connections between children's explicit understanding of scientific thinking skills and children's implicit abilities to use those skills.

Outline of the Book

Our overarching goal in this book is to examine how scientific thinking develops. We specifically focus on bridging the gap between children's early causal reasoning abilities and later-developing, more mature reasoning abilities, which involve further development of these causal reasoning abilities and include aspects of metacognitive reflection. To do so, in chapter 2, we first describe the rational constructivist view of development and introduce a formal framework (causal graphical models), which provides a good description of children's causal reasoning. In chapter 3, we address some of the challenges associated with using this framework to describe how children engage in causal learning and children's development more generally. These chapters lay out a theoretical position that sets the stage for drawing connections between causal reasoning and scientific thinking. These chapters also provide an overview of what we think young children's causal reasoning capacities are and how they might form part of the foundation for scientific thinking.

We then consider links between the literature on causal reasoning that emerges from rational constructivism and the literature on the development of scientific thinking, focusing particularly on one crucial aspect of scientific thinking: the control of variables strategy. Throughout the book, we aim to take seriously the idea that there is a difference between causal reasoning and children's capacities for scientific thinking. Chapter 4 identifies several possible reasons for why research on causal reasoning and scientific thinking have outlined different developmental trajectories for and have drawn different conclusions about children's abilities.

The crux of our argument is this: Children develop sophisticated causal reasoning abilities early in life, and those causal reasoning capacities are the foundation of their scientific thinking. Although this groundwork is laid early, the emergence of mature scientific thinking abilities requires a series of developments beyond the preschool years, which allow children to more fully express their thinking skills in formal learning contexts and in real-world scenarios. Children's causal reasoning is thus not a monolithic capacity, but one that undergoes a great deal of development. As that capacity develops, so too does children's scientific thinking. For instance, children must integrate evidence into existing theories and recognize explicitly that they sometimes have to change their beliefs given new data. They also have to be able to represent and process uncertainty and possibilities. Moreover, children have to understand how to engage in systematic investigations and draw the right conclusions from those investigations. This set of more explicit, metacognitive capacities has a prolonged developmental trajectory. The development of these cognitive and metacognitive capacities allows children to move from causal reasoning to scientific thinking.

Following this identification of these key variables, section II of the book (chapters 5 to 7) present empirical work that aims to build a bridge between aspects of children's early causal reasoning abilities and the more mature set of scientific thinking abilities that they will eventually need to acquire. Critically, we focus on just one problem that relates causal reasoning to scientific thinking. These chapters are thus meant as a set of case studies, not as a complete explication of the only way that the two literatures can be linked. Specifically, chapter 5 introduces a diagnostic reasoning task that incorporates elements from both research areas as a way to test the emergence of scientific thinking in young childhood through a causal reasoning lens. Chapters 6 and 7 then use this task to investigate two potential factors that could affect children's scientific thinking abilities: the type of scientific content that the problem presents, which we call its *contextualization* (chapter 6) and the presence of *metacognitive thinking* abilities (chapter 7).

Engaging in causal reasoning and systematically using causal reasoning abilities to think about sets of data are necessary components of mature scientific thinking, but, as we argued above, so is an explicit understanding of what science is. In section III, we thus turn to the issue of how children conceptualize science and related concepts like learning. In chapter 8, we present data from a series of studies asking children to define "science" and to think about what kinds of activities are scientific. This chapter also considers the hypothesized connection between understanding what science is and children's abilities to engage in certain aspects of scientific thinking. Chapter 9 extends these investigations of children's explicit definitions to the concepts of learning, play, and teaching. Chapter 10 describes similar investigations on pretending and explores the relations between imagination and scientific thinking.

Our concluding chapter (chapter 11) revisits the arguments made throughout the book about what scientific thinking is and how it develops. We then use these arguments to provide recommendations for formal and informal science education and to outline broad directions for future empirical research.

Finally, as a note to the reader, throughout the book we report research findings from papers that our labs and our collaborators' labs have already published. In such cases, we indicate this and provide references to the published paper for the statistical details. In other places, we report the results of experiments that are not published elsewhere, or we conduct analyses beyond our published work. In those cases, we report actual statistical analyses in endnotes for the interested reader. Additionally, the data sets that we report on for the first time in this book are publicly available on the Open Science Framework (osf.io/dtp78/), and recordings of some testing sessions are available on Databrary (databrary.com). Finally, as we have already done, we use the pronoun "we" to describe that the work comes from either one of our labs or from our collaborative efforts. This should not be taken to negate the role of the many mentors, collaborators, and students we have worked with over the years, whom we gratefully acknowledge as helping us to construct our science.

2 The Evolution of Rational Constructivism

Theory Theory (or the Child-as-Scientist Metaphor)

Developmental psychologists, observing how rapidly children are able to learn about the world around them, have sometimes claimed that children are little scientists (e.g., Gopnik et al., 1999). But what does it mean to say that a child is (or is like) a scientist? As we noted in chapter 1, the relevant dimension of similarity between children and scientists is that both are faced with the task of figuring out how the world works. Scientists go about this task in a systematic fashion, by making observations, designing experiments, collecting data, and drawing conclusions. The metaphorical view of child-as-scientist argues that children do the same thing. In their own way, children also make observations, design experiments, collect data, and draw conclusions.

A classic example of this can be found in children's play. When Dave's son was a toddler, he had a set of stacking cups in the kitchen, which he liked to stack up and knock down, over and over. Such toys—and such stacking activities—are common in many households, at least in our culture. The child-as-scientist metaphor argues that toddlers are not doing this randomly. Rather, their behavior is allowing them to learn about the properties of objects: They fall down when they are unsupported. At early ages, such stacking activities provide children with first-person experiences of building and with information about objects' mass and center of gravity that is necessary for appreciating physical relations involving support. This particular behavior even has a two-for-one bonus. At later ages, once concepts of support might be mastered, such stacking activities often become more of psychological intervention on the part of the child: How many times can I make this loud banging noise before dad stops cooking and plays with me (or takes these toys away)?[1]

The idea of children as little scientists has been formalized into one of the major modern theories of cognitive development, known as the *theory theory*. Theory theory states that babies have a set of initial theories about the world. These theories are not exactly the same as the ones that scientists use, but they have a similar structure. Like full-blown scientific theories, babies' theories are abstract, coherent representations of causal structure (Gopnik & Meltzoff, 1997; Gopnik & Wellman, 1994). For example, babies might have a theory of the physical world that includes a concept of gravity: Objects fall down when they are not fully supported by a surface. Although these theories are simple, they allow babies and children to make predictions about the world and to interpret the information they observe. As children's language capacities develop, they become able to use what other people are saying as another form of evidence for their theories. Eventually, they become able to generate their own explanations.

One of the most important features of children's theories is that, like scientific theories, they can change over time. This is good, because the initial theories that children hold usually are not fully accurate. For example, babies initially behave as though any amount of support under an object will prevent it from falling, no matter how tiny of a sliver of the object's bottom is resting on the surface underneath (Baillargeon et al., 1992). There is something correct about this theory—objects that are entirely unsupported do fall—but this theory also needs to change to account for different types of scenarios. For example, some objects are asymmetrical and will fall unless they are balanced properly on the supporting surface. As children observe or hear about new evidence, they are able to incorporate this new information into their theories and adjust these theories to be a better fit to how the world actually is (e.g., Karmiloff-Smith & Inhelder, 1974).

To take another example, children's initial observations generally lead them to believe—incorrectly, but understandably—that the world is flat. As they grow up, they hear testimony from trusted adults that the world is actually round, and they may observe pictures of the Earth from space or encounter globes. Through a slow process of weighing this new observation against their initial belief, children come to learn that the world is actually round; they change their theory (e.g., Vosniadou, 1994). This kind of theory change can be described as directly analogous to scientific reasoning: Children's theories change through the acquisition of new evidence and through a process of integration of old with new information, similar to how scientific theories change as new evidence is generated or discovered.

This example raises an important question to consider before we more fully discuss the development of scientific thinking: If children engage in this process of theory change, why do some erroneous, pseudoscientific beliefs persist? Consider again the process of learning that the Earth is round. Centuries of scientific data back up this assertion, yet some individuals continue to espouse the belief that the Earth is flat. These Flat Earthers have societies and conventions. Although there are many fascinating aspects to this phenomenon, we want to highlight that Flat Earthers do not seem to hold other beliefs that are incommensurate with reality. They believe that unsupported objects fall and that solid objects cannot pass through one another. Neither do they deny that science exists, nor do they say that scientific thinking is invalid. Indeed, perhaps surprisingly, they often try to use scientific methods to support their claims. To quote a report on one of their conferences, "While flat earthers seem to trust and support scientific methods, what they don't trust is scientists" (Dyer, 2018). That is, what seems to promote Flat Earth beliefs (and, presumably, many of the other beliefs that adults hold that are incommensurate with scientific evidence) is a misunderstanding about the relation between truth and power. Demonstrating that something (particularly something nonobvious) is true has the potential to be misinterpreted as a form of scientific elitism, which seems to be a big part of the Flat Earth movement. Some Flat Earthers potentially dislike the idea that there are truths about the world they did not (and potentially cannot) discover through own direct observations, possibly because they perceive that they lose some power in this process.

We suspect (although to our knowledge this has never been investigated) that individuals who hold Flat Earth beliefs (or who deny climate change or the safety of vaccines, or who hold other erroneous, pseudoscientific beliefs) developed their scientific thinking skills in the same way as individuals who do not hold these beliefs as adults. One of the reasons we had for writing this book was to articulate the idea that scientific thinking is not exclusive to a particular group of individuals (i.e., scientists), and that the power of scientific thinking is available to everyone. All individuals have the potential to engage in the kinds of causal reasoning and scientific thinking that leads to our understanding that the Earth is round, that climate change is real, and that vaccines are safe.

A Problem with the Child-as-Scientist Metaphor

Theory theory has a lot to offer. Most notably, it can explain both the origin and the development of children's knowledge. But this theory, and the child-as-scientist metaphor in general, nevertheless suffers from a serious problem: It's vague.

Theory theory can be stated in two sentences: *Children have theories about how the world works. Their theories change as they gain more information.* But how, exactly, are children's theories represented? And how do these theories change as children observe information in the world?

In fairness, vagueness is a challenge faced by all theories of development. For example, to explain how children advance from one development stage to the next, Jean Piaget posited two mechanisms of development: *assimilation* and *accommodation.* As any developmental textbook will describe, children move from one state of equilibrium to another via these processes. Information that fits with a child's existing concepts is assimilated to that concept, while information that does not fit with the concept can be used to transform (accommodate) the concept to a new one. To take the example used above, as children are told that the Earth is round instead of flat, they come to accommodate their ideas about the shape of the Earth to the standard concept. To use the terms of theory theory, children's astronomical theories undergo change. But both of these descriptions, while sensible, are too vague to really be of much use in describing children's conceptual development. How do children's concepts change on the basis of new information that they receive? Changes in children's responses and behaviors can be observed, but what is actually happening in children's minds as they revise their theories? And how can we access these cognitive processes?

Speaking about the development of children's knowledge as a process of theory change can function as a productive representation of development only if there is some way to cash out the details of how this happens. Otherwise, it is only a metaphor—one that some people might be tempted to take too literally. For example, Dave was once challenged by a well-known researcher who argued that the child-as-scientist metaphor could not be right because children do not actually wear lab coats and carry around clipboards. We're pretty sure that this is an apt observation: Children do not wear lab coats or carry clipboards that often.[2] But the fact that children do not do these things does not invalidate the child-as-scientist metaphor. The broader point of the metaphor is that children are actively testing their beliefs about the world and changing their knowledge based on their observations and their interactions with the world, much like the process of science.

Similarly, there is a strong and important tradition in studies of scientific thinking to observe scientists in their laboratories or other natural environments to see how they approach problem-solving (e.g., Dunbar, 1995, 2000; see also Latour & Woolgar, 1979). We think that this is also a valid way to study how children might learn science, reason scientifically, and, perhaps most importantly, become engaged with science. Knowing how scientists do their work also allows educators to emulate these processes with children. But analyzing the practices of adult scientists as a way of telling us how children should think scientifically misses a crucial point. There are certain practices that adult scientists engage in that children would never do. For example, as practicing scientists, we keep notebooks detailing our findings and experimental ideas. We use those notes to help design new experiments or to integrate findings together (and, sometimes, to remember what groceries to get). No 4-year-old would ever make such notes or engage in such abstract reasoning practices. But the absence of this behavior, just like the absence of lab coats and clipboards, does not indicate an inability to engage in any kind of scientific thinking; it just suggests the possibility that scientific thinking looks different in young children than in adult scientists. If we take the approach that scientific thinking is only and exactly what adult scientists do, then we will miss situations in which children (and adults) engage in scientific thinking in the course of their everyday activities.

Nevertheless, obviously, to make the child-as-scientist approach do real work for developmental science, we must go beyond the metaphor. We must specify what theories are, how they are represented, and the process by which one representation of a theory changes to another. Luckily, a response to this challenge comes from the definition of theory theory laid out above: Theories are abstract, coherent representations of *causal structure*. To understand how theories are represented and how theories change, we have to understand how causal structures are represented, how they change, and how they can allow for abstraction and coherence. This is something that researchers in psychology, philosophy, and computer science have been working on for a long time.

Causal Reasoning as Associative Learning

Following Piaget's (1929) description of young children as "precausal," many early theories of children's causal reasoning posited that they made inferences based on associations, hence took their cues from the long literature on associative learning in comparative psychology.[3] In this view, over the course of many observations, human beings (and nonhuman animals) build up stronger and stronger links between events that tend to co-occur, which are generally causes and effects. Nonhuman animals can learn various kinds of associative relations in order to make predictions about their environments. For example, nonhuman animals can learn over the course of many trials that a particular type of action on their part will tend to be accompanied by a reward. The more these animals perform this action, the stronger this link between action and reward becomes. This can be interpreted as understanding the relation between a cause and an effect—action causes reward. Based on these sorts of experiments, classic psychological theories of associative learning formalized how animals relate conditioned and unconditioned stimuli to each other in such a way as to make better predictions about their environments (e.g., Mackintosh, 1974; Pearce & Hall, 1980; Rescorla & Wagner, 1972).

These frameworks provide one way of thinking about causal reasoning in human beings, including in children, because human beings can use associative mechanisms similar to the ones that are available to nonhuman animals (see e.g., Cramer et al., 2002; Dickinson, 2001; Dickinson & Shanks, 1995). This process manifests in the ability to pick out associations among events in the world. Indeed, infants and young children appear to possess sophisticated capacities for noticing statistical regularities in the environment (e.g., Fiser & Aslin, 2002; Goldstein et al., 2009; Haith et al., 1993; Kirkham et al., 2002; Saffran et al., 1996), making such a hypothesis developmentally viable.

The ability to learn individual associations is a powerful mechanism for interpreting the world. However, this mechanism is limited. One limitation is the computational complexity that this system would require. For example, most of the research cited above focuses on how children learn that events unfold over time (e.g., in a given language, syllable A is more likely to be followed by syllable B than by syllable C, and is never followed by syllable D) or how events are paired in space (e.g., members of category X have both features A and B, while nonmembers of category X lack both features A and B). Massive demands on memory and reasoning are needed in order to learn all the associations necessary to make such inferences. Given the sheer volume of data necessary to make these inferences, such reasoning requires a good amount of exposure to data. Moreover, many attentional resources are involved. The world has many statistical regularities, and many of those associations fail to indicate causal relations. To avoid being swamped by thousands of spurious associations, learning from association must also involve attentional or arousal systems to know what events to process.

Finally, for associative learning to be a good representation of causal structure, one also needs to be able to stack associations on top of one another to create higher-level correlations. Knowing that event A correlates with event B, and that event A correlates with event C, does not indicate that event B will correlate with event C without the knowledge that A is also present. For example, a simple associative learning system will associate objects that have hands together with objects that have legs (i.e., bodies). Infants can register these kinds of correlations among object features (Younger & Cohen, 1983). Infants also understand that hands (but not other objects like sticks) produce goal-directed actions (Sommerville et al., 2005; Woodward, 1998). Registering the kind of higher-order correlations we are describing means that infants will additionally be able to infer that objects with legs will engage in goal-directed actions, even if the infant never sees the object's hands.

Although it may seem implausible that children or babies could engage in such complicated inferences, several studies have demonstrated that young children can indeed detect such second-order correlations among static features of objects (Cuevas et al., 2006; Yermolayeva & Rakison, 2016) as well as among dynamic features of objects (Rakison & Benton, 2019). Some of our work demonstrated that children can use second-order correlations to make causal inferences about nonobvious object properties (Benton, Rakison & Sobel, 2021). We introduced 2- and 3-year-olds to two objects that differed in shape, color, and size (objects A and B). Object A had a unique feature (X), while object B had a different unique feature (Y). The two objects just sat on a table in front of the children for about 10 seconds, so that children could see them but not explore them. Then the objects were put away.

Children were then introduced to a novel machine (a blicket detector, which we describe below in more detail, but for now, we can just describe it as a small box) and two new objects. One object was identical to object A, but did not have the unique X feature. The other object was identical to object B, but did not have the unique Y feature. These two objects were placed on the machine individually; one of them activated it (caused it to light up and play music when it came into contact with it) and the other did not.

Finally, children were introduced to a third pair of objects. These objects looked completely different from A and B, except that one of them had feature X and the other had feature Y from the original pair of objects that children saw. Children were asked to make the machine go with these new objects. Children tended to choose the object that had the second-order correlation with the machine's activation. That is, if the object that looked like object A had activated the machine, then children chose the object in the test pair with feature X, whereas if the object that looked like object B had activated the machine, they chose the object with feature Y. These data suggest that even 2-year-olds can register second-order causal relations. In turn, these results support the idea that children's (and adults') causal reasoning could be the result of layers of associative reasoning, which in turn keep track of different kinds of correlations.

Although children may have access to the computational abilities to represent a wide variety of complex associative relations, we believe that an associationist framework is not the best way to model children's causal reasoning, because this framework makes a series of predictions that are not borne out by other data. For example, in cases like the study described in the last paragraph, higher-order associative reasoning may lead children to simply put both objects on the machine. If they do not know which one makes it go, or if they have even the slightest uncertainty, putting both objects on the machine is a reasonable thing to do, because this would reflect the highest possible correlational structure. Critically, children never did this in our study, and they almost never do this in other studies where they could solve a problem by using this kind of associative reasoning (see e.g., Gopnik et al., 2001; Sobel, 2020, for similar results). We need to look elsewhere for a full answer to the question of how children construct and represent theories.

Constraining Causal Inferences

Associative reasoning is a good first step, but learning causal structure involves more than just understanding that events are associated. It involves appreciating the hows and the whys of that association. The hows involve appreciating *mechanism*, that is, the ways that events relate to one another; this allows for *interventions*, that is, the ability to act on events in the world to produce outcomes. The whys involve appreciating the reasons behind those mechanisms—understanding when mechanisms generalize to novel events.

Empirical work demonstrates that children do appreciate more than just associations between events. For example, children can integrate various pieces of causal knowledge they possess with their associative reasoning and statistical learning capacities (e.g., Denison & Xu, 2010; Madole & Cohen, 1995). In one study of this ability, Madole and Cohen (1995) showed 14- and 18-month-olds a set of trucks. The trucks had either small black wheels or large yellow ones, and the top parts of the truck looked like either a small person or a green tree. The wheels could either roll or were fixed, and the top part could make a whistling sound. These researchers initially showed that infants could learn the relation between an object's parts and those parts' functions. Specifically, they habituated infants to trucks on which the black wheels rolled, but the yellow ones did not, regardless of what the top part of the truck was. At the same time, they habituated infants to trucks on which the top part that looked like a tree whistled, but the part that looked like a person did not. Infants in both age groups could learn these relations, understanding that the perceptual features of the parts (black or yellow wheels; top that looked like a tree or a person) predicted the function of those parts (rolling vs. stationary; whistling vs. not).The main idea of their experiment, however, was not to show that infants could learn these correlations, but to show that infants did not learn all kinds of correlations. There are a lot of correlations in the world, and most of them have little to do with the actual causal structure of the world. For instance, there is a strong inverse correlation between the number of pirates on Earth and the Earth's temperature, but it's hard to find a causal mechanism for how fewer pirates directly causes an increase in the Earth's temperature. We naturally reject this correlation as indicating causal structure because there is no mechanism by which it could work. Nor do we think that popularizing "Talk Like a Pirate Day" (a real thing, apparently—it's September 19) will effectively combat climate change.

To show that infants understand this as well, Madole and Cohen (1995) presented infants with a different kind of correlation. Now, the perceptual appearance of one part predicted the function of the *other* part. So if a truck had black wheels, the top part made a whistling sound, regardless of whether it was shaped like a person or a tree. Similarly, if the top part was a tree, then the truck's wheels rolled, regardless of whether they were black or yellow. The nature of the correlations in this condition were the same as in the other one. The amount of data that the babies were given to learn these correlations was the same. But these correlations just do not make sense. Like in the case of pirates and climate change, there does not tend to be real-world situations in which the shape of one part of an object correlates with the function of another part of the object, which can differ in shape. Eighteen-month-olds did not learn this correlation. That is, they did not dishabituate to the trucks that violated the correlation. However, 14-month-olds did. We interpret these findings to show that the older babies did not pay attention to this correlation in the first place, or that they did not think it was worth learning. And why should they? By this age, they have other aspects of causal knowledge that allowed them to register this correlation as unimportant and to reject it, or to not attend to it at all in the first place.

This study shows that, by 18 months, children can use aspects of their real-world knowledge to constrain which correlations they will learn. This implies that human causal learning is not purely driven by correlations. Human beings have access to other information that guides their learning of particular types of correlations but not others.

We drew the same conclusion using a different approach (Tummeltshammer, Wu, Sobel & Kirkham, 2014). In this experiment, we were interested in understanding how well infants could track the reliability of others as a source of knowledge. We showed 8-month-olds videos of a person in the middle of a screen. The person on the screen got the baby's attention, and then turned to a location in space (one of the four corners of the screen). An interesting cartoon with a weird sound effect then appeared in one of the four corners of the screen. The trick of the experiment is that one set of videos showed a person who was always accurate; the person looked to different locations four times, and, all four times, that's where the cartoon appeared. The other set of videos showed a different person who was accurate only 25% of the time, only looking at the corner where the video appeared one out of four times. We wanted to determine whether the infants would learn to follow these two people's gaze differently from this relatively brief exposure. And infants did. When the accurate informant looked to a novel location on the screen, 8-month-olds followed their gaze. But they did not do so when that same location was cued by the inaccurate speaker.

These two conditions of the study show that even 8-month-olds can track the accuracy with which other people generate information (a point that we raised in chapter 1 and that we return to in chapter 3). But the more important point for the present purposes is the control condition. In that condition, the faces were replaced by two different blobs that morphed into arrow-like shapes to cue the location of the cartoon. Eight-month-olds did not learn in this condition; they did not look in the direction that the blob was "pointing" at test. Like the Madole and Cohen (1995) study described above, this suggests that babies are not just always associating information together to make inferences. Instead, they are using higher-level information—that people can be sources of knowledge, but objects are less likely to be—to figure out when to track the accuracy of someone or something.

What is important about all of these studies is that infants can integrate their existing knowledge about how the world works into their ability to register and reason about statistical regularities they observe. Critically, there might not be a specific age when they can do this generally; the two studies we reviewed examined infants of different ages. This is because the specific pieces of causal knowledge that are necessary to constrain particular causal inferences develop at different times. The important point is that the causal knowledge children possess at any given time constrains how they process new information and has cascading effects on the development of subsequent knowledge. In turn, this strongly implies that children's reasoning abilities are supported by something more than mere association.

Bridging Statistical Learning to Causal Models

The fact that some associations are stronger or more salient than others is not a novel idea. This is even the case for nonhuman animals, and it is consistent with classic models of associative reasoning.[4] For example, rodents learn taste aversion in a single trial, rather than needing several trials to build up the negative association (e.g., Garcia et al., 1955). This suggests that there are other forces (perhaps evolutionary ones) that prime or bias rodents' learning of the relation between taste and food avoidance, or that make this kind of relation particularly salient.

Based on results like this one, many models that started as purely associative began incorporating a way to calculate the causal strength of known causal relations between events or properties, allowing them to integrate top-down knowledge with associative learning principles (e.g., Krushke & Blair, 2000; McClelland & Thompson, 2007; Rogers & McClelland, 2004; Van Hamme & Wasserman, 1994; Wasserman & Berglan, 1998). A related approach relies on estimating causal parameters based on the frequency with which events co-occur, such as in the ΔP model (Allan, 1980; Jenkins & Ward, 1965; Shanks, 1995) and the Power PC model (Cheng, 1997; Novick & Cheng, 2004).

Although many published papers validate each of these models (and many more papers challenge them), a problem for our purposes is that these models were designed with adult reasoners in mind, and the evidence that has been generated in support of these models comes from adult participants. Empirically, children's causal reasoning does not match adults' in many cases, and there is evidence that children's inferences are not well-explained by any of these models (e.g., Griffiths et al., 2011; Lucas et al., 2014; Sobel et al., 2004).

There are also theoretical concerns with using these models in development. For example, in some of the models mentioned above, learners are assumed to have separate processes for identifying events in the world that could be causes and effects and for deciding how those events fit together in a causal structure. That is, they first identify the causal structure, and then separately determine the strength of individual causal relations. It seems reasonable to keep these processes separate; describing which events in the world might be potential causes and effects can be done independently of determining how these causes and effects might be related to one another. If I ask you to determine the relation between flipping a switch and a light turning on, for example, it is relatively straightforward to posit that the switch causes the light to turn on, and not the opposite. This process could be independent from determining the strength of the relation between the switch and the light.[5] The trouble is that it's not clear whether these processes actually are separate for young children.

Given these complexities, researchers in cognitive science started to look for alternative computational frameworks to describe how children represent causal knowledge and engage in causal inference. One successful approach has relied on a computational framework called *causal graphical models* (Glymour, 2001; Pearl, 1988, 2000; Spirites et al., 1993; Woodward, 2003). These models came out of a research program at the intersection of philosophy and computer science and are now a popular tool in cognitive and developmental psychology. In particular, Gopnik et al. (2004; Gopnik & Wellman, 2012) argued that causal graphical models could solve the vagueness of the theory theory by specifying how causes and effects could be represented and at what level of abstraction. Moreover, well-established algorithms have been developed for these models to describe how relations between causes and effects could be learned, particularly from observed data. Further, these models provide a description of how inference and counterfactual reasoning occur, supporting the coherence of these representations.

Overall, then, causal graphical models seem like they could be an especially good fit for describing children's behavior. But are young children's inferences and representations of causal structure really consistent with this framework? In the next two sections, we describe a body of work that shows that they are, establishing the causal graphical model framework as a viable and productive way of understanding causal reasoning, hence theory representation and theory change, in development.

What Are Causal Graphical Models?

The first step is to explain what a causal graphical model is.[6] To explain that, we have to explain what a plain graphical model is. Briefly, a graphical model is a formal (mathematical) way to represent a joint probability distribution.

So what's a joint probability distribution? Imagine a list of all possible combinations of the events under consideration and the probability that each combination occurs—that's a joint probability distribution. This can be represented by a graph. In this formalism, the nodes in the graph are used to represent events, objects, or their properties. The vertices in the model are used to represent particular types of dependencies between such objects or events, such as their relational structure. Figure 2.1 shows a simple example: Variable X is related to variable Y, and both X and Y are independent of Z.

Importantly, information about conditional probabilities (that is, which events are likely to occur given that other events have occurred) can be extracted from these structures. For the simple example in figure 2.1, if you know that X has occurred, you will think that it's likely that Y has occurred too (or, more precisely, it affects the probability that Y occurs). But if you know that Z has occurred, you do not have any information about either X or Y.

Given that graphical models can be used to represent events and the probabilities that they occurred, it starts to become clearer how these kinds of models can be used to talk about causal reasoning. However, before we can formally interpret these models as psychological representations of causal knowledge, we must make two further assumptions about the underlying structure of the connections between events (nodes) and the relations among them (vertices): the *faithfulness assumption* and the *Markov assumption*. Importantly, these are assumptions, not empirical claims. While there is some empirical evidence that suggests that these assumptions are being used, they are part of the necessary background for reasoning with graphical models.

Faithfulness

The faithfulness assumption is the assumption that the data we observe are indicative of the actual, ontological structure of the world. The broad assumption here is that we see the world the way it really is.[7] For our purposes, what faithfulness means is that there is never a situation in which there would be a direct contradiction between the causal data that we observe and the causal structure that actually exists in the world.

For example, suppose that we observe that two events are independent, like X and Z in the example described above (figure 2.1, or the top panel of

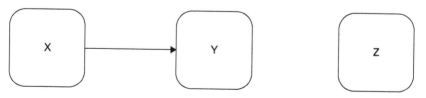

Figure 2.1
A representation of a (not very interesting) graphical model.

figure 2.2). The faithfulness assumption says that there is never going to be another event (like the unobservable naughty monkey, shown in the bottom panel of figure 2.2) that inhibits the occurrence of Z with exactly the same causal efficacy as event X causes the presence of Z whenever X occurs, thus masking the presence of a causal relation between X and Z (represented by the dashed line in the bottom panel of figure 2.2). Put another way, in order to get causal reasoning off the ground, we have to assume that there are no unobservable naughty monkeys playing with our perceptions and inferences about the world.

Of course, no specific empirical evidence fully supports or invalidates the faithfulness assumption. How can we prove that there are no unobservable naughty monkeys? Wouldn't the unobservable naughty monkeys just

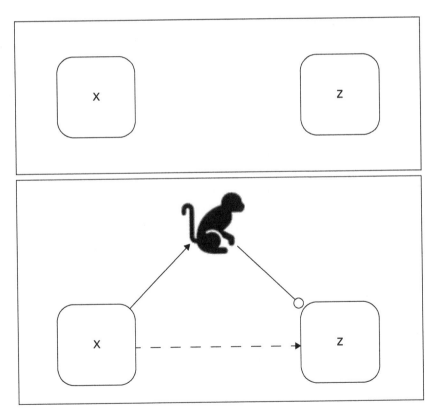

Figure 2.2
Illustrating the faithfulness assumption.

manipulate any experiment we do to result in evidence that does not suggest their presence? This is correct; we can't prove that there are not unobservable naughty monkeys (as Descartes noted in his *Meditations*). Faithfulness is an assumption. But once we accept it, we can make enormous progress in conceptualizing how the world works.[8]

The Markov Assumption

The Markov assumption states that the value of an event (i.e., a node in the graph) is independent of all other events, with the exception of its children (i.e., its direct effects) and its parents (i.e., its direct causes). To take a concrete example, let's look at an incredibly simple model of the weather, shown in figure 2.3.

For the graph shown in figure 2.3, all else being equal, the Markov assumption states that the event of raining yesterday and the event of raining tomorrow are independent; there is no direct relation between them. Whether it rained yesterday and whether it rains today are related, as are whether it rains today and whether it will rain tomorrow. That is, the event of raining yesterday and the event of raining tomorrow are dependent on each other, because they are related through the event of raining today. However, the only influence that raining yesterday has on raining tomorrow depends on whether it rains today; these two events are otherwise independent. If /rɔθ/ ("roTh," the backwards doG of Thunder[9]) comes along and makes it rain today, it no longer matters whether it rained yesterday when predicting whether it will rain tomorrow. Knowing that it rained today is all you need.

Do children reason about the relations among events using the Markov assumption? And, if they do, how could we test this? To answer this question, our test would need to include a way to examine whether children can recognize that some events are independent of each other. But, as reviewed

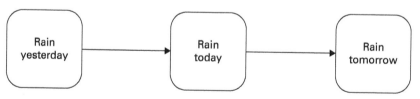

Figure 2.3
A simple model of the weather.

in chapter 1, even young children have a good deal of knowledge of how events are causally related to each other and could bring that prior knowledge to bear when responding in our studies. This means that any test of the Markov assumption in children would need to avoid using any causal relations about which children could have prior knowledge.

That's what blicket detectors are for.

Blicket Detectors

As a graduate student, Dave was part of a team[10] that developed a novel paradigm for testing children's causal knowledge: the *blicket detector*. An example is shown in figure 2.4. The blicket detector is a box that can light up or play music when certain kinds of objects are placed on top of it, making it look as though these objects have activated the machine. The machine actually works via an enabler switch, which is controlled by an experimenter and kept hidden

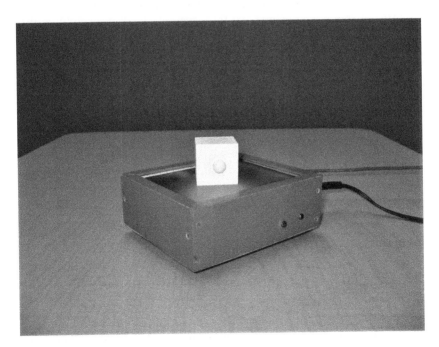

Figure 2.4
The original blicket detector with an object on it. This box lit up red and played a MIDI recording of *Fur Elise* when it was active.

from the participant. When the switch is in the "on" position, anything that is placed on the box activates it; when the switch is in the "off" position, nothing activates the machine. The top of the box has a pressure-sensitive plate, which is connected to this enabling switch and which makes it appear as though the objects that are placed on top of the machine are making it turn on.[11] Later versions of the detector added more features, like turning different colors or being activated by remote control, which could allow researchers to present different examples of causality at a distance (e.g., Kushnir & Gopnik, 2007; Sobel & Buchanan, 2009). The crucial aspect of these machines, however, always stays the same: Some objects placed on them make them activate while others do not, controlled by the experimenter. This machine, though simple, is a powerful tool for studying children's causal reasoning abilities.

The first studies to use the blicket detector did not focus on children's causal reasoning, but rather on the extent to which causal features of objects were important for categorical inferences. We (Gopnik & Sobel, 2000) presented 2- to 4-year-olds with a set of nondescript objects (wooden blocks of different shapes and colors) and the machine. We showed children that some objects made the machine activate, while others did not. For example, we placed four objects on the table with the machine and then placed the objects on the machine one at a time. Two activated the machine and two did not. We then labeled one of the objects that made the machine go a "blicket" and asked children to show us the other "blicket." Children—certainly by the age of 4—picked the second object that had activated the machine (see also Nazzi & Gopnik, 2000).

Most of these studies also ran a condition in which the experimenter would hold each object over the machine in one hand and use their other hand to press the panel on the machine so that it activated. This way, the object was associated with the machine's activation, but there was clearly another cause for the machine's activation (i.e., the experimenter's hand). In this case, children responded at chance when asked to find the other blicket. That is, children were differentially sensitive to cases where an object seemed to cause the machine's activation and cases where an object was merely associated with the machine's activation.

The crucial point of connection between the blicket detector and the causal graphical model framework introduced above is that the detector presents a novel causal system. Although children may have some prior knowledge about machines and some basic knowledge about causal relations (like the fact that causes temporally precede their effects), children have never seen this machine before and hence do not have any expectations about how it works. That is part of what makes the blicket detector such a powerful tool for testing children's causal reasoning abilities: It allows researchers to flexibly present new causal systems without having to worry about whether different children are approaching these systems with different levels of domain-specific causal knowledge. In this way, it allows for a good test of whether children are using the Markov assumption, because some events can be presented as dependent on each other, and some can be presented as independent. Children can then be tested to determine how they interpret these new events.

This is what we did in Gopnik, Sobel, Schulz, and Glymour (2001). We started by noting the importance of the control condition in Gopnik and Sobel (2000), in which the experimenter holds the object over the machine and presses the panel of the machine down with his hand to activate the machine. In this case, even though an object is associated with the machine's activation, there is another candidate cause (the experimenter's hand) that is a better explanation for the machine's activation. The assumption that children seemed to make is that the experimenter's hand explains the activation of the machine, and the presence of the object over the machine is independent of the machine's activation (or of the experimenter's decision to activate the machine). That is, children appeared to reason according to the Markov assumption.

To test this more directly, this study contrasted children's inferences about objects that activate the machine by themselves with objects that only activate the machine when another object is present. We told 3- and 4-year-olds that certain objects, called "blickets," would activate the detector. Then, children observed one of two types of trials. On the *one-cause* trials (figure 2.5), children were shown two objects (A and B). Each object was placed by itself onto the detector. One object (A) activated the detector by itself. The other object (B) did not. Children then saw objects A and B placed on the detector together (twice), and the detector activated (both times). This means that object B was associated with the activation of the detector two out of the three times it was placed on it. If children were keeping track of only associations between causes and effects, they might reasonably conclude that B was a blicket. But if children were reasoning about the events that they observed

Figure 2.5
Illustration of the one-cause trial from Gopnik et al. (2001).

using the Markov assumption, they should realize that B only activated the detector dependent on the presence of object A. They should not use the positive association between object B and the machine's activation to infer that B alone could make the machine activate.

This is what children did. To test children's understanding of these events, at the end of the experiment, they were asked whether each object was a blicket. Children in the experiment generally labeled only object A as a blicket. That is, they recognized that B's association with the detector's activation (the effect) was dependent on the presence of object A. To put it the other way around, children understood that object B lacked causal efficacy when object A was not in the equation.

On the analogous *two-cause* trials (figure 2.6), everything was the same except that B activated the machine independently two of the three times it was placed on the machine. Even though B was associated with the effect (machine activating) with the same frequency as in the one-cause trials, children tended to label B as a blicket in this case. In these trials, object B activated the machine independently from object A, and children used this fact to conclude that it was also a blicket.

Various investigations since the publication of this paper have extended these findings to infants (e.g., Sobel & Kirkham, 2006), to other domains of knowledge (Schulz & Gopnik, 2004), and to other kinds of inferences supported by the Markov assumption (i.e., those involving causal chains or common causes, Sobel & Sommerville, 2009). These data all suggest that children adhere to the Markov assumption in their causal inferences. In turn, we have good reason to think that the causal graphical model framework provides a productive way to describe and explain young children's causal reasoning.

Figure 2.6
Illustration of the two-cause trial from Gopnik et al. (2001).

Back to Associative Reasoning

The Gopnik et al. (2001) findings suggest that children are reasoning according to the Markov assumption. However, what that work did not consider is *how* they might be doing that. One worrying possibility is that other reasoning mechanisms could simulate the Markov assumption. That is, it could appear as though children were reasoning in a manner consistent with the Markov assumption but were not representing causal knowledge via the causal graphical model framework; rather, a simpler mechanism could account for their inferences. A candidate for that simpler mechanism is associative reasoning. Indeed, the findings on second-order conditioning and the findings on statistical learning described earlier in this chapter are both consistent with the idea that children's causal reasoning is associations all the way down. Although some of these associative models of causal reasoning include top-down information, at base, they still calculate the associations between events to make causal judgments. This has led numerous researchers across different domains of psychology to argue for the importance of associative mechanisms in causal reasoning (e.g., Blaisdell, 2008; Cramer et al., 2002; Dickenson & Shanks, 1995; Heyes, 2012; see also Hanus, 2016, which provides a good review of the comparative literature but does not consider developmental work).

To illustrate how this might work, consider again the procedure from Gopnik et al. (2001). In the one-cause condition, object A activates the machine by itself (we'll represent this as A+), then B fails to activate the machine by itself (B–), then they both activate the machine together twice (AB+, AB+). In the two-cause condition, the associations are the same—it's just that the objects are never presented together (A+, A+, A+, B–, B+, B+). On a simple version of associative reasoning, in which you just count the number of times the stimulus (the object A or B) is paired with the effect (+), these two conditions are equivalent. But children treated these two conditions differently: In the one-cause trials, B was not considered to be a blicket, but in the two-cause trials, it was. Because of this, we can conclude that children are not reasoning according to simple association; something more complicated is going on.

If you are familiar with associative reasoning, you should recognize that the pattern of data presented to children in the one-cause condition is similar to the pattern of data presented in nonhuman animals in studies on *blocking* (Kamin, 1969). In basic Pavlovian conditioning experiments, if a conditioned stimulus (A) is paired with an outcome, and then that stimulus A is presented in compound with a novel stimulus (AB) and also paired with that outcome, animals will learn the association between A and the outcome more than between B and the outcome. These data suggest that a simple model of associative reasoning cannot account for the animal's behavior. But more complex models of reinforcement learning could.

One such model is the Rescorla-Wagner (1972) model, which was an attempt to describe how associative learning worked; one of its earliest successes was that it provided an explanation of the phenomenon of blocking. We are going to step through the calculations of this model, because the associative challenge is significant not only here, but also to arguments we want to make in the chapter 3. So it's important to appreciate how it (and more contemporary models like it) describe the process of making inferences from data.[12]

The Rescorla-Wagner model is a way of updating the associative relation between conditioned and unconditioned stimuli, though for our purposes, we will just say a stimulus and an outcome. It is defined with the following formula:

$$\Delta V_{n+1} = K(\lambda - \Sigma V_n)$$

To unpack this, start with the first term, ΔV_{n+1}. This represents the change in the associative strength of a relation between a stimulus and an outcome on the next trial (i.e., exposure to the pairing). Here, V is a variable that represents associative strength, and the Greek letter Δ is a symbol that means "change in." So the change in the associative strength of a stimulus is a function of three things: (1) the salience of the stimulus and the outcome (represented here by K),[13] (2) the relation between the associative strength of all the stimuli that are present on trial n (ΣV_n), and (3) how much can be learned on a given trial (λ). This last variable (λ) represents how much of an associative relation that stimulus can have with this outcome. It's the largest amount of associative strength that can be learned about this outcome. Finally, ΣV_n represents the sum (Σ) of the associative strength (V) of all stimuli that are present on this trial (trial number n). Taken together, the formula is a way to represent what an organism learns from a particular event, given a stimulus, an outcome, and the organism's prior knowledge about the strength of the association between and the stimulus and the outcome.

There is one more important point here. The units used in doing calculations with this model are arbitrary, because there is not a unit of measurement for the strength of an associative relation. Rather, this formula provides a mathematical way of contrasting differences among different learning scenarios. As long as you keep your numbers constant within a modeling framework, what matters is the difference in values, not the values themselves.

Let's apply this formula to the Gopnik et al. (2001) procedure, starting with the two-cause condition. To do so, we will make up some values. We have no reason to assume that object A's relation with the machine is more salient than object B's relation, so we can assign $K_A = K_B$. For this example, let's assign both of these variables the value of 0.2. Let's also posit that the most associative strength you can learn when the effect is present is 100 units. So $\lambda = 100$. Now, when you run the numbers:

Initially: $V_A = V_B = 0$
Trial 1 (A+): $\Delta V_A = .2(100 - 0) = 20$; $V_A = 20$; $V_B = 0$
Trial 2 (A+): $\Delta V_A = .2(100 - 20) = 16$; $V_A = 36$; $V_B = 0$
Trial 3 (A+): $\Delta V_A = .2(100 - 36) = 12.8$; $V_A = 48.8$; $V_B = 0$
Trial 4 (B–): $\Delta V_B = .2(0 - 0) = 0$; $V_A = 48.8$; $V_B = 0$
Trial 5 (B+): $\Delta V_B = .2(100 - 0) = 20$; $V_A = 48.8$; $V_B = 20$
Trial 6 (B+): $\Delta V_B = .2(100 - 20) = 16$; $V_A = 48.8$; $V_B = 36$

Let's unpack this. On Trial 1, the change in associative strength of A is the K value (0.2) multiplied by the λ value (100, because the effect was present) minus the strength of all the stimuli that were present on this trial. The only stimulus that was present was A, and its associative strength was 0. Hence $\Delta V_A = 20$. Trials 2 and 3 follow that same logic—the only thing that changes is the associative strength of the stimuli present.

At this point, we want to point out something really cool about this model, which hopefully reflects an intuition you have about the world: As you are exposed to the same association more and more, you learn from it less and less each time. On Trial 1, the change in the associative strength of A was 20, but on Trial 2, it was 16. Even if you pair a stimulus with an effect thousands of times individually, you can't ever achieve an associative strength beyond the λ value. This was an important point of the model; it was focused on the idea that the slope of a learning curve changed as the animal was given more trials. An analogy in human learning is automaticity: Some things are learned so well that further exposure does little to strengthen the relation. In the human case, that can make these associations hard to unlearn (e.g., Shiffrin & Schneider, 1977).

On Trial 4, you should notice a pretty big change to the calculations. On this trial, the λ value is 0. That is because the effect *did not occur* on this trial. In this case, B does not accrue any associative strength, because there is nothing to associate with its presence. This feature of the model illustrates another important part of the learning process: Associations can also be unlearned, a process known as *extinction*. But the more strength an association accrues, the harder it is to unlearn.

So at the end of this process, the associative strength between object A and the machine's activation is 48.8 and the associative strength between object B and the machine's activation is 36. Children are then asked, "Is this (A/B) a blicket?" Presumably, to answer that question, the associative strength has to exceed a certain value. Because the numbers are arbitrary, we can choose that value, as long as it stays the same throughout the demonstration. So let's pick a value of 30. In this scenario, then, A and B are both blickets because their associative strength exceeds that threshold. This matches nicely with the empirical data from Gopnik et al. (2001).

But what about the one-cause trial? Again, let's run the numbers:

Initially: $V_A = V_B = 0$

Trial 1 (A+): $\Delta V_A = .2(100 - 0) = 20$; $V_A = 20$; $V_B = 0$

Trial 2 (B−): $\Delta V_B = .2(0 - 0) = 0$; $V_A = 20$; $V_B = 0$

Trial 3 (AB+), $\Delta V_A = \Delta V_B = .2(100 - 20) = 16$; $V_A = 36$; $V_B = 16$

Trial 4 (AB+), $\Delta V_A = \Delta V_B = .2(100 - 52) = 9.6$; $V_A = 45.6$; $V_B = 25.6$

Notice a few things here: First, Trials 1 and 2 result in the same effects as the first times A and B were put on the machine in the two-cause case. On

Trial 3, the change in associative strength for A and B is the same, but it is lessened by the fact that A already has some associative strength. It's lessened even more on Trial 4 because A and B are both present, and both have some associative strength. If we use the same arbitrary threshold for "blicketness" as we used above (an object is a blicket if its V is greater than 30), then A is a blicket in this example, and B is not. Again, this result matches the data from Gopnik et al. (2001). This conclusion lends credence to the idea that children's performance on that task (and on similar tasks) could be the result of this kind of associative learning, hence might not be due to children's abilities to reason according to the tenets of the causal graphical model framework or to use the Markov assumption.

An important objection to this example is the arbitrariness of the values that we assigned, particularly our threshold for deciding that something is a blicket. This is true; these numbers are arbitrary. But instead of thinking of it as a bug, think of it as a feature. For one thing, these numbers are constants, so the results will replicate regardless of their exact value. But more importantly, we can choose these numbers to be as generous to this associative modeling framework as possible, so that it has the best chance of explaining children's performance (although we are going to show in the next section that it does not work all the time). The point right now is that, based on the modeling, the associative strength of object B in the two-cause condition exceeds that of the strength of object B in the one-cause condition. As long as that is the case, there is a concern that simple statistical learning mechanisms—and not the more sophisticated graphical model approach that we favor—can underlie children's causal inferences.

Bayesian Inference

As discussed above, one of the main motivations behind adopting the causal graphical model framework as a description of children's learning is that it brings precision to the otherwise vague description of children's learning provided by theory theory. These models allow us to specify how children represent causal structures and the connections among events that they observe in the world. But our review of the Rescorla-Wagner model in the last section suggests that these processes might not require such a sophisticated computational framework. It might be that young children are just engaging in a form of *blocking*, where they are discounting one potential cause in favor of another event they already know is a cause.

Although it is well-established that this kind of discounting or blocking can be explained by associative models, other results are more challenging for these models. One example is "backward blocking," which just reverses the order of the blocking paradigm. In studies on backward blocking, similar to the blicket detector studies described above, adult participants observed two potential causes (A and B) produce an outcome. Then, one of those causes alone (A) produced the same outcome. Participants were less likely to judge that B was a cause than when they only observed A and B together (Shanks, 1985; Shanks & Dickinson, 1988). The Rescorla-Wagner model has difficulty explaining these data, because the associative strength of B is the same in both conditions.

To examine this developmentally, we (Sobel, Tenenbaum & Gopnik, 2004) implemented a backward blocking procedure with preschoolers. We first showed 3- and 4-year-olds two objects (A and B), which activated the machine together. Then we placed object A on the machine alone. In some conditions, A did not activate the machine by itself. Here, children should make a similar inference as they did in the Gopnik et al. (2001) study described above: object A was just "along for the ride" in this case, and only object B was efficacious. Unsurprisingly, given the earlier results, this was exactly the inference that they made. (It is worth noting that if you just show children that objects A and B together make the machine go, and then ask them about object A, children usually say that it is a blicket.)

The more interesting case is when A did activate the machine by itself; this is a version of the backward blocking paradigm. So A and B activate the machine together, and then A activates the machine by itself. In this case, what is the efficacy of object B? The correct answer is, "I have no idea." This is because object B's causal efficacy is ambiguous. But given that children now have an explanation for why the machine activated when A and B were on it together (i.e., object A unambiguously makes the machine go), they might explain away object B as a potential cause. And this is basically how children responded. Four-year-olds, for example, stated that object B had efficacy only 13% of the time. Other researchers (e.g., McCormack et al., 2009) have generated similar findings on slightly older children, further demonstrating that children are retrospectively reevaluating the probability that objects have causal efficacy.[14]

Although these data are a challenge for the Rescorla-Wagner model, many other theories of causal inference, particularly those from the literature on adult cognition, can account for them (e.g., Krushke & Blair, 2000; Van Hamme & Wasserman, 1994; Wasserman & Berglan, 1998). Similarly, other investigations have proposed that causal learning relies on the estimation of causal parameters, again based on the frequency with which events co-occur (Allan, 1980; Cheng, 1997; Jenkins & Ward, 1965; Shanks, 1995). What these models all have in common is that they use statistical learning principles to make a calculation about the causal strength between stimuli. What differs among these accounts is the math, not necessarily the way in which causal inferences are made. Further, most of these models could be categorized as making a lower-level kind of inference, as opposed to using the kinds of graphical representations we are advocating for (although see Glymour & Cheng, 1998).

Even if backward blocking can be accommodated on these models, we believe that there are other patterns in children's reasoning that cannot straightforwardly be explained within an associative learning framework. The example we work through here involves children's use of *base rates*—the frequency with which an event tends to occur in the environment. Some work with the blicket detector shows that even young children can track the frequency with which objects activate the machine and use that information when making inferences about ambiguous cases. That is, instead of just explaining away ambiguous data, children assume that they should default to the base rate of blickets when they do not know whether something is a blicket.

In one study on children's use of base rate information, we (Sobel, Tenenbaum & Gopnik, 2004, Experiment 3) showed 3- and 4-year-olds a set of identical objects and put 12 of those objects on the machine, one at a time. Either 2 of the objects or 10 of the objects activated the machine; there were different base rates of efficacious objects across the two conditions. Then, we showed children that the 13th and 14th objects (we'll call these objects A and B) together made the machine go, and then that object A made the machine go by itself. When the base rate of blickets was low (2 out of 12 objects had activated the machine in the initial demonstration), 4-year-olds judged object B to be a blicket approximately 16% of the time. When the base rate of blickets was high (10 out of 12 objects had activated the machine in the initial demonstration), 4-year-olds judged object B to be a blicket approximately 83% of the time. This was almost identical to the base rates.

These results cannot straightforwardly be explained via association because the associative relation between object B and the machine is the same in both conditions, yet children treat them quite differently. These results are more parsimoniously explained by algorithms that involve the causal graphical model framework. Specifically, the modeling approach that Sobel et al. (2004) suggested was based on Bayesian inference (following Griffiths & Tenenbaum, 2009; Tenenbaum & Griffiths, 2001), which takes us back to theory theory. According to theory theory, children have some knowledge about a situation (i.e., their theories). In modeling terms, this knowledge allows them to construct a set of hypotheses about the world, in which each hypothesis has a probability value attached to it (*priors*). Then, they observe new data, and they use those data to update their priors so that each hypothesis now has a new probability value (*posteriors*). This allows children to construct new representations of the world—that is, to learn. Bayesian inference provides a formal approach for describing exactly how this learning happens, specifically, how individuals update their priors based on data. It is expressed using this formula:

$$P(H \mid D) = (P(D \mid H) * P(H)) / P(D)$$

Here, H stands for "hypothesis" and D for "data." The initial term, $P(H \mid D)$, expresses the probability that the hypothesis H is true given the current set of data (D). Bayes' theorem allows us to calculate this probability as a function of three other terms: the probability that we would observe the data that we have observed if the hypothesis were true, $P(D|H)$, the probability of observing the data, $P(D)$, and the probability that the hypothesis is true, $P(H)$. Bayes' theorem provides a formal description of how different hypotheses are weighed against each other given the data from the world and given how existing knowledge can be rationally integrated with observed data to make novel inferences. Soon after this idea of Bayesian inference was incorporated into theory theory, the theory was rebranded as a new form of constructivism we will refer to as *rational constructivism*. The goal of rational constructivism is to focus on the ways that children use their existing knowledge to make new inferences about the world based on the data they observe in this rational way, which allows for a formal definition of how theories can change (Gopnik & Wellman, 2012; Xu & Kushnir, 2012; Xu & Tenenbaum, 2007[15]).

Algorithms that use Bayesian inference can provide a good description of how children reason. Specifically, they can explain how young children's existing knowledge (their priors) can constrain future inferences. Children's use of base rates in their reasoning provides a particularly powerful example of this, because this procedure makes it clear how different sets of prior information lead children to make different inferences at the end of the procedure.

Interestingly, although the 4-year-olds tested in the base rates study described above were able to do this, the 3-year-olds were not. They just said that B was a blicket in both conditions. Why were only the older children in this procedure able to attend to the base rate with which objects have causal efficacy? One possibility is that the 3-year-olds did not have the same priors (i.e., knowledge) as the older children.[16]

To succeed at the base rates task (and any causal reasoning task involving the blicket detector), children must register that there are at least two types of entities in the environment: objects and detectors. Blickets are objects that have the ability to activate a detector. Children are told all of this information in the beginning of the experiment. What they have to infer is an unobserved attribute of the objects—that some of the objects in front of them are blickets. To do so from the evidence, they must possess three pieces of prior knowledge. The first two are *temporal priority* and *spatial independence*. Temporal priority allows children to understand that certain objects being placed on the detector are responsible for the detector's activation, as opposed to the idea that the detector's activation causes the experimenter to place an object on it. Spatial independence allows children to understand that the identity of an object is independent of its spatial location and the spatial location of all other objects. That is, placing one object on the detector does not cause another object to become a blicket or cause another object to be moved in space. These are fairly basic assumptions about how causality works in general; as discussed earlier, there is some evidence that young children and even infants make such assumptions (e.g., Bullock et al., 1982; Leslie & Keeble, 1987; Oakes & Cohen, 1990; Sophian & Huber, 1984).

The third assumption is what Sobel et al. (2004, following Tenenbaum & Griffiths, 2001) called the *activation law*: the machine activates if and only if at least one blicket is placed on top of it, and only blickets make the machine activate. Children had to recognize that there was something about the object that connected its being placed on the machine to the machine's activation. A candidate cause for that "something" was a nonobvious property that the object possessed.

To test whether children made this assumption, we (Sobel, Yoachim, Gopnik, Meltzoff & Blumenthal, 2007) ran a new blicket detector study where we presented 3- and 4-year-olds with three objects. Two were identical; one was unique. One of the two identical objects and the unique object activated the machine. The other object did not. Children were then shown that the member of the identical pair that had activated the machine also had an internal property (e.g., a piece of hard plastic in its center). They were asked which among the two other objects had the same inside. Four-year-olds in this study responded based on the causal efficacy and chose the unique object. However, the 3-year-olds responded based on perceptual similarity. These data suggest that, between the ages of 3 and 4, the children are developing the understanding that there is something inherent about objects that make the machine go. This in turn can explain the difference in their performance in Sobel et al. (2004): Only 4-year-olds could attend to the base rates because only they could apply their understanding of the hidden properties of blickets to the objects.

To make this case more strongly, one of our later experiments (Sobel & Munro, 2009) connected these two ideas. In this experiment, we replicated the Sobel et al. (2004) procedure in which causes were rare and the Sobel, Yoachim et al. (2007) "insides" procedure and found strong correlations, controlling for age. It did not matter whether children were 3 or 4; what mattered was their understanding of the relation between causes and insides. If children understood that an object's causal efficacy related to its internal properties, then they were likely to use the base rate information and say that object B was not efficacious in the final trial. This experiment also manipulated the way the machine's activation was described to children. Sometimes, it was presented as a machine; other times it was presented as an agent ("Mr. Blicket") who liked things. Desire is a mental state that 3-year-olds understand well (e.g., Repacholi & Gopnik, 1997; Wellman & Liu, 2004). In this condition, 3-year-olds were almost always able to correctly reason about the relation between the activation and the internal properties of the objects, as well as use base rates in their reasoning.

This study provides a clear illustration of how content knowledge constrains causal reasoning, as discussed in chapter 1. It also continues to show how the causal graphical model framework and the Bayesian inference procedure for updating beliefs on the basis of new evidence provide a better match to children's behavior than do models of associative learning. It also illustrates another feature of this framework: the idea that children represent a *hypothesis space* of potential causal models. Under the causal graphical model framework, one looks at the data that one has available and constructs a set of possible explanations for how those data could have come about. Bayesian inference can then be used to select which explanation (that is, which hypothesis in the space of possibilities) is the best fit to the data. This process can be used to describe children's reasoning behavior in tasks like the ones described above. The exact details of Bayesian inference are not important for the argument we want to make here (see Gopnik & Bonawitz, 2015, for a good introduction to this topic). What is important is the question of how a hypothesis space is formed in the first place. We propose that this mechanism relies on our imagination: We must extrapolate away from the observed data in order to think about all of the *possible* mechanisms that could have caused it. This proposal illustrates the tight links between imagination and causal reasoning, which we explore further in chapter 10.[17]

In general, though, these data all suggest that children's causal reasoning is well-described by Bayesian inference, which allows children to use their existing knowledge (e.g., about base rates) to construct hypotheses about what they will observe in a new situation and to update those hypotheses on the basis of new information. In addition, numerous researchers have applied these ideas to other domains (e.g., Schulz et al., 2007) and to more domain-general learning problems (e.g., Goodman et al., 2011). More recently, Bayesian approaches have offered interesting and important contrasts (as well as supplements) to approaches in machine learning and artificial intelligence (e.g., Lake et al., 2017). They thus constitute a promising way for describing how children learn in general.

A Concern about Mechanisms

There is one thing that we have left out of our discussion of causal graphical models and Bayesian inference so far (actually, there are quite a number of things that we have left out, but we are trying to restrict our discussion to what will be useful for our later arguments about scientific thinking). This is the issue of *mechanisms*, which are important for using causal graphical models as representation of human causal knowledge (see Goodman et al., 2011). A mechanism is a description of how the nodes in a graph are related to one another. Knowing something about mechanisms helps you relate the data that you observe to the model that you represent.

More specifically, in a graphical model, a vertex (the arrow that goes from one node to another) represents a dependency: a probabilistic relation between one object, event or property and another object, event or property. Although this seems simple, it raises a problem: Even the simplest causal graph $(X \rightarrow Y)$ can be consistent with an infinite set of interpretations because we do not necessarily know what the probability of that relation is. The dependency between Y and X might be near deterministic (e.g., the probability that Y occurs given that X occurs might be .99), or it could be .63, or .17, or any real number between 0 and 1. But in all of these cases, the graph itself is the same. If graphical models are meant to be representations of our causal knowledge, how do we know how to interpret them, given that any particular representation is compatible with so many different possible structures?

It gets worse. Think about a graph in which two events each cause a third (see figure 2.7). Here, events X and Y are independent, but each cause Z. But what is the nature of the relations among X, Y, and Z? A natural way to think about this graph is that Z occurs either if X occurs or if Y occurs, with some probability. That is, this graph could represent a set of disjunctive relations, called a "noisy-or" parameterization when the relations among the variables are not deterministic. Such relations are fairly common in everyday causal reasoning, which might be why it is so easy to think this way. But this is only one possible interpretation of this graph. It could also be the case that X and Y both have to occur to cause Z (i.e., a conjunctive parameterization). Or it might be the case that only either X or Y is necessary for Z to occur, but the other is probabilistically related to Z.

Yet another issue with understanding the mechanisms represented by a graphical model is the question of abstraction. To illustrate this issue, consider two possible causal models. In one, you posit that diseases cause symptoms. In another, you posit that having a cold causes you to have a runny nose. When you observe that individuals who have colds also have runny noses, which model do you learn? This distinction between learning "specific" theories and "framework" theories (Wellman & Gelman, 1992) is one of the most difficult to describe in causal learning.

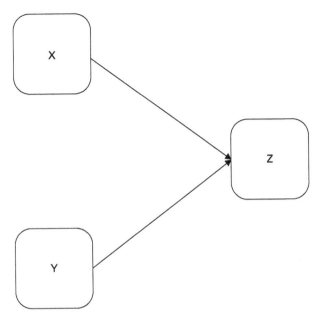

Figure 2.7
A common effect model. Even simple models like this one can represent a wide array of possible causal structures.

The answer might be that you learn both of these models; as you observe data in the world, you can formulate a representation not only of the specific causal structure (between runny noses and colds), but also an abstraction of broader frameworks or types of causal relations (between diseases and symptoms). Goodman et al. (2011) refer to this as the "blessing of abstraction."[18] They suggest that hierarchical modeling can allow learners to do both at once: to reconstruct both the specific causal model specified by the data and also the general principles (which they call "theories") that govern how such representations of causal structure should be constructed. Once those general principles are in place, they can also constrain subsequent causal learning given new data. Although human reasoners can do this, at least to a certain extent, it remains a challenge to understand which level(s) of abstraction people are thinking about in any given circumstance.

These problems occur largely because causal graphical models were not originally designed to represent human cognition; they are just formal tools.

But to make them do useful work for psychology, we need to sort out exactly what kinds of relations people are representing. The good news is that, at least with respect to the issue of different types of parameterizations as illustrated in figure 2.7, it turns out that young children can recognize different forms that these relations can take. For example, when shown evidence for disjunctive or conjunctive causal relations, 4- and 5-year-olds have no difficulty telling the difference (Lucas et al., 2014). They also do not have much difficulty discerning the difference between deterministic and probabilistic environments (Griffiths et al. 2011).

Because of this, we think that even young children posit that there are some kinds of mechanisms that relate different events to each other, and they do so with reasonable sophistication even from a very young age (similar to how the infants in the Madole and Cohen (1995) experiment we described above recognized which regularities to attend to). Although this is not a formal part of the causal graphical model framework (at least not in the computations), it provides a good starting point for thinking about the relation between a representation of causal knowledge and the world. But this is merely a starting point, and it does not fully resolve all the complexities of using causal graphical models to describe human learning that we outlined above. We return to these issues in chapter 3, where we talk about some potential challenges to the causal graphical modeling framework as a description of human causal reasoning.

More generally, what this discussion of mechanism tells us is that children are faced with a challenge, which is how they instantiate the models that they represent. In blicket detector tasks, this is basically done for the children. They can see that there is a machine that activates and that there are objects that are potential causes of that activation. Moreover, they are often told by a (presumably) knowledgeable experimenter that the machine is a blicket machine and that objects that activate the machine are blickets. The pedagogical situation between the experimenter and the child establishes that there are blickets in the world and that this machine will help figure out whether objects get that label. In situations that involve richer real-world contexts, however, there might be an additional problem for children: They not only have to learn how the variables relate to one another, they also have to grapple with the information presented by the context. We return to this discussion when we explore the role of contextualization in children's scientific thinking in chapter 6.

Levels of Explanation

The causal graphical model framework outlined above provides a precise way to discuss children's (and adults') causal reasoning abilities and how causal relations between objects and events can be represented and updated (i.e., learned). The blicket detector was used to test whether children's causal inferences indeed match with this formalism, and results from studies using the detector indicate that they generally do. In later chapters, we examine how more complex causal reasoning tasks can be presented using blicket detectors, in order to test other aspects of children's causal reasoning. For now, the important point is that this reasoning can be well represented with causal graphical models, providing a more precise language for describing how children learn about the causal structure of the world.

We (along with many other researchers) are in favor of the causal graphical model description of causal reasoning because these models—in combination with algorithms like Bayesian inference that describe how one might build up these models from observed data and prior beliefs—allow us to accurately and formally represent children's reasoning processes. But it is important to recognize what these models are supposed to be doing in our theory of cognitive development.

In his classic book on vision, Marr (1982) noted that it was possible to speak about a symbolic system at different levels. One could focus on exactly how that system is instantiated in the substrate that makes it up (e.g., neurons for humans; silicon chips for computers); this is the implementation level. Or one could focus on the steps that the system takes to perform its computations; this is the algorithmic level. Or one could focus on the general processes that the system performs and what it does; this is the computational level. To take a concrete example, both a smartphone and an abacus can be used to do addition, but they have different algorithms for doing so, which are implemented in different ways. But at the computational level, they are similar. They take representations of two numbers and produce a sum.

We take the causal graphical model framework to be describing children's reasoning processes at the computational level. These models offer a description of how children learn causal knowledge and make causal inferences, but they are neutral about exactly what steps children go through in order to implement this process and also about exactly how those steps are represented and carried out in the brain.

We discuss these distinctions among levels of explanation in order to note that the causal graphical model framework is not meant to describe exactly how children go through their reasoning processes (i.e., at the algorithmic level). The framework is meant to provide observations about children's reasoning processes, and it is meant to be used to make predictions about how children will behave in certain circumstances, given certain kinds of data. But even though it describes those processes precisely and formally, it is not meant to say anything further about exactly *how* those processes are carried out in human minds. That is, it can be a successful representation of children's reasoning processes (at the computational level) without being a fully detailed description of all the processes children undergo when reasoning (at the algorithmic level).

This discussion makes it clear that we do not yet have an algorithmic-level or an implementation-level description of children's reasoning. That is, we do not yet know exactly how the causal graphical models are instantiated in children's neural structures (or even if they are—it is probable that the instantiation is quite different and only simulates such a modeling framework). This leaves open the question of what is happening at the algorithmic level. If our brains are not genuinely representing causal graphs, then what underlies our abilities to use these computational tools?

Some researchers are currently investigating this issue (e.g., Bonawitz et al., 2014). One likely possibility is that the brain is sampling information in a way that mimics the representational structures and algorithms described by causal graphical models (e.g., Sanborn & Chater, 2016). Although this view is satisfying to some, others are not convinced. The dissatisfaction has mostly centered on the Bayesian inference part of the framework (Jones & Love, 2011; Marcus & Davis, 2013). Without reviewing all of these arguments, our goal for this section is simply to say that not all researchers believe that the causal graphical model framework is the be-all and end-all of human causal reasoning. However, this framework does present an abstract, coherent way to represent causal knowledge and describe causal inference, one that can be useful in constructing explanations and making predictions. For developmental psychologists interested in looking at the relation between children's causal reasoning capacities and other facets of cognitive development, this suffices.

As an example, one of our studies (Yang, Bushnell, Buchanan & Sobel, 2013) presented 15-month-olds with a novel object (X) that had a manipulandum, such as a bright red button (call this A). In this study, we showed the infants that activating the manipulandum (pushing the button) caused a nonobvious effect from the toy (e.g., it would make a mooing sound). Infants were allowed to imitate this novel effect on the object, and most did so. We then showed the infants an identical object (X) with a different manipulandum, such as a blue lever (call this B), which did nothing when manipulated. Again, infants imitated the action on this toy. The critical part of the study was when we showed infants a third identical object (X) that had both manipulanda on it (A and B; see figure 2.8). Which one would infants go for? Could they generalize what they had observed in imitation to produce a causal action on a novel object?

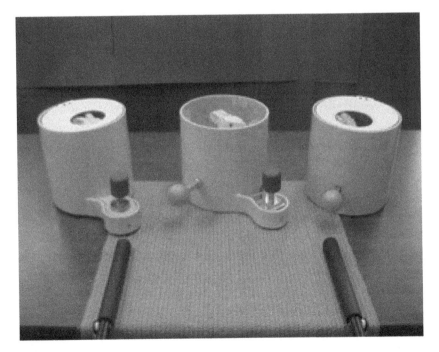

Figure 2.8
Stimuli from Yang et al. (2013). The objects used in the demonstration phase are on the left and right, and the object used in the test phase is in the center.

The answer is yes—infants acted on whichever manipulandum they had previously seen make the toy do something interesting (in this example, the button). But they only did this when the third toy—the one with both manipulanda—was identical to the first two toys. In a second experiment, we showed infants the same demonstration, but now the test object (Z) looked completely different, although it had the same two manipulanda (A and B). In this case, infants did not generalize—they went for either the button or the lever at random. These data suggest that infants' generalization capacities are relatively weak, and they were not able to identify that their actions on the manipulanda were the cause of the effect. When they saw the novel stimulus, they resorted to an irrational basis for deciding what to do.

But perhaps infants were behaving rationally in both cases. In both experiments, infants had observed a presumably knowledgeable adult model the efficacy of two objects that looked quite similar but behaved differently. Infants also imitated both of those actions, indicating that they could perform both actions and that they had some motivation to interact with both objects. When the test object with both manipulanda was the same, they generalized because they recognized that the action related to that object. But when the test object was different, there was no reason to generalize. Because the base object was different, this could plausibly indicate that it operated according to a completely different set of causal relations. It is reasonable to assume that pushing a button might work on toys of type X, but not all toys in general. There is potentially also a Bayesian analysis of this explanation: The demonstration phase of the experiment does not provide enough information to support the posterior hypothesis that one should generalize when the test object changes shape.

This line of reasoning suggested a third experiment. Instead of presenting two similar objects with different manipulanda during the demonstration and then a novel object with both manipulanda at test, this new study presented three different objects: First, children saw object X with manipulandum A, which caused an effect, and object Y with manipulandum B, which did not do anything. Then, children were shown the test object, Z, with both A and B on it. Fifteen-month-olds generalized in this case; they chose to intervene first on the manipulandum that was previously shown to be efficacious.

We also presented a formal computational model of these data, based on Bayesian inference over causal graphical models. The model wound up explaining the data nicely, including one completely unanticipated aspect of the results. In the first experiment, in which all three manipulanda had the same base shape, the model predicts that the likelihood of the previously seen efficacious manipulandum being efficacious on the test object is about 80%. But in the third experiment, in which all three objects have different bases, the model predicts that this likelihood is only 65%. In both cases, given that they have two choices at test, children should still attempt to act on the manipulandum that was previously shown to be efficacious; in both cases, it is more than 50% likely to work.

But in both of these studies, the test object—the one with two manipulanda—was inert, so that even when the infant manipulated the previously seen efficacious manipulandum, the object did not do anything. The question is then what children should do when they observe that this manipulandum is not efficacious on the test object. The model suggests that they should be quicker to switch to acting on the other manipulandum in the third experiment (where the likelihood of the first manipulandum working is 65%), as opposed to in the first experiment (where the likelihood of the first manipulandum working is 80%). This is exactly what the infants did. They persisted more with the first manipulandum when all of the objects were the same than when all of the objects were different.

We do not know if this is a good computational model or a bad one, or a good set of experiments or a bad set. That's not the point. Rather, the point is that the model helped explain some data we had collected and made novel predictions about new data. It even explained something completely unexpected in the data—something that we did not even know we were looking for.

Computational modeling is now a large part of cognitive science. Explaining behavior via such models is useful because it provides a formal account of what kinds of inferences children are making—not necessarily how they are making them, but what they are and are not capable of. We have presented a particular framework, but that does not mean that other computational frameworks are not also good descriptions. Regardless of whether causal graphical models or Bayesian inference mechanisms are instantiated as such in our brains, or whether other computational models provide a better description of children's reasoning, our goal is to use these frameworks productively in developmental science to better explain children's behavior and learning.

One of the things that we are going to do throughout the rest of this book is use this framework to describe aspects of the development of children's causal reasoning, particularly as it relates to scientific thinking. This does not mean that one has to know about these computational frameworks or even pay attention to them in order to understand our arguments. Rather, we articulate this framework here because we find it helpful in guiding the theoretical arguments we are trying to construct.

3 Beyond Rational Constructivism

We argued in chapter 2 that young children can integrate their existing knowledge with new data in order to learn about the world, developing and changing representations of a set of naïve theories in ways analogous to the process of science. We also argued that this process of learning and theory formation can be described by the causal graphical model framework, because this process is essentially one of making a series of causal inferences. Finally, we suggested that Bayesian inference serves as a good computational-level description of how children (and potentially adults) engage in causal reasoning.

These theoretical constructs were introduced to solve a problem with the way that theory theory was initially formulated. Theory theory described cognitive development by stating that children are born with impoverished theories, which they refine through observation and interaction with the world. This description, though generally accurate (we believe), was vague. And while we left a great many of the computational details out of chapter 2 (so maybe *we're* vague), the computational framework of Bayesian inference over causal graphical models is not a vague description of how learning works (see Gopnik & Schulz, 2007; Gopnik & Tenenbaum, 2007; Tenenbaum et al., 2006; Woodward, 2003, for many more computationally oriented examples).

However, the rational constructivist framework is not without its challenges. In this chapter, we identify several of them and attempt to respond to them. It is important to note that our goal in this chapter is not to present a full description and defense of this framework, or a full accounting of its faults and flaws. The theoretical dissatisfactions we want to highlight here are ones primarily related to the main focus of this book: how children's causal reasoning underpins their scientific thinking. For that reason, we will not exhaustively cover every objection to the causal graphical modeling framework or to rational constructivism more generally. Rather, we outline a few key problems as a way of beginning to build toward a reconciliation of this work (which has primarily been influential within the field of cognitive development) with work on children's scientific thinking. These analyses serve as background for the empirical studies presented in part II of this book, which provides several case studies for how this reconciliation could proceed.

Nonindependence

One objection to the causal graphical model framework as a description of children's causal inference is that most of the studies in this area (including Gopnik et al., 2001, and many of the other studies that we mentioned in chapter 2) only tested children's inferences about one kind of causal structure: a *common effect* structure in which two objects could independently be potential causes of the machine's activation. But this is not the only way to test the Markov assumption. One could also use a *causal chain* structure (as illustrated by the example of the backwards doG of Thunder in chapter 2) or a *common cause* structure. Interestingly, when presented with some of these structures, adults do not always reason according to the Markov assumption (e.g., Rehder & Burnett, 2005; Walsh & Sloman, 2004). But if adults do not reason according to this assumption in simple paradigms where they can be asked fine-grained questions, it seems even less likely that children do reason according to this assumption. In that case, why should we believe that the causal graphical model framework serves as a good starting point for describing children's causal reasoning?

At the outset of our response to this challenge, we want to say that some investigations with preschoolers have shown that they reason according to the Markov assumption for all three of these different kinds of causal structures, not just common effect structures (e.g., Schulz et al., 2007; Sobel & Sommerville, 2009).[1] But the broader challenge from the adult literature has important implications for our understanding of children's causal reasoning, so it is worth discussing in detail.

Consider, for example, a common cause model, such as the one shown in figure 3.1. In this model, event A causes two other events (B and C).

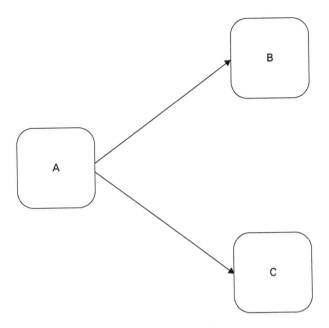

Figure 3.1
A causal graphical model with a common cause structure.

There is a question about how people make predictions when told about this model. Specifically, if we observe event A occur, but then event B fails to occur, should this change our judgment about whether event C occurred? Is our judgment in this case different from the case where we observe only event A occurring with no knowledge of event B? Under the Markov assumption, causes produce their effects independently of each other. More precisely, the Markov assumption says that the probability that C happens given that A happened is the same as the probability that C happens given that both A and B happened. This means that the answer to the questions above should be that these cases are the same; we should make judgements about C's occurrence based only on A, not on B.

However, that prediction does not match how many adults think about these two situations. Several experiments indicate that adults show these robust *nonindependence* effects: Adults change their judgments about whether C should occur based on what they know about B (e.g., Rehder & Burnett, 2005). Nonindependence has been interpreted either as a fallacy of the modeling framework or as evidence that the modeling framework does not provide a good description of causal reasoning.

Leaving aside the possibility that adults and children might engage in causal reasoning differently (see e.g., Liquin & Gopnik, 2022; Lucas et al., 2014, for good evidence for this claim), there is another explanation for nonindependence effects. When researchers tell adult participants about a causal model, either by describing it verbally or by instantiating it in the real world, these participants might not be reasoning about the specific causal model that the researchers intend. Because these models are descriptions of the way children and adults might represent causal knowledge, they are subject to other domain-general processes that could influence those representations. One such process, which we discuss in detail in chapter 10, is imagination. Simply put, when told about or shown causal evidence, reasoners might represent not only those events, but many other possible ways in which those events could be related. This relates to the issue of mechanisms that we discussed in chapter 2: Because any given model could be compatible with many different mechanisms, more information is required to narrow down the set of possibilities, even though both adults and children understand that there must be some mechanism or other by which these events are related.

So let's consider again the common cause model introduced above (figure 3.1). We can instantiate this model with an example (adapted from Buchanan & Sobel, 2014): Alice is using her cell phone to try to call Bob and Charlie. Alice dials Charlie. What's the likelihood that she connects with Charlie? Cellular technology being what it is, the answer is probably not 100%. But it is also not 0%, and (hopefully) it is closer to the former than the latter (otherwise Alice should really switch her service provider). Now what if you know that Alice tried dialing Bob first but could not connect. Do you revise your estimate about Alice's chances of successfully calling Charlie?

You probably said yes. If Alice can't get through to Bob, then there is probably something wrong, and whatever is wrong might affect Alice's ability to connect with Charlie, or possibly anyone. Maybe she just misdialed the numbers. But maybe the cellular signal is weak where Alice is, or her battery is dead, or someone is jamming her signal. It is also possible that something is wrong with Bob's phone—he could be in a dead zone, or his battery died, or he's the target of jammers, or his phone is turned off.

Some of these lines of thinking could lead you (and participants in the experiments cited above) to violate the Markov assumption and to consider events that should be independent as not. That is, maybe people don't reason according to this principle after all.

We can respond to this objection by noticing that, once you start imagining these *somethings* that could be affecting Alice's or Bob's phones, you are no longer reasoning about the simple common cause model depicted in figure 3.1. Instead, you're reasoning about a different model, one with a placeholder for the *somethings* that you assumed were happening with Alice's and Bob's phones. For example, some of the situations that we described would be better represented with a model like that in figure 3.2.

If this is the model that you are using to think about the situation, then you are not necessarily violating the Markov assumption. You have just constructed a different model. In a model that specifies an extra "something" (i.e., another node or set of nodes) between event A and events B and C, the probability of C given A and B is not the same as the probability of C given just A. This point, made by a number of researchers (e.g., Rehder & Burnett, 2005), provides a good way of thinking about how to solve the problem

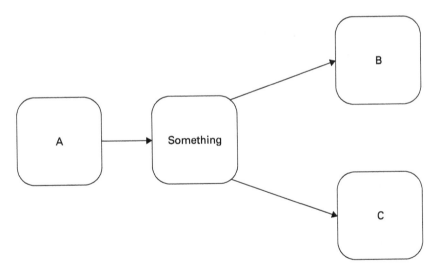

Figure 3.2
Constructing a common cause model with an additional node can explain why people sometimes appear to violate the Markov assumption.

of nonindependence. On the other side of the same coin, Park and Sloman (2013) showed that adult reasoners show nonindependence effects only in certain circumstances. If adults have a good representation of a potential hidden mechanism that might relate observed variables together, they are much more likely to reason according to the Markov assumption.

To address these issues, we extended the causal graphical modeling framework to account for the possibility that children and adults might imagine possible mechanisms for the relations they observe (Buchanan & Sobel, 2014). We called this extension *edge replacement*. In the edge replacement framework, people reason about a set of causal graphical models that posit different representations of mechanisms between each causal relation. That is, for each graph, reasoners simulate a set of possible mechanisms and try to match those mechanisms to the data. In this view, nonindependence is a feature, not a bug. It suggests that people are not reasoning about a simple model that contains only Alice, Bob, and Charlie (or at least, only their phones). Rather, people are reasoning about the mechanisms through which these causal relations might function (i.e., how their phones work). Critically, edge replacement does not say that there's necessarily a single "something" that stands for causal mechanisms. Rather, it provides a rational way that people could generate and reason about a hypothesis space of possible mechanisms—possible "somethings," which may relate to each other in complex ways. An important facet of this framework is that it allows a learner to posit nonobvious causes or side effects that are not part of the initial representation of events. This is also a way to scale up reasoning from the toy examples that we (and others) have often provided to more real-world contexts. We refer the interested reader to Buchanan and Sobel (2014; see also Buchanan et al., 2010) for the computational details and to Buchanan and Sobel (2011), Erb, Buchanan, and Sobel (2013), and Ahl and Keil (2017) for empirical data consistent with this framework.[2]

The important point for our purposes is that the edge replacement framework provides a computational-level description of how to generate possible hypotheses for how events are causally related to each other and a way of evaluating the likelihood of those possible hypotheses. So although psychologists build incredibly simple causal graphical models in their studies, these models could stand for much more complex and more contextualized systems. This framework also addresses the concern that we as human beings almost never construct models that precisely represent the actual causal structure of the world. For example, we tend to like simple models, like the one that states that having a cold causes sneezing. But the actual causal structure that describes this relation is *really* complicated. The edge replacement framework provides a way of explaining how the shorthand of Cold → Sneezing could stand for this actual causal structure.

Besides being of theoretical interest as a potential objection to the causal graphical model framework, discussions of nonindependence are important because they force us to consider the context in which people are reasoning about causal relations. As mentioned in chapter 2, and as illustrated by the cellphone example, context may affect how both adults and children reason about the data they are given. This is relevant to our discussions of the development of scientific thinking because one of the big differences between measures of young children's causal reasoning abilities and measures of their scientific thinking abilities is their use of context (or their *contextualization*). As we discuss in more detail in chapter 4, causal reasoning studies often use blicket detectors or other machines or vignettes in which the mechanisms are stripped down and in which relatively few hidden causes are assumed or implied. Scientific thinking studies, in contrast, can have many hidden causes or can imply many different types of hidden structures. We believe that these differences may explain some of the perceived conflicts between the results using these different types of tasks.

As an aside, the idea that reasoners are constantly thinking about potential mechanisms, particularly when they observe probabilistic events, could lead to interesting explanations for various cognitive phenomena. One example of such a case is the *gambler's fallacy* (Tune, 1964; Tverksy & Kahneman, 1971), in which reasoners misunderstand the independence of random events. People tend to think that, because a certain number has not hit on a roulette wheel for a while, it's "due," hence more likely to come up on the next spin. Similarly, if you hear that a friend won the lottery, you assume it's the friend who has been playing each week for ten years as opposed to the friend who played for the first time. In both cases, as in the cellphone case above, people posit (faulty, incorrect, improper) mechanisms that lead to these inferences.

Another example is the *illusion of explanatory depth* (Rozenblit & Keil, 2002), in which adults believe that they understand a concept better than they actually do. In studies of this phenomenon, adults are asked how well they understand how common objects work, like bicycles or toilets. Most people rate their knowledge fairly high. But these ratings drop precipitously after they're asked to actually explain how a bicycle or a toilet works, which most adults are unable to do. Second graders also show this effect (using modified procedures, see Mills & Keil, 2004). The tendency to posit extra "somethings" in causal models as stand-ins for our mechanistic knowledge can also explain these results, because our constant thinking about mechanisms might lead us to believe that we understand causal structures more than we actually do (see Alter et al., 2010, who make a similar argument, although not relying on this computational framework).

Where Does a Concept of "Cause" Come From?

The previous section was concerned about objections to the causal graphical model framework from a line of research in adult cognitive psychology. We suggested that there are ways of integrating these objections into rational constructivism, which uses this framework to formalize its representation of children's learning. An equally important concern comes from studies on the causal reasoning capacities of very young infants, which raise the question of where a concept of "cause" comes from. Are infants born perceiving the world in terms of causal structure? Or are the reasoning mechanisms available to preschoolers and school-age children different from the ones available to infants? And if the latter view is correct, then how should we describe the nature of this discontinuity?

Rational constructivism answers these questions by suggesting a developmental continuity. In this view, children are born with some aspects of theories—representations of causal structure—which means that the inherently causal nature of these representations is available to children from the earliest ages. Some versions of this theory have suggested that, while the nature of these representations changes with development, the basic concept of what a cause is stays the same. This aspect of this theoretical approach is similar to contemporary nativist theories, which suggest that causality is piece of core cognition (e.g., Carey, 2009).[3]

The evidence for this position comes from studies on the inferences that infants make about causal relations among events. Babies around 6 or 7 months old can recognize various aspects of contact causality, including registering certain configurations of perceptual features as causal (e.g., Leslie & Keeble, 1987; Oakes & Cohen, 1990) and understanding the relations between the speed and force of objects (e.g., Kominsky et al., 2017). They can also recognize that there are differences between self-moving agents and inanimate objects, namely that the former can cause the latter to move but not the reverse (e.g., Saxe et al., 2005; see also Muentener & Carey, 2010, and Newman et al., 2010, for related findings on slightly older infants).

In summarizing these findings, Carey (2009) writes that the fact that "infants take into account the causally relevant properties of the participants in events in their representation of these events provides evidence that they are making causal attributions" (p. 488). And we agree. By about 6 months of age, infants have a variety of sophisticated causal reasoning capacities. But where we disagree is that the concept of a cause is innate in Carey's sense of the word: shared with other animals,[4] not learned, and unchanging throughout the process of development.

We would like to suggest a different possibility. We question whether infants at very early ages (i.e., newborns) have the capacity to engage in causal inferences, as opposed to merely registering statistical regularities among events (following Haith et al., 1993; Kirkham et al., 2002). As we suggested in chapter 2, one of the big challenges for a description of causal reasoning is that learning the associative relations among events gets you pretty far.[5] That is, it is possible that infants represent causal relations first by recognizing and using associative information. In the second half of the first year of life, they move to appreciating those statistical regularities in terms of more genuinely causal (not merely associative) structure. Those representations now support the interventions that are part of the causal graphical model framework.

This hypothesis is consistent with a multitude of findings that suggest that, while infant causal perception undergoes development related to information processing (e.g., Cohen et al., 1999), it is mostly in place by the second half of the first year of life (Leslie & Keeble, 1987; Oakes & Cohen, 1990; Saxe et al., 2005, 2007). Indeed, most of the citations that Carey (2009) uses to argue that causality is an innate concept come from infants in the second half of the first year of life.

What kind of evidence would differentiate between the claim that infants register relations among events in terms of their statistical regularity and the claim that they register events in terms of causality? One is their use of the Markov assumption. Some work that we did on infants' causal inference suggests that 8-month-olds respond to the backward blocking sequences we described in chapter 2 in the same manner as 4-year-olds (Sobel & Kirkham, 2006), but 5-month-olds treat these sequences in terms of their associative relations (Sobel & Kirkham, 2007).

Another way to differentiate between these claims is with reference to the fact that the causal graphical model framework supports interventions. In this framework, interventions force variables in the network to take on a particular value. Psychologically, interventions are typically actions on the world that bring about a situation. It is important to consider what role the development of action abilities might play on the emergence of a causal reasoning system that places such an emphasis on actions.

During the second half of the first year of life, infants develop the capacity to coordinate their actions in complex ways. For instance, at 7 months old, infants can reach for an object directly in front of them, but they struggle when they have to coordinate their actions with barriers or obstacles in the environment to make that reach. As an example, Diamond and Gilbert (1989) showed that infants develop the capacity to reach for objects in a coordinated fashion between the ages of 7 and 11 months. As part of this study, they placed a LEGO brick in an open, transparent box. When the brick was placed in the middle of the box, infants across all of the ages that they tested could reach for it without difficulty. Similarly, when the brick was placed outside of the box, adjacent to it, infants at all ages could reach for it without difficulty. However, when the brick was inside the box, adjacent to the side of the box closest to the infants, the youngest infants struggled with their reaching. There was a clear developmental trajectory in their ability to reach for the brick in this configuration between 7 and 11 months. The trouble with this particular setup is that infants have to coordinate their movement—they have to reach into the box from the top and then change the direction of their reach to get to the brick, which requires inhibitory demands that are developing during this time (Diamond, 1991a, 1991b). Critically, even the youngest infants had no difficulty when the brick was in the same position, but the box was tilted, such that they did not have to change the direction of their reach.

These data suggest that, during the second half of the first year of life, infants are developing domain-general capacities, which facilitate their ability to coordinate their actions to accomplish a goal. We speculate that, in addition to this developing inhibitory control, the concurrent maturation of the supplemental motor cortex and dorsolateral prefrontal cortex (the neural systems that Diamond, 1991a, hypothesized were responsible for pattern of results described above) might serve to integrate children's representation of the structure of their own actions with their understanding of statistical regularity, allowing them to construct causal representations. That is, prior to infants being able to coordinate the ways in which they intervene on the world (e.g., by reaching and grasping), they may see the world in terms of statistical regularity. Being able to generate actions on the world may then afford the infant more of a fully causal interpretation (i.e., I can change the state of the world through my actions).

An analogy for this idea comes from an interpretation of a body of research on children's capacity to understand that actions are goal-directed. As an example of this kind of research, a study by Woodward (1998) habituated 5- to 9-month-old infants to a display in which a hand reached to one of two objects. After habituation, the locations of the two objects were switched. Infants saw the hand reach for either the same object (via a new path) or the other object (via the same path). Overall, infants looked longer when the hand reached for the other object (that is, when it reached to the same location as before). Woodward interpreted these results as demonstrating infants' abilities to encode the actions of others as being goal-directed: The agent (represented by the hand) intended their action to reach a particular object and not simply to reach toward a particular point in space.

What is interesting about this capacity is that younger infants—3-month-olds—do not interpret this experiment in the same manner. Instead, they look equally to the same object (new path) and same path (new object) events. This suggests that they do not encode the relation between the actor and the goal of the action (see Sommerville et al., 2005, footnote 1). But Sommerville et al. (2005) argued that these very young infants could recognize this relation, they just needed the experience to do so. Their argument is similar to the one we made above regarding the link between self-generated actions and an understanding of causality. Specifically, 5-month-olds have enough coordinated motor development to reach for objects and grasp them themselves. Their reaching ability is not adult-like, but they can begin to manipulate objects in their environment. Three-month-olds? Not so much. They can flail their arms about, but they cannot make a coordinated reach for an object. This difference in motor abilities could potentially explain why 5-month-olds in Woodward's experiment interpret the hand's reach as goal-driven, but 3-month-olds do not: They themselves cannot coordinate reaches for objects in a goal-directed way, so they do not appreciate goal-directedness in others.

This argument implies that, if 3-month-olds were given experience with manipulating objects, they would become able to register that reaches are goal-directed. That is, the infants' own reaching actions could help them to appreciate that others' actions have goal-directed structure. To test this, Sommerville et al. (2005) sewed Velcro strips onto teeny-tiny mittens and gave the 3-month-olds two minutes of experience manipulating a Velcro block world with the mittens on. When using these "sticky mittens," as mentioned in chapter 1, these babies' flailing attempts to pick up objects would actually successfully allow them to manipulate the block—or at least to move it around in space. After the mittens experience, the infants were presented with a version of Woodward's experiment described above, in which the actor was wearing the same kind of Velcro mitten as the infants had worn and reached for a Velcro block that looked similar to the one that the infant participants had themselves played with. These infants picked up on the goal-directedness of the reach, and looked longer after habituation when the actor reached for the new object than for the same object on the new path—just like the 5-month-olds did.

One can interpret these data as suggesting that the motor experience of being able to reach for objects provides infants with the ability to register that those actions are goal-directed, and thus to infer that similar actions by others are goal-directed. While that understanding naturally develops between the ages of 3 and 5 months as infants learn to reach and grasp objects, it can be accelerated artificially through experience.[6] In turn, this suggests that a concept of causality might develop out of infants' growing understanding that their actions can affect the world in a variety of ways.[7] Given these findings, we suggest that infants are not necessarily born with the concept of "cause"; they develop this concept during the first year of life, based on their observations of the world and especially on their experiences with acting on the world.

Although we think that the issue of where a concept of "cause" comes from is interesting in its own right, we highlighted this issue here for another reason. In chapter 4, we suggest that the causal reasoning system that children have during the preschool years develops during the early elementary-school years; it changes in fundamental ways that might facilitate success on some scientific thinking measures. As such, it is important to conceptualize children's causal reasoning as a system that itself undergoes significant change throughout the lifespan, starting in infancy.

Active Learning

An interesting set of findings in from the psychological literature shows that people overemphasize data that they themselves generate. Much like the infants in the sticky mitten experiment described above, both adults (Lagnado & Sloman, 2004; Sobel & Kushnir, 2006; Steyvers et al., 2003) and children (Sobel & Sommerville, 2010) benefit from engaging in actions themselves when learning causal structure, particularly when compared to cases in which they simply observe others' actions. But young children, in particular, can be overinfluenced by their own actions; they have been shown to rely on a single data point that they themselves have generated as opposed to considering the aggregate data that they observed (Kushnir & Gopnik, 2005; see also Sobel & Letourneau, 2018). Moreover, numerous studies suggest that various aspects of young children's cognition are influenced by their movement and actions (see Lillard, 2004, for a review).

This tendency to favor data that one has generated oneself poses a problem for the causal graphical model framework as a description of children's inferences. As we suggested in the previous section, an important aspect of this framework is intervention—the idea that actions on the world can support different inferences than simply observing data. Interventions are important because data from interventions can provide conditional independence information that may be difficult to detect by observation alone. For example, we can observe the correlation between a rooster crowing and the sun rising, but making the rooster crow at 2 am will not cause the sun to rise (example taken from a 1913 manuscript of *Theory of Knowledge* by Bertrand Russell, as cited in Pearl, 2000). However, the framework does not distinguish between data that come from one's own interventions and data that come from someone else's. Given that learners do seem to process these different types of data differently, this tendency could undermine the ability of the framework to serve as a model for how learning works.

Some hints for how to address this problem comes from work on adults, who—although they do tend to be influenced by self-generated data during causal learning—might be doing so for rational reasons. For example, Sobel and Kushnir (2006) suggested that the reason adults learn better from their own actions than from observing the same data generated by others is that, when one generates actions, one can test one's own hypotheses. This study showed that adults were better learners when they generated their own interventions on a causal system than when they observed another adult generate the same interventions, in support of the work cited above. But this study also showed that the number of critical interventions that learners made (interventions that generated the relevant conditional probability information necessary to learn the particular causal structure) predicted adults' learning, but only when they generated their own actions. When adult participants observed the same sequence of data, there was no relation between the quality of the data sequence and adults' learning. These results suggest that, while adults may appear to simply favor data generated by their own interventions, these data may be genuinely more informative for them, hence it may not be an error to rely more heavily on them.

To test this hypothesis with children, Sobel and Sommerville (2010) presented 4-year-olds with a situation in which they could learn causal structures from their own actions as opposed to observing an adult play with the machine. Like the adults in Sobel and Kushnir (2006), children were better at describing the causal structure in the former condition, and were especially better when those actions discovered novel information. We'll discuss findings like this more in chapter 4, but for now we want to note that there is also neuroscientific evidence that suggests that self-generated action recruits different neural systems for learning in children. James and Swain (2011) introduced 5-year-olds to novel actions on novel objects, which the children either performed themselves or observed another person perform. Each action was given a novel label. When these children were exposed to the stimuli and novel labels, they showed greater recruitment of motor regions for the labels that named actions that they had performed themselves as opposed to the ones they had observed. These data suggest that children conceptualize their own actions differently than the actions of others (see also Sommerville & Hammond, 2007).

This work has close ties to a body of literature examining the role of play in development, much of which asks a similar question: whether children learn better when adults provide them with direct instruction or when they are allowed to explore the world on their own (Klahr, 2000; Klahr & Nigam, 2004; Kuhn, 1989; Kuhn & Dean, 2005). With respect to this question, there is a general notion that *guided play* is a good compromise between these two approaches to learning (Mayer, 2004; Weisberg, Hirsh-Pasek, et al., 2013, Weisberg et al., 2016). For example, in a large-scale meta-analysis of learning outcomes, Alfieri et al. (2011) argued that unguided ("free") play typically resulted in worse learning outcomes than direct instruction, although adding some kind of adult guidance to the play improved learning.

Similarly, research on *scaffolding* (Wood et al., 1976) suggests that parents (or adults more generally) can teach children to solve problems by presenting them with the parts of the task that they believe children are capable of doing successfully. The adults then build them up to engage in more and more complex behaviors (e.g., Connor & Cross, 2003; Meins, 1997; Pratt et al., 1988; Wood & Middleton, 1975; see Mermelshtine, 2017, for a review). Of particular importance is the contingent interaction that adults generate during these interactions. For scaffolding to be effective, adults must react to children's actions to facilitate learning. Scaffolding allows children to take an active role in completing aspects of the problem that they understand, while also being exposed to more parts of the problem that they might be able to tackle the next time it is encountered. Research on both guided play and scaffolding assert a crucial role for the child's own self-guided actions in learning.

We investigated this issue as part of a large-scale study of parent-child interaction while families played at a gear exhibit at three children's museums around the United States (Callanan, Legare, Sobel, et al. 2020). After families played together, children were asked to participate in a set of activities that measured their causal knowledge about gears, while parents completed demographic questionnaires and other measures about their beliefs toward science and play.

While this work examined many facets of children's interaction and learning, here we want to highlight one set of findings that is pertinent to this discussion: How did children explore when they played? To examine children's actions in detail, we looked for a set of behaviors that we thought a priori would be important to learning. We called these behaviors *systematic exploration*. These behaviors primarily included building a gear machine and then testing that machine by spinning it. Perhaps unsurprisingly, we found that the proportion of systematic exploration behaviors that children engaged in increased with age. Independent of age and of many other social factors like household income and parental education levels, this proportion was related to children's persistence in problem-solving when things went wrong during their play. This proportion was also related to children's abilities to successfully engage in causal reasoning about gears, as measured by our post-exploration activities. Systematic exploration generated by children themselves therefore seems to be an important set of behaviors for children's learning.

On the surface, there is a clear interpretation of this result (and of many of the results that show a benefit for children's learning from their own interventions) that fits with the rational constructivist framework: How children explore the exhibit relates to the way in which they interpret the exhibit content. Because rational constructivism is committed to children's learning (i.e., updating the causal representations of their theories), it should not be problematic that children can do so through their active engagement with the world, or even that such exploration is an effective way for them to learn. So although it may be difficult for the causal graphical model framework to take this into account, we believe that this framework could be amended to include information about where the intervention data is coming from and to reflect more accurately children's (and adults') biases for their own actions.

But this is only half of the story. The idea that self-generated action is an effective way to learn could potentially be accommodated by rational constructivism, but it is more difficult for rational constructivism to account for the ways in which parents' and children's interactions at this exhibit affected the children's learning. Specifically, our analyses focused on the explanations that parents provided while their children were playing. The overall proportion of parents' causal language had no bearing on how children played; it didn't relate to the proportion of systematic exploration that children engaged in. Nor did parents' proportion of causal language relate to children's causal reasoning when they were tested on their understanding of gears. But what did relate to children's systematic exploration was the *timing* of when parents generated causal language. Specifically, if parents began their causal utterance while children were connecting gears together in their exploration, children were more likely to complete this systematic exploration sequence than if parents began the causal utterance after the gears were already connected, or before or after this entire sequence. Although parental language in the aggregate might not relate directly to children's causal reasoning, parents do influence whether children engage in behaviors that might be related to their causal learning.

To accommodate findings like these, rational constructivism needs to be integrated with the idea that children are not just internally constructing meaning, representations, and concepts through their actions on the world. Rather, learning from self-generated action occurs within the context of an interactive social framework (e.g., Vygotsky, 1978; Wood et al., 1976). We suspect that the best way to reconcile these findings with the tenets of rational constructivism is to suggest that children have internal mechanisms for learning, but that those mechanisms are affected by various different kinds of social and cultural interactions.

Unfortunately for this suggestion, sociocultural theories have typically been cast in opposition to constructivist approaches (see e.g., Cobb, 1994). There are two general reasons for this. The first is theoretical; the two theories posit different mechanisms for how learning is accomplished. Specifically, there is disagreement as to whether knowledge is represented internally by the mind or co-constructed by the interaction (see e.g., Rogoff, 1990); we discuss this idea in more detail below. The second reason for the opposition between these theories is more practical, in that the views often lead to different educational approaches, which emphasize different cultural practices (e.g., Cobb, 1994; Van Oers, 1990). Nevertheless, because both theoretical approaches have important insights to offer, we believe that constructivist approaches can—and should—be integrated with sociocultural approaches to cognitive development (see Callanan & Valle, 2008; Legare et al., 2017). We begin to take up this challenge in the next section.

As an important note, we think it is unlikely that the parents in our museum study explicitly timed their causal utterances to match children's behaviors during their interaction. This study is correlational and does not imply that parents controlled when they generated causal language in an attempt to facilitate children's causal learning. No demographic factor that we measured related to whether parents generated causal language at that particular point in time. In contrast, parental education level related to whether they generated causal language *after* children completed a systematic exploration behavior. This suggests that some parents might notice the behavior itself and react to it with causal language, but this reaction did not relate to children's exploration. Instead, parental interaction with children during play might create opportunities for parents to help children generate co-constructed interpretations of their own actions.

The Social Nature of Learning

As noted in the previous section, another source of challenges to rational constructivism comes from a large literature on the *social* nature of learning, noting that learning is largely (perhaps essentially) a social process. As such, blicket detector experiments and similar bodies of work, on their own, do not provide a complete picture of children's learning, hence the causal graphical model framework may not be able to fully account for this process. Here, we review two bodies of work that examine children's learning in social contexts. In both cases, we attempt to synthesize these results together with rational constructivism as well as try to explain how rational constructivism needs to evolve to take these results into account.

Trust in Testimony

For the past fifteen years, in response to research on children's causal learning, a body of work within cognitive development has argued that learning from intervention and observation is not particularly efficient. While one can (re)discover causal concepts from observation and intervention, many of those discoveries take quite a bit of time and data. As noted in chapter 1, preschoolers have a great deal of knowledge about the physical, psychological, and biological worlds, and it is unlikely that they learn all of this information from observation and intervention alone. More to the point, children possess many beliefs for which they have no evidence. Children have never seen a germ or a ghost, and yet they tend to believe that both of these entities are real. Further, they have different kinds of beliefs about these two particular entities, and they have differential beliefs about scientific and religious entities in general (e.g., Harris & Koenig, 2006).

The thrust of this argument is that children cannot gather enough evidence about the world simply through observation or solitary play (Harris et al., 2006; see also Keil, 2010); they are not "stubborn autodidacts" (Faucher et al., 2002, p. 341; Harris, 2002). Rather, they need social transmission of knowledge for learning to occur. Specifically, children's beliefs about the world are acquired not just from observation and interaction with the environment, but from social transmission and communication (e.g., Csibra & Gergely, 2009; Harris, 2002, 2012; Vygotsky, 1978). This process is often referred to as children's "trust in testimony" (Harris, 2002), and studies in this area have shown how children learn directly from others' explicit verbal statements (see Sobel & Finiasz, 2020, and Tong et al., 2020, for two meta-analyses of this literature).

Children are also influenced by the pedagogical nature of the environment they are in, and they learn more when others are teaching them directly and presenting them with data in a way that ensures their attention (e.g., Butler & Markman, 2012, 2014). Such pedagogical inferences can have a cost, however. In particular, children assume that teachers tell learners all they need to know, so when teachers commit sins of omission in their teaching, children's exploration is negatively affected (Bonawitz et al., 2011), as are their beliefs about the teacher's credibility (Gweon et al., 2014). These data generally suggest that, while children can learn from observation and interaction with the world, they are also learning from the social interactions that they have (Harris et al., 2006).

These studies illustrate how this research tradition not only emphasizes communicative norms, like word meanings, but also constraints on the acquisition of social knowledge, like cultural conventions and norms. Knowing from whom to learn and under what circumstances to learn, and then actually learning from others, is the basis of cultural knowledge and transmission (e.g., Bergstrom et al., 2006; Harris & Koenig, 2006; Kline, 2015; Mascaro & Sperber, 2009).

Given this, an important question is whether children simply believe what others tell them, or whether they learn from others in a more judicial manner. That is, are children credulous or skeptical, and does their credulity or skepticism change with development and experience with informants? Traditionally, young children have been thought to be overly credulous, trusting what others tell them (e.g., Piaget, 1930). This assumption, however, has not been widely supported. Across many experiments, it has been shown that preschoolers can track the reliability of other informants and use that information to make judgments about their epistemic competence and about whether to learn from them. For example, preschoolers will not learn a label for a new object if that label is said by a speaker who has previously labeled objects inaccurately (e.g., Clément et al., 2004; Koenig et al., 2004; Koenig & Harris, 2005). *Selective learning* capacities are an important part of what makes children such good learners.

But how do children's selective learning capacities develop? One major theory about children's trust in testimony posits that children are credulous early in development—they have what Jaswal and colleagues describe as a "default bias to trust" (Jaswal et al., 2010, p. 1541). In this view, what develops is the capacity to discount inaccurate testimony in favor of other sources of information, such as one's own direct observations or testimony from a more reliable source. That is, learning from others requires assessing whether another person is providing accurate information and discounting that information when one judges the other person's statements to be false. Children's developing cognitive control thus underlies their capacity to inhibit others' inaccurate testimony (Harris et al., 2018; Jaswal et al., 2014). A similar—although separate—proposal was put forward by Mills (2013), who suggested that the developing capacity to track others' reliability involves the development of the metacognitive capacity to register skepticism. This latter proposal also suggests that children have an initial default bias to trust, but what develops is a broader and more metacognitive conception of skepticism—that others' statements can be false, regardless of their intention.

Both of these proposals have a great deal of merit. While 3- and 4-year-olds are capable of discerning between reliable and unreliable sources of knowledge, a study by Krogh-Jespersen and Echols (2012) suggested that 2-year-olds trusted an accurate, inaccurate, and ignorant informant equally. The 2-year-olds did not discount the inaccurate informants' information, although older children did. This study is important because it supports the argument that children have a default bias to trust.

But it is also important because it used a somewhat different method for studying this topic. So far, the work we have discussed that examines whether children can learn selectively introduced children to multiple informants who differed in some way. In some cases, they differed on epistemic competence (e.g., one was an accurate source of knowledge and the other was not), and in other cases they differed on a social or physical cue that could indicate differences in epistemic knowledge (e.g., one was an adult while the other was a child, or one spoke in the child's native language with a native accent while the other spoke with a nonnative accent). But this is a not a particularly ecologically valid paradigm. How often do children observe two people generating different labels for the same object, particularly for familiar objects (that's a stapler, no, that's a cup)? A more ecologically valid method might be to introduce children to a single informant and vary the epistemic or social information that that informant generated between groups of children.

That is the method used in Krogh-Jespersen and Echols (2012). Children in that study were asked about the testimony of only one informant, who in the past had been accurate (or inaccurate or ignorant). These researchers found that 2-year-olds tended to use the labels generated by an interlocutor, regardless of that individual's epistemic status. Sabbagh and Baldwin (2001) also used a single-informant paradigm and found that the 3- and 4-year-olds they studied learned word meanings better from a knowledgeable than from an ignorant informant. So 2-year-olds might have more of a default bias to trust, but there is development away from that default bias during the preschool years.

That very young children can learn selectively from conflicting informants, but not from a single informant, speaks to the possibility that selective learning is indeed governed by inhibitory control factors. It is more difficult for young children to integrate past information with an informant's current statement than it is to judge between two informants who provide this information at roughly the same time. In support of this argument, some studies suggested that children were more likely to respond selectively when they observed two informants disagree about the objects' labels, as opposed to when they observed one informant simply generate inaccurate labels (Koenig & Woodward, 2010; Vanderbilt et al., 2014). More directly, Jaswal et al. (2014) showed that preschoolers' capacity to discount inaccurate information—particularly when it conflicted with their own observations of a situation—correlated with their inhibitory control capacity, measured by a separate battery of tasks. Inhibitory control (or cognitive control more generally) does seem necessary to learn from others selectively.

We agree that inhibitory control or other executive function abilities are necessary to make accurate judgments about others' epistemic competence; in fact, this makes a nice parallel to our arguments above about motor skills being necessary prerequisites to infants' abilities to understand others' goal-directed actions. However, it is questionable to us whether inhibitory control is *sufficient* for explaining these findings. In particular, all of these studies ask children to make a choice, for instance between the labels provided by two speakers, or between two novel words as the label for a novel object. These measures seem reasonable for older preschoolers, but they might introduce demand characteristics, particularly for children at the younger end of the age range, like the 2-year-olds discussed above.

There are other techniques for eliciting responses from children this young, many of which rely on measuring the amount of time children spend looking at different displays (called looking time paradigms). To illustrate how these methods can be used to determine how toddlers decide whom to trust, we (Luchkina, Sobel & Morgan, 2018) presented 18-month-olds with speakers who were either accurate or inaccurate. Then, one of those speakers labeled a set of novel objects with novel labels. This speaker then showed the child one of those labeled novel objects together with another novel object, which the child had not yet seen. Children were asked "Where's the [label]?" where the label used was the novel label that the speaker had previously generated for the first object in the test set. When that speaker had been accurate, the 18-month-olds stared more at that object as opposed to the other one, suggesting that they had accepted this label. When that speaker had been inaccurate, they stared equally at the two objects, suggesting that they did not know which object the speaker was labeling. That is, children seem able to track the reliability of others at very early ages. They are not just biased to trust, but rather can be evaluative. However, they may have difficulty displaying these abilities in all cases, illustrating that children's developing cognitive control abilities still have an important role to play.

There is an important control in this study, which further illustrates these children's abilities to think critically about the information they receive. In this control condition, we ran essentially the same procedure with a new group of 18-month-olds but changed one aspect of the design. Usually, in measures of selective learning, children are shown familiar objects, and the informant says "That's an X," where X is either an accurate or inaccurate label. That's what we did in the study described above. But in our second experiment, each informant asked "Is that an X?," where the label that the informant generated was either accurate or inaccurate. The critical idea is that questions do not convey information about epistemic competence, so using an inaccurate label in a question is not a sign of incompetence, like it is in a statement. Children registered this, and they did not learn from either informant in this case. This is important because, when the speaker generated the accurate label in question form, they have generated the same associative information as in the statement case; the speaker has said the correct label in both cases. The fact that children learn differently from statements than from questions suggests that even very young children are not simply using the associative properties of the language they hear. Rather, they undergo a more sophisticated process of evaluating the pragmatics of the situation in order to judge a speaker's accuracy.

This conclusion is related to a set of findings generated by Sabbagh and Shafman (2009), using a single-speaker paradigm. The 4- and 5-year-olds in this study could both remember a knowledgeable speaker's label for a novel object and also use that information to make an inference about meaning; they could state that the object had that label. When a speaker expressed more ignorance about the label, the children in this study could remember the speaker's label, but did not necessarily endorse the object as having that label. As in Luchkina et al. (2018), these children did not simply use a speaker-object-label association to learn the meaning of words (see also Mangardich & Sabbagh, 2018). In addition to inhibitory control, which plays a role in helping children consider a variety of information in their decisions about whom to trust, there are also important situational constraints on learning. Causal knowledge is necessary for judging whether to use the information that others generate.

To appreciate this more integrative perspective, consider work by Kushnir et al. (2008), who varied whether an informant was knowledgeable or ignorant about a novel toy (an epistemic constraint) and also whether the source was permitted to use that knowledge in performing an action (a situational constraint). Specifically, they showed 3- and 4-year-olds two puppets and a machine (a blicket detector, although it was never called that) and a set of blocks. Children were told that one of the puppets knew which blocks activated the machine and the other puppets did not. Across their experiments, children observed the puppets either intentionally pick blocks to make the machine go, observed the puppets pick blocks while blindfolded, or picked two blocks themselves for the puppets to put on the machine. In all three cases, the two blocks were placed on the machine together and the machine activated. Children were asked which block was more likely to be efficacious. Only when the puppets picked the blocks intentionally did children say that the knowledgeable puppet's block was the efficacious one; in the other two cases, they responded at chance.

This study demonstrates that children do not seem to be relying on a single cognitive capacity to make decisions about the value of information. Rather they are integrating both situational and epistemic knowledge when tracking and making inferences about others' accuracy. In turn, this implies that both developing cognitive control abilities and developing knowledge about epistemic states (in this case, the knowledge that one has to have perceptual access in order to intentionally choose an efficacious object) are necessary to explain children's behavior.

Similarly, other work has shown that preschoolers still have access to low-level associative learning mechanisms underlying certain aspects of their ability to track others' reliability (Luchkina et al., 2020), which can be described by dual process accounts of selective learning (e.g., Hermes et al., 2018). Just as children have early-emerging capacities for statistical learning and for appreciating regularity in data, children also have early-emerging capacities to track others' accuracy. Those capacities become integrated with new information and can be constrained by other domain-general cognitive processes like children's developing cognitive control. As children's knowledge changes, it has cascading effects on how they evaluate and learn from the information generated by others.

In general, then, this body of work on children's understanding of testimony and the relative accuracy of different speakers illustrates one way in which social factors influence children's learning. Although children do learn from playing alone, they also learn from playing with others (e.g., Gauvain & Rogoff, 1989). They learn from observing others (e.g., Williamson et al., 2010), but they also ask questions and seek explanations (e.g., Callanan & Oakes, 1992, Chouinard, 2007; Frazier et al., 2009; Mills et al., 2010; Mills et al., 2011). And while children learn some ideas, concepts, and skills readily from direct instruction (e.g., Klahr & Nigam, 2004; Markson & Bloom, 1997), other ideas, concepts, and skills take significantly more time, particularly when that instruction runs contrary to direct observations that children themselves make (e.g., DiSessa, 1993; Vosniadou & Brewer, 1992).

It is unlikely that such learning is governed by two separate processes— one for (relatively solitary) causal learning and the other for social learning. Understanding children's learning thus requires a framework that can encompass both of these situations. Importantly, such an account should make systematic predictions regarding when and from whom children learn and when and from whom they do not. Moreover, such an account should predict what happens when multiple sources of information interact or conflict.

We believe that rational constructivism could serve as such an account because it can be extended to capture children's selective social learning

(Sobel & Kushnir, 2013). Much like children have causal reasoning systems dedicated to integrating their existing knowledge and beliefs with observed data, children also have domain-general capacities to track statistical information generated by other people (e.g., accuracy information). They can integrate that accuracy information into what they know about others and about the situation to make inferences about the reliability of a person as a source of novel information. Rational constructivism should thus be expanded to consider not only the kind of information that children receive, but also the social contexts in which that information is presented to them.

Integrating Cultural and Sociocultural Approaches with Causal Learning

An appeal of rational constructivism is that it emphasizes how learning mechanisms are part of children's cognitive development: Children observe, interact with, and interpret the world around them. Children also process social information as they observe and interact with the world. In this theoretical perspective, learning is often reduced to information processing, following metaphors from symbolic AI that have dominated cognitive science (e.g., the mind is a computer; Miller, 1956; Turing, 1950). Turing actually considered this, suggested that "instead of trying to produce a programme to simulate the adult mind, why not rather try to produce one which simulates the child? . . . Presumably the child-brain is something like a note-book as one buys it from the stationers. Rather little mechanism, and lots of blank sheets" (1950, p. 456). Turing did not use causal graphical models or Bayesian inference (although some suggest that Turing used Bayesian methods in his code-breaking, see Good, 2000), and the algorithms that researchers use to model learning today are different. But this idea—that we can model learning as the internal workings of the mind, particularly of a child's mind—has not fundamentally changed, at least in certain parts of cognitive science.

In contrast, sociocultural theories of cognitive development suggest that children's learning is not mainly about information processing. Rather, learning should be understood as being contextualized in social activities and cultural practices (Gauvain & Perez, 2015; Vygotsky, 1978), meaning that there is not a specific algorithm that results in particular outputs when given particular inputs. For example, Rogoff (2003) describes children as developing "within cultural communities. . . . Their development can be understood only in light of the cultural practices" (pp. 3–4). In this account, developmental processes are dynamic (both nonlinear and multidimensional)[8] and are dependent on cultural practices and norms, and therefore may not be able to be captured by the modeling frameworks that we have been espousing.

Given this, a fundamental challenge for rational constructivism is that children's learning is embedded in their culture, which may bring about profound individual differences based on what Gutiérrez and Rogoff (2003) call a "repertoire of practice." For instance, we argued above that children and adults learned better from their own actions. But the studies that we used to draw that conclusion used samples of convenience composed mainly of white, upper-middle-class families in so-called Western cultures. Learning from self-generated actions might have been encouraged by those participants' repertoires or by the researchers' beliefs that their own cultural practices are the norm (Medin et al., 2010).

We do not yet know whether these findings generalize across other practices that do not emphasize self-generated action. Rogoff and colleagues, for example, studied Mexican communities in which children are included in a range of activities from which children in European-American communities would be excluded. The cultural practices of such communities might promote different observational practices, attentional practices, or information-seeking practices on the part of children, which are all related to their learning (e.g., Mejía-Arauz et al., 2005; Rogoff et al., 2003; see also Correa-Chávez & Rogoff, 2009).

In general, work on children's causal reasoning and scientific thinking has a markedly WEIRD focus, tending to study cultures that are Western, Educated, Industrialized, Rich, and Democratic (Henrich et al., 2010). One exception is work by Shultz (1982), who tested a group of children in Mali on two of the same experiments he presented to children in Montreal—though we note that it is potentially problematic to assume that a task developed within one cultural context will straightforwardly translate to a different context (see Gutiérrez & Rogoff, 2003). Shultz's experiments tested children's understanding of what he called "generative transmission," the idea that causal relations require some form of mechanistic force to transfer from cause to effect. This work—and the broader idea that there should not be many cross-cultural differences in certain kinds of causal reasoning—was formed as a response to archaic beliefs advocated particularly by Levy-Bruhl (1926) about the absence of causal reasoning capacities within various non-Western cultures.[9] Critically, in direct opposition to this view, Shultz (1982) found numerous similarities between the two samples and emphasized the importance of this similarity: "This provides considerable support for the notion of psychic unity as applied to causal reasoning, and is consistent with the hypothesis that causal reasoning is a very basic mental function not easily influenced by even substantial cultural variations" (p. 39).

Wente et al. (2019) used a similar approach to support this conclusion. These researchers tested a group of children and college-age adults in Peru on whether they could infer the causal structure of a system that presented either conjunctive or disjunctive causality (similar measures as used in Lucas et al., 2014). They found similar results in their Peruvian sample as in their Western sample: Children and adults in both cultural contexts made similar inferences about the disjunctive causal relations, but both groups of children were more likely than both groups of adults to succeed at making inferences about the conjunctive ones.

Although this study and the one done by Shultz (1982) show some cross-cultural consistency in causal reasoning, we should be cautious about drawing that conclusion broadly. Reasoning tasks developed within a particular cultural frame in order to assess a particular group of individuals might be understood differently in a different frame and by different groups of individuals (see Novaes, 2013; Scribner, 1975).

Other research studies that document the effects of cultural practices on children's learning have investigated science learning in the classroom (e.g., Bang et al., 2012; Hudicourt-Barnes, 2003), classroom-based practices more generally (Chavajay, 2006; Davis et al., 2021; Gurven et al., 2017; Heyneman & Loxley, 1983), and the ways that parents and children interact with each other based on individual differences within a culture. For example, Solis and Callanan (2016) examined two groups of Mexican-heritage families, where the parents differed in their level of formal schooling. Parents and children engaged in a science-related activity about floating and sinking. The children of the parents with higher levels of formal schooling asked more questions about the procedure, while the children of parents with lower levels of formal schooling asked more questions about the conceptual structure of the task. These researchers' explanation of this difference is that the parents with higher levels of schooling took on a "teacher-like role and focused on asking children known-answer questions and evaluating children's performance. . . . In contrast, the parents with basic schooling seemed to engage in the task as co-learners with their children" (Callanan, Solis, et al., 2020, p. 84). Parents had different approaches to this task, which in turn focused their children on different facets of what they could learn.

Although one can imagine computational models that might begin to account for this kind of causal learning across cultures in a manner consistent with rational constructivism (e.g., Werchan et al., 2016[10]), some sociocultural approaches to cognitive development reject such information-processing accounts of development in favor of representing knowledge more within social frameworks (e.g., Rogoff, 2003). We want to suggest a compromise between these positions. Cultural factors change dynamically, which relate to how children learn. And because we generally agree that learning can be fruitfully described within the context of an information-processing approach, rational constructivism must endeavor to integrate more deeply with sociocultural perspectives and to take seriously the ways in which cultural and subcultural contexts may not straightforwardly translate to information processing.

For example, ojalehto and Medin (2015) suggest that different cultural frameworks can synthesize together to create unique causal models of explanation. To illustrate this process, they describe studies of adults in South Africa who were asked to provide explanations about the cause of AIDS. Within these explanations, different explanatory frameworks overlapped (e.g., supernatural entities like witches can use natural causes like viruses and other germs to make one sick; Legare & Gelman, 2008; see also Legare et al., 2012).

Similarly, in the work on gear exhibits in museums described previously, we found that parents' causal talk, particularly the timing of such talk, influenced children's causal actions (Callanan, Legare, Sobel, et al., 2020). How parents feel about children's play and their beliefs about how children might learn through play might relate to the ways that children learn through play as well as the ways they interact with their children during play.[11] For instance, Lancy writes, "Unlike the Euro-American cultural model of childhood where parent-child play may be considered 'essential,' elsewhere adults do not play with children (Lancy, 2007), in large part because it violates the child's independence and takes the adult away from more important activity" (2016, p. 658). That is, different cultures might value learning from play differently, which in turn might affect how children learn from play, or what beliefs they have about learning from play, or the process of learning from parent-child interaction and scaffolding.

Even within the WEIRD culture, in which we personally work, most studies of children's causal reasoning involve children participating in a novel activity setting in a social context like a lab or a museum, which is not necessarily a reflection of their authentic experiences. Children are often brought to a strange place, where a strange adult asks them strange questions, or shows them strange stories, or places them in strange situations.[12] Different children (and their parents) construct different meanings from those social contexts, and these differences should be investigated and documented more systematically. In the absence of a robust body of empirical work on this issue, we support the conclusion drawn by ojalehto and Medin (2015): "Investigating causal cognition from the perspective of complex systems may afford new insights into conceptual behavior. Given that humans are surrounded by complex systems (ecologies, societies, consciousness), this is an area of study that deserves more investigation" (p. 265).

These arguments apply directly to the central theme of the book: the development of children's scientific thinking abilities. Scientific thinking, about both the content and the process of science, is often done in a social context, such as a classroom or a museum. Children may inherently possess some causal reasoning and scientific thinking abilities, but these abilities may not exist independently of the social nature of the world in which they acquired them or of the social nature of the tasks in which they are being asked to express them. For example, Packer and Goicoechea (2000) argued that "any social context—a classroom, for example—is itself the product of human language and social practice, not fixed but dynamic, changing over time, in what we call history" (p. 232). How children are taught science, and what children believe that science is, likely influences the way they learn it, and both of these factors are embedded within children's social and cultural contexts. To be a productive theory of children's scientific thinking, rational constructivism must take these factors into account.

Questions of Explanations

Questions about cultural differences and about different kinds of adult-child interactions illustrate one set of ways in which the contexts in which children develop affect how they learn and how they interpret their experiences. But these contexts are not simply provided to children, nor are they static; developmental contexts are actively and continually co-constructed by children, their parents, and others in the environment. One place in which this process can be observed in children's construction of and requests for explanations—their active seeking for meaning within their environments and their abilities to talk about their causal knowledge.

Prior work has shown that young children are able to generate explanations about a number of different concepts, and these explanations are often argued to reflect their causal knowledge (e.g., Hickling & Wellman, 2001; Legare et al., 2009; Schult & Wellman, 1997; Sobel, 2004b; Wellman et al., 1997). Children also generate various kinds of causal utterances at relatively early ages (e.g., Hood & Bloom 1979). These utterances are often quite sophisticated, particularly in contrast to their predictive inferences. For example, Bartsch and Wellman (1995) documented that young children learning English in the US generated what they called *false belief contrastives*—utterances in which children indicated that they had a false belief about a state of events—during their third year, well before they solved explicit measures of representational change (e.g., Gopnik & Astington, 1988; Perner et al., 1987; see also Sabbagh & Callanan, 1998, for similar findings).[13]

But children's explanations and causal utterances do more than reveal what they already know; they allow children to interpret information within a social context. Based on research on school-age children's problem-solving (e.g., Chi et al., 1994), several studies suggest that encouraging preschoolers to generate explanations of data they observe helps them make causal inferences (e.g., Bonawitz & Lombrozo, 2012; Macris & Sobel, 2017; Walker et al., 2017). Indeed, in their meta-analysis of play-based approaches to learning, Alfieri et al. (2011) showed that the clearest benefit to children's learning outcomes was when children generated their own explanations of the situation. Explanations also allow young children to learn novel pieces of information, which in turn allow them to use systematic strategies in requesting explanations from others.

For example, children begin to ask "why" questions around 30 months old (Hood & Bloom, 1979) and 3- to 5-year-olds use such questions spontaneously (Callanan & Oakes, 1992). Similar work found that preschoolers used "why" questions—and questions more generally—in systematic ways to gain information (Frazier et al., 2009). Children in this study continued to ask the same kind of question when given a nonexplanatory reply, but shifted to a new topic or a different question when they received an explanation.[14] This strategy clearly develops between the ages of 3 and 5; for example, Callanan et al. (1995) showed that children shifted from simply asking single-word "why" questions to elicit explanations to jointly constructing such explanations with their parents.

Various studies also examine children's likelihood to seek or generate explanations in order to address gaps in their knowledge—an expression of their curiosity (Jirout & Klahr, 2012). Legare (2012), for example, showed that children are more likely to generate novel explanations when they observe outcomes that are inconsistent with their existing knowledge. Stahl and Feigenson (2015) demonstrated a similar finding in infants' exploratory behavior: Infants explore objects for a longer amount of time when these objects violate their expectations. Exploratory behavior and the search for explanations potentially share information-seeking mechanisms (see e.g., Legare et al., 2017).

These information-seeking mechanisms might be in place before children's linguistic capacities allow them to articulate a "drive" for explanations (Gopnik, 1998), meaning that some of the increased complexity of young children's abilities to generate explanations over the course of the preschool years reflects their more sophisticated language capacities. This is another case in which children's performance on a task is constrained by development in their domain-general abilities. For example, Ruggeri et al. (2017) showed that a group of 5-year-olds were more likely to choose which of two questions would correctly provide them with more information than a group of 3-year-olds. When children were given the opportunity to ask questions that they designed themselves to learn about the cause of an event, there was a shift between what Ruggeri and Lombrozo (2015) called "hypothesis scanning" questions, which reduced the hypothesis space by considering one particular hypothesis out of many possible ones, to "constraint seeking" questions, which eliminated large sets of possibilities. Seven- and 8-year-olds in this study tended to ask many hypothesis scanning questions, but the frequency of these questions decreased as children got older. The younger children could choose which of two questions (designed by another person) provided them with more information, but there is a developmental lag before children can design and ask those questions themselves in this specific context.

Of importance is the trajectory of children's metacognitive development with respect to understanding how best to seek information. For example, children not only use explanations for learning, they also evaluate others' explanations and learn judiciously. Corriveau and Kurkul (2014) found that 5-year-olds preferred to learn from informants who generated noncircular explanations rather than circular ones, while the younger children they investigated did not (although see Mercier et al., 2014, for findings that suggest that even 3-year-olds are sensitive to others' circular explanations). However, the ability to probe further when given a weak explanation in a more challenging context seems to have a longer developmental trajectory, particularly between the ages of 7 and 10 years (e.g., Danovitch et al., 2021; Mills et al., 2019). Attending to unexpected events and inconsistencies often leads children to seek explanations, which in turn may help them to progress along the path toward knowledge acquisition or hypothesis revision, particularly as they develop the understanding that the weaker explanations that result from their own or others' gaps in knowledge may motivate further exploration.

With respect to the theme of this book, there is potentially an important connection between the drive to generate explanations and scientific thinking. Scientific thinking does not involve ignoring or rejecting inconsistent evidence, nor does it involve declaring inconsistencies or mysteries to be beyond the scope of the investigation (Chinn & Brewer, 1993). Engaging in explanation and exploration thus may serve as a critical mechanism within a cultural context for integrating and reconciling discordant or ambiguous information with existing theories. Further, this may reduce engagement in heuristics like confirmation bias, which preserve children's existing causal knowledge and prevent learning. Frankly, this kind of critical thinking and curious seeking after explanations could help adults as well.

4 Variables Relating Causal Reasoning to Scientific Thinking

In chapter 3, we suggested that rational constructivism, especially as instantiated with the causal graphical model framework, faced certain challenges and had certain theoretical insularities. Throughout the rest of the book, we want to address one issue that illustrates an aspect of that insularity—the relation between causal reasoning and scientific thinking. If children are so good at causal reasoning, as our views about cognitive development and learning suggest, why is scientific thinking challenging, even for adults? This is one of the key questions of this book.

Although not all scientific thinking involves causality, we want to suggest that causal reasoning is the basis of many aspects of the scientific process. Children's abilities to learn about the world through understanding causal relations could potentially be used when they need to interpret scientific content or predict the results of both formal experiments and informal actions on the world. While preschoolers may possess sophisticated causal reasoning abilities, as illustrated by the experiments we reviewed in chapters 2 and 3, it is more of a challenge for them to make inferences about experimental design, or, more generally, to solve scientific thinking problems in the classroom or in real life. Simply put, the fact that children can make inferences about blicket detector systems in a developmental laboratory does not mean that they can engage in the kind of systematic, content-rich inference-making required for scientific investigations.

But why should this be the case? Given that children are so good at causal reasoning, which underpins much of the scientific thinking they are asked to do, why do they struggle with explicit scientific thinking? Our main goal in this chapter is to begin to address this question. We review the similarities and differences between children's causal reasoning capacities and children's scientific thinking abilities, with the goal of beginning to reconcile these two bodies of work. We begin this reconciliation by accepting a few premises:

(1) Young children have sophisticated causal reasoning abilities by the time they enter preschool, and likely earlier. Those inferential abilities are evident in children's reasoning during laboratory-based tasks but are also evident in children's learning from play and social interactions, perhaps particularly in informal learning environments.

(2) Causal graphical models provide an initial description of how young children represent their causal knowledge. That is, algorithms from this modeling framework provide a useful (computational-level) description of how children engage in causal inference.

(3) These same children struggle with the scientific thinking problems presented to them both in the classroom and in the laboratory (although maybe less so in their play). They may lack the ability to demonstrate these kinds of reasoning abilities until late in elementary school or later.

Taken together, these premises imply that making scientific inferences is not the same process as making causal inferences. As reviewed in chapter 1, scientific thinking is a set of abilities that allows one to generate hypotheses, solve problems, and explain aspects of the world. Scientific thinking involves distinguishing between a hypothesis and evidence, an appreciation that theories that generate hypotheses can be wrong and can be revised with the appropriate evidence, and the metacognitive capacity to reflect on this process. All of these capacities have unique developmental trajectories (e.g., Bullock et al., 2009; Kuhn, 1989).

To narrow the scope of our investigation somewhat, here we focus on one crucial ability within the scope of scientific thinking: the ability to use the *control of variables strategy*. Briefly, as the name suggests, the control of variables strategy involves selecting or conducting unconfounded experiments so that one can learn a causal structure by changing exactly one variable and holding all else the same (e.g., Chen & Klahr, 1999; Inhelder & Piaget, 1958; Tschirgi, 1980). This is the method that underlies much of modern experimental practice because it allows for the isolation of particular events, providing a definitive answer to which factors are genuinely causal and which are merely correlated with a particular outcome.

For example, consider the slopes task used in Chen and Klahr (1999). Participants are introduced to two ramps and two different kinds of balls.

The ramps each have a starting gate where a ball can be placed. The angle of each ramp can be adjusted (making the slope steeper or gentler), as can the location of the starting gate (so that the ball can start at a higher or lower point on the ramp). Also, each ramp can be covered with one of two different surfaces, changing the amount of friction between the ramp and the ball as it rolls. Participants are told to learn which factors are important to determining how far a ball will roll down the ramp. Participants can set the angle of the slope, the starting position of each ramp, and the surface, and then can observe the two different balls roll down the ramps from the starting gate. The key idea to solving this task is to manipulate each variable independently in order to consider its efficacy in isolation.

Because of its importance, a large body of work has investigated whether children understand the control of variables strategy and whether and how they can learn this strategy (e.g., Chen & Klahr, 1999; Klahr & Nigam, 2004; Kuhn et al., 1992, 1995). In general, the young children tested in these studies often do not use this strategy without instruction, and even with instruction, they do not use it all the time. Even worse, they tend to misremember the evidence they generate and to misunderstand the relation between that evidence and the theories they might have (e.g., Amsel & Brock, 1996; Croker & Buchanan, 2011; Schauble et al., 1995). In a meta-analysis of studies on children's understanding of this strategy, age did not significantly moderate the outcomes of these kinds of studies (Schwichow et al., 2016). But the youngest children tested in that analysis were age 6, older than the participants in most of the research on causal reasoning discussed in chapter 2.[1]

Although the control of variables strategy is an important aspect of scientific thinking, as noted above, understanding this strategy (and other aspects of scientific thinking) has different conceptual requirements than understanding causality more generally, making direct comparisons between these reasoning processes difficult, if not impossible (see discussions in Kuhn, 2007a; Kuhn & Pearsall, 2000; Ruffman et al., 1993; Sobel et al., 2017; Sodian et al., 1991). Rather than try to align these two abilities directly, then, our approach in this chapter is to identify the main variables on which causal reasoning and scientific thinking—particularly as expressed in the control of variables strategy—differ.

As we make our arguments, we focus on two key example tasks, one from the literature on children's causal reasoning and one from the literature on children's scientific thinking. To represent the causal reasoning literature, we use the blicket detector study as reported in Gopnik et al. (2001), described in chapter 2. In this study, 3- and 4-year-olds observed the effects of two objects on the detector and recognized the difference between an object that activated the machine directly and one that only did so conditionally, dependent on the other object.

To represent the scientific thinking literature, we use computer-based inquiry tasks, such as Earthquake Forecaster as described by Kuhn and Dean (2005; see also Dean & Kuhn, 2007; Kuhn et al., 2009). In this task, children (usually 4th or 6th graders) play a computer game in which they try to predict the risk of earthquakes given particular features. For example, Kuhn (2007b; Dean & Kuhn, 2007) showed children that, on each turn of the game, they could select to observe cases that combined five potential causes of earthquake risk: soil type, S-wave rate, water quality, snake activity, and gas level (although other versions use other features, such as water pollution, water temperature, soil depth, and elevation). Each of these variables had two levels; for example, soil type could be either sedimentary or igneous. Students could query a database and find out how each combination corresponded to a geographic area that had one of four levels of earthquake risk: low, medium, high, or extreme. Students were charged with finding out whether and how each variable contributed to the outcome.

In general, the children tested in these studies rarely used the control of variables strategy when just given the program (e.g., see the "pilot assessment" reported by Dean & Kuhn, 2007 or the pretest performance in Kuhn, 2007b). After being exposed to the program (or a very similar one) over several sessions, children were able to make valid inferences about the potential causes of earthquake risk less than half of the time (Dean & Kuhn, 2007). Some children (19 of the 30 children tested) did show improvement in their reasoning with practice (Kuhn, 2007b), and children who had been given prompts to focus on only one variable at a time also performed better (Dean & Kuhn, 2007; see also Kuhn & Dean, 2005).

Throughout the rest of the chapter, we use the blicket detector and Earthquake Forecaster tasks to illustrate what we take to be the main factors that differ between the literatures on causal reasoning and scientific thinking, acknowledging that these tasks are merely examples of their respective literatures and do not represent the entire state of the field.

Age

One of the clearest differences between most of the research on children's causal reasoning and most of the research on children's scientific thinking is the age of the participants. Studies of causal inference tend to recruit young participants, often toddlers and preschoolers. For instance, in chapter 3, we talked about the debate over whether the concept of "cause" was innate or developed from interactions with the environment. This dialogue places emphasis on research with infants and toddlers, which for the most part has demonstrated that these very young children have sophisticated causal reasoning abilities, at least after about 6 months of age (e.g., Denison et al., 2013; Saxe et al., 2005; Sobel & Kirkham, 2006, 2007). By contrast, studies of children's scientific thinking tend to focus on students in elementary school or even older (e.g., Chen & Klahr, 1999; Kuhn & Dean, 2005; Schwichow et al., 2016).

This difference in the age of the target populations is driven by the different (and equally valid) foci of the two literatures. Studies of causal reasoning have typically sought to find the basic building blocks for children's abilities, hence they have focused on young children and infants. By contrast, studies of scientific thinking have tended to be more concerned with understanding children's thinking within formal, educational contexts, aiming to connect this research to the practical problem of having to teach children science in schools.

A few studies have presented the same measures to young children and adults, or to children across a wide age range. Some of this research has argued for continuity in the mechanisms children and adults use to make inferences. For example, one of our studies presented adults with causal systems similar to the blicket detector (e.g., the "superpencil detector," which detects "super-lead," a nonobvious property of golf pencils[2]) (Griffiths, Sobel, Tenenbaum & Gopnik, 2011). In this study, we replicated experiments previously done on preschoolers with adult samples and found similar (although more nuanced) results. We also generated new causal reasoning problems with this task and showed that both adults and children solved them in similar ways, again suggesting continuity between 4-year-olds and adults.

Other research, however, has found discontinuities in development. Lucas et al. (2014) presented causal reasoning tasks using the blicket detector to both adults and children, and they demonstrated that reasoning was relatively similar in some cases. But in other cases, children were better at learning causal structures than adults, particularly cases in which causal relations were conjunctive. The conclusion from this work could be that children are better than adults at causal reasoning. However, a more likely explanation is that this difference can be explained by different degrees or types of prior knowledge in these two populations. Specifically, although conjunctive causality is not rare in everyday life, conjunctive mechanisms are rarely discussed and are usually limited to enabling cases. For example, oxygen must be present with a spark to start a fire, but the absence of oxygen is not a counterfactual most will think about in order to explain how a fire could have been prevented. As a result, when a conjunctive cause is presented outside of an enabling paradigm, adults might be less open to registering this relation. Children, on the other hand, simply accept that this is the mechanism of this strange, novel machine. What changes in children's cognition is not the reasoning mechanism, but the prior knowledge that children use to instantiate initial hypotheses (analogous to the ideas in Bayesian inference that we discussed in chapter 2).

These studies suggest that preschoolers, older children, and adults may possess roughly equivalent abilities with respect to causal reasoning (just different amounts of prior knowledge). But there are some reasons to believe that there is more to the story. For one thing, these studies use tasks with similar (low) amounts of real-world scientific content, and that similarity might have led to the similarity in results. More importantly, the fact that studies in causal reasoning and scientific thinking have tended to recruit children of different ages has led to a variety of other design choices that then co-vary with age, including the issues that we discuss below: the causal systems' levels and types of complexity, whether these systems present scientific content, and whether children are asked to observe data about the system or generate their own. These confounds across the two literatures make it difficult to determine why young children often tend to succeed in causal reasoning tasks using methods like the blicket detector, while older children often respond differently on scientific thinking tasks like Earthquake Forecaster. One possibility is that children's early-developing causal reasoning skills are maintained as they grow older, but are masked by the increased task demands of systems like those in Earthquake Forecaster. A second possibility is that children's early-developing reasoning skills can be used only for simple systems or only for systems that lack recognizably scientific content, and children require different or additional skills in order to think appropriately about the kinds of systems presented by scientific thinking measures.

Complexity

Studies of children's causal reasoning capacities often focus on preschoolers, toddlers, or even infants. One reason for this is historical, because one of the fundamental questions behind these studies was whether young children could engage in causal inference, particularly in response to Piaget's claims that young children were "precausal." Because of this emphasis, the demonstrations and causal systems used in this work tend to be simple, and the procedures are designed to walk children through the demands of the study. These simplifications are designed to eliminate any extraneous factors that might get in the way of children's abilities to demonstrate their causal reasoning.

Just as choosing to test young participants introduces certain limitations to studies' designs and materials, choosing to test older children also guides the choice of methods and procedures. While blicket detectors are fascinating to 3-year-olds (and strange curiosities to adults), 9-year-olds or middle-school students are often bored by the simple causal systems presented in blicket detector tasks. Moreover, older children and adults (at least in Western, educated, industrialized cultures) tend to possess a great deal of existing causal knowledge about electronics, which allows them to make certain assumptions about how the machine works; these assumptions can lead them to do better on certain kinds of causal inference measures, and worse on others (as in the Lucas et al., 2014, example we described above). Tasks like Earthquake Forecaster (or the slopes task [e.g., Chen & Klahr, 1999], or playing with spring tension [e.g., Schauble, 1990], or experimenting with sinking and floating [e.g., van Schaik et al., 2020], or making inferences about planets [e.g., Panagiotaki et al., 2009[3]], or many other tasks in which scientific thinking has been tested) can be more engaging to children in this older group precisely because they are more complex. For the blicket detector, once you get over the initial surprise that the machine lights up, it's just not that interesting. Using more complex tasks provides a richer real-world context for children to explore, which makes them more interesting to do and which increases their resemblance to real tasks involving scientific thinking.

For these reasons, studies on children's causal inference and studies on children's scientific thinking tend to vary in the amount and type of complexity instantiated in the systems that these studies present. To make these differences more concrete, consider again the contrast between the blicket detector and Earthquake Forecaster. In the original blicket detector task, children saw two potential causes (objects A and B), each of which could be in one of two states (on or off the machine). These potential causes could have one of two effects: turning the machine on or failing to do so. By contrast, Earthquake Forecaster presented five potential causes (e.g., soil type, snake activity), each of which had two levels (e.g., igneous or sedimentary, high or low). This system also presented four possible effects (i.e., four levels of earthquake risk). The number of factors that a participant must keep track of is thus much larger in Earthquake Forecaster; this alone may contribute to older children's struggles in reasoning about such systems.

In addition, and perhaps more importantly, the causal structure of the system presented by Earthquake Forecaster is more complex than the one presented by the original blicket detector task. This task presents an interactive (specifically, an additive) system, in which some of the five causes individually can be diagnostic of one level of risk for earthquakes, but jointly combine to produce different levels of risk. Previous work clearly shows that thinking about interactive systems is difficult, even for older children (e.g., Schauble, 1990, 1996; see also Novick & Cheng, 2004). By contrast, most studies in causal reasoning use disjunctive causality, in which the presence of any cause would lead to the same effect.

Finally, a crucial difference between the two types of task is the presence of causes whose efficacy is unknown. The blicket detector task showed children what happened when each object was placed on the detector by itself, and no other objects were present. This task thus presents no uncertainty about which things could potentially be causes or about what the effects of all potential causes are. Earthquake Forecaster (and most measures of scientific thinking), presents situations where the efficacy of individual causes is unknown. In many cases, this is because of the number of variables about which children have to reason.

Some of our previous work has shown that the presence of potential causes whose efficacy is unknown can compromise children's abilities to reason about a causal system (e.g., Erb & Sobel, 2014; Fernbach, Macris & Sobel, 2012; Sobel, Erb, Tassin & Weisberg, 2017). For example, Sobel et al. (2017,

Study 1) presented 3- to 7-year-olds with a blicket detector and a set of four objects. In the first trial, children saw each object placed on the machine by itself. The first object activated the machine, making it light up and play music. The second did not. The third and fourth objects also activated the machine. The experimenter then brought out a large piece of cardboard and hid the machine and objects from the child. He told the child he was putting one of the objects on the machine, and that the machine lit up. Even though children could not see this, they could confirm it because they heard the music that went with activation. The experimenter then said he was taking the object off the machine and putting it back. (In reality, the experimenter never moved any of the four objects—he just mimicked doing so and activated the machine remotely, so that there would be no visual clues as to which object had been moved.)

The experimenter then removed the cardboard occluder and asked children a set of diagnostic reasoning questions. He first asked which object he had used to make the machine go. Whichever object children chose, the experimenter said, "That's a good guess, but it's not right." He removed that object and asked children to try again. Regardless of what they chose the second time around, he also indicated the choice was wrong, and asked for a third guess, this time with only two objects left on the table.

Responding correctly to these questions is straightforward: Just don't pick the object that you saw fail to activate the machine. However, 3- and 4-year-olds were equally likely to choose an object that had previously activated the machine as to choose the object that had not. What is important here is that they could (in a separate trial) correctly make predictions about what would happen if each object was placed on the machine, showing that they had remembered each one's effects (i.e., they could predict that the object that had failed to activate the machine would fail to activate it again, suggesting that their performance on the diagnostic reasoning question was not the result of poor memory). What was more of a challenge was applying that knowledge across numerous requests to revise their guess. Five-, 6-, and 7-year-olds, however, almost never got this trial wrong.

The more complicated finding comes from comparing performance on this trial with performance on two others (shown in figure 4.1). In the second trial, everything was same as the first, except at the initial demonstration phase. In this case, the first object activated the machine (same as above), the second object did not (also the same), the third did (also the

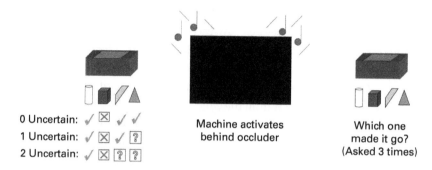

0 Uncertain: ✓ ☒ ✓ ✓
1 Uncertain: ✓ ☒ ✓ ?
2 Uncertain: ✓ ☒ ? ?

Machine activates
behind occluder

Which one
made it go?
(Asked 3 times)

Figure 4.1
Schematic of procedure used by Sobel et al. (2017, Experiment 1).

same), but the fourth object was never placed on the machine (different from above). Now there is an object that could be a cause, but the child simply does not know. In the third trial, the first object made the machine go, the second did not, and the other two were never placed on the machine. In this case, there were two objects with unknown efficacy.

The critical finding of this study was that there was a significant correlation between age and the number of unknown causes children could accurately reason about diagnostically. Five-year-olds had no problem when all of the causes were known but fell to chance levels of performance in the other two trials. That is, children at this age sometimes mistakenly chose the object that had previously failed to activate the machine as a potential cause when they had not been able to observe the efficacy of all of the objects. As children got older, specifically between the ages of 5 and 7, they were able to handle to more and more uncertainty in their diagnostic inferences, becoming less and less likely to choose the inefficacious object, even in the presence of some uncertainty.

These results match up well with some of the differences that we have been emphasizing between young children's performance on tasks involving the blicket detector and older children's performance with more complex scientific thinking tasks like Earthquake Forecaster. The number of causes and causal relations, the complexity of these causal relations, and the amount of uncertainty presented by the system tend to be much greater for studies on older children's scientific thinking abilities than for studies on younger children's causal reasoning abilities. Again, this makes comparing them difficult;

it is not clear whether children perform less well on scientific thinking tasks because they lack some basic reasoning skills or because some aspect of the complexity of these tasks simply takes them out of their grasp.

Use of Scientific Content

As noted in chapter 2, causal reasoning measures like those that use the blicket detector do not reflect the structure of any particular real-world system. They present causal reasoning problems in an abstract way. Although some have described these tasks as *decontextualized*, lacking any reference to real-world causal structures (e.g., Kuhn, 2007a), it would be more accurate to say that such tasks are just *less* contextualized than other problems, in that they do not reflect any particular piece of real-world content knowledge. Making inferences about causal relations among objects and a machine requires children to possess knowledge about general causal factors such as temporal priority (causes precede their effects) and spatial contact (the machine probably works by having the blocks touch it, not merely come near it). We cannot imagine a task that is purely context-free (see Sobel & Munro, 2009).

Still, beyond these basic factors, causal reasoning measures like those that use the blicket detector do not require children to draw on much prior experience with blocks or machines or other real-world causal systems. This is a strength of the paradigm; children do not need much prior knowledge to make causal inferences about these systems. This is also a strength of the causal graphical model framework; it is a domain-general reasoning system. At an abstract level, a graphical model does not care what it is about; the nonindependence findings that we discussed with the cellphone example in chapter 3 potentially emanate from the fact that *participants* care. They might have been reasoning about models that were not intended by the researchers. Hierarchical modeling (e.g., as discussed by Goodman et al., 2011) as well as the edge replacement findings we discussed in chapter 3 may provide a way to conceptualize how this prior knowledge constrains children's reasoning and thinking, allowing the framework to take participants' tendencies into account.

However, a lack of context is also a weakness of the blicket detector paradigm. Although the causal systems instantiated in the blicket detector could be abstract representations of real-world causal systems, that fact is likely opaque to young children. More importantly, children's responses to such systems, in which their reasoning abilities can be neither helped nor hindered by their prior knowledge about specific scientific domains, may not provide a true reflection of children's scientific thinking abilities. These tasks might show how children think about minimally contextualized systems, but they could have little or no bearing on our understanding of children's thinking about real-world causal systems. That is, the abilities that children need in order to solve a causal reasoning problem with a blicket detector might be different from the reasoning processes used in scientific investigations in the lab, the classroom, and everyday life. In support of this argument, Dunbar (2002) suggests that "causal reasoning in science is not a unitary cognitive process . . . but a combination of very specific cognitive processes that are coordinated to achieve a causal explanation such as an unexpected finding" (p. 157).

To expand on this statement, designing experiments, integrating data with existing theory, revising beliefs, and communicating both the conclusions and the process by which one arrives at those conclusions are all part of causal reasoning in science. So far, the blicket detector and other measures of children's causal reasoning have really focused on only one cognitive process: making appropriate causal inferences. In part II of the book, we explore several ways that blicket detectors can be used to probe different aspects of children's abilities. But, for now, we still must deal with the objection that research using novel and artificial causal systems like the blicket detector have no bearing on children's scientific thinking in real-life contexts because they do not adequately resemble the work that real scientists do. If this is the case, then it becomes vitally important to understand why tasks that do resemble the work that real scientists do tend to be more difficult. We can again use tasks like Earthquake Forecaster, which present more richly contextualized systems that instantiate their variables in real-world problems, to identify the source of these problems. Instead of blocks that can be either on or off the machine, children are asked to think about potential causes that might conceptually relate to causes of earthquakes (e.g., what type of rock is more abundant, whether there is water pollution, the elevation of the land mass).

This choice to contextualize the causal system is both a strength and a weakness. As noted above, a major strength of this task is its realism. When children think about the structure of the real world, they need to grapple with data that come with meaningful labels and that produce effects in a real-world context, because most reasoning tasks that children will have to solve will involve some kind of real-world content. Plausibly, in order to solve such tasks, children would need to strip away the specific labels and effects and construct more abstract representations of the variables that they represent. The blicket detector and other laboratory-based measures eliminate the need for this step. As a result, they might be overestimating children's abilities. Children will need to develop the skills to deal with both the real-world content and the abstract causal structures that underlie it to become fully mature scientific thinkers.

But richly contextualized systems could be making the underlying reasoning task too difficult for children, and hence may be underestimating their abilities. These tasks not only ask children to solve the abstract problem of how different variables contribute to an effect, they also require children to navigate potentially unfamiliar vocabulary, like "igneous" and "sedimentary," or to think carefully about whether a general variable like elevation relates to earthquakes. Additionally, different children may have come to this task with more prior knowledge about earthquakes and their causes, which might put them at an unfair advantage. For instance, there is a complex relation between the activity of nonhuman animals (including snakes) and the probability of earthquakes (Woith et al., 2018). Despite this, many laypersons seem to believe that animal activity is a risk factor. However, we doubt whether young children believe this initially, and thus might dismiss this variable as potentially irrelevant, much like infants ignore certain correlations (e.g., Madole & Cohen, 1995, as discussed in chapter 2).[4] Prior knowledge may work against them in this case, and their performance on this task may be poor not because they were unable to think about how combinations of potential causes lead to an effect, but because the surface features of the task threw them off the trail.

This distinction relates to the idea of the "seductive allure" of particular types of scientific content that we have investigated in a series of studies on adults' understanding of science (e.g., Hopkins, Weisberg & Taylor 2016; Weisberg et al., 2008; Weisberg, Hopkins & Taylor, 2018). Briefly, this effect occurs because adults tend to prefer explanations of scientific phenomena that involve reducing those phenomena to a more fundamental science, such as an explanation of a chemical phenomenon in terms of physics. This is especially the case for psychological explanations, which seem more satisfying when they include neuroscientific terms, even when the neuroscience information does not add any explanatory value.

One potential explanation for this preference is that adults believe that the language in reductionist explanations sounds smarter and more complex, hence more "scientific." But that's adults. For children, those "smarter" explanations—the ones that use more complex vocabulary—might sound too smart. Children might get so caught up in trying to understand these new concepts that they fail to reason appropriately, or they might get so turned off by needing to understand the strange vocabulary that they do not try to reason appropriately at all.

More generally, the way in which a measure of scientific thinking is designed might affect children differently based on what prior knowledge they bring to the reasoning task. Parents who talk to their children endlessly about sinking and floating while in the bathtub might have children who approach sinking and floating tasks with a better (or at least a different) understanding of the role of density than parents who talk to their children more about the importance of hygiene. But these latter children might perform better in laboratory tasks about the role of germs in disease transmission (e.g., Conrad et al., 2020).

This is a potentially important explanation for why studies have found sophisticated causal reasoning when using the blicket detector and other less contextualized measures, and less robust reasoning when tasks are more contextualized: Asking children about blicket detectors may make it easier for them to demonstrate their reasoning about causal systems per se, while asking children about more contextualized problems with scientific content, like predicting the risk of earthquakes, may make it harder for them to understand how to bring those capacities to bear on the problem they are presented with. Similarly, this might be why scientific inference in the laboratory or the classroom is hard to demonstrate, but seems more evident in free play either in the lab (e.g., Cook et al., 2011) or in museum settings (e.g., Callanan et al., 2020), as we documented in chapter 3.

In the end, the extent to which one agrees with the findings of one set of studies over the other may be a matter of perspective: Is it more important to emphasize the thinking skills, regardless of context? Or is it more important to test children's abilities in the context of real-world information, because that is where they will tend to need to display these abilities? We address these issues directly in chapter 6, where we compare children's performance with more and less contextualized causal systems, and in chapter 11, where we discuss children's understanding of learning in different learning environments.

Observing versus Generating Data

As noted above, scientific thinking often involves experimentation—doing things and observing their results. For example, the Earthquake Forecaster program asks children to figure out how five two-level variables combine to reflect four different levels of earthquake risk. The program provides them with a database of cases that they can query. Their main task is to use the program to design controlled experiments—changing only a single parameter while keeping everything else fixed—because only this method of testing will allow them to draw conclusions about which variables are causally relevant and how the different variables are causally related.

However, results with the Earthquake Forecaster program and similar systems show that children do not always apply the control of variables strategy, even late in elementary school. They tend to not test hypotheses explicitly and generally do not design unconfounded interventions to learn new causal relations. Instead, they tend to change many variables at once. Naturally, because their experiments are generally ineffective at disentangling different causal factors, children in these studies often do not discover the causal structure of the system—at least not without practice or direct instruction, and even then, this thinking is still difficult.

Although this is partially due to the other variables discussed above, the fact that children have to generate their own data may be a major contributor to their poor performance with this program, particularly when they are not given instructions to focus on individual variables. Indeed, paradigms in scientific thinking, contextualized with rich scientific content, present children with two distinct challenges. First, as noted above, these tasks ask children to construct appropriate experiments using this system to tease apart which combinations of variables lead to the effects. This poses a challenge because most children either do not know how to use the control of variables strategy or do not know how to apply the strategy to this situation. Second, children must draw conclusions from a set of data. This may be a difficult task in itself, even when the set of data has been properly constructed (Kuhn, 2007b).

Causal inference measures that use less contextualized paradigms, on the other hand, almost always present children with sets of data and do not ask children to design their own experiments. These sets of data are constructed to lead children to discover the underlying causal structure of the system.

Indeed, children tend to view these situations pedagogically, understanding the experimenter's presentation of the data as being designed to teach them something about the system (e.g., Gweon et al., 2010; Yang et al., 2013). This removes the requirement for children to be able to generate productive data, allowing them to focus only on the second task: drawing conclusions from the data they observe.

This is not to say that children cannot learn from their own actions, particularly when those actions are set up so that they do not conflict with the efficacy of the objects. Preschoolers do engage in productive exploratory behavior when faced with ambiguous evidence, and that exploratory behavior can lead to them generating information that allows them to resolve the ambiguity (e.g., Cook et al., 2011; Legare, 2012; Schulz & Bonawitz, 2007). And preschoolers also can learn a novel set of causal relations better from exploring their environment than from observing the actions of another, at least under some circumstances (e.g., Sobel & Sommerville, 2010). Further, as we discussed in chapter 3, there are cases of children learning from their own actions in museum settings, particularly when those actions are construed as play (e.g., Callanan, Legare et al., 2020; Sobel, Letourneau et al., 2021). However, the ability to produce a genuine control of variables strategy emerges slowly over the course of elementary school, and even adults find aspects of such reasoning hard (e.g., Klahr et al., 1993; Schauble, 1996). Small wonder, then, that children struggle more to solve tasks like Earthquake Forecaster than tasks like the blicket detector: Even with some degree of understanding of how to explore a causal system, it is far less taxing to their cognitive resources to merely draw conclusions from a set of data that has been designed to help them draw the correct conclusions than to have to additionally generate this set of data.

There's one other important difference. Most measures of causal reasoning, like those using the blicket detector, involve generative causality. Children are shown a set of potential causes (usually objects), which individually or together have the efficacy to activate the machine. During the experiment, those objects are placed on the machine in an active intervention. In contrast, Earthquake Forecaster is more diagnostic. Children select a certain combination of features and then observe a case that had those features. But selecting an observation is not the same as making an intervention. That is, when children choose to observe a case in which there are more igneous rocks and lower snake activity, they are not creating a new case with that combination of variables. Selecting observations of data that have occurred in the past does not necessarily provide the same conditional probability information necessary for causal learning as conducting interventions on a generative system (Lagnado & Sloman, 2004). While many measures of scientific thinking, like the slopes task, involve children generating their own novel data from their interventions, other measures rely on selecting observations, which could interfere with children's learning.

Recognizing and Using the Control of Variables Strategy

So far, we have suggested that an important finding of a variety of studies of children's scientific thinking is that the control of variables strategy has a prolonged developmental trajectory, extending beyond the preschool years (e.g., Klahr & Nigam, 2004; Kuhn et al., 1995). Second graders can learn the strategy, particularly through direct instruction (Chen & Klahr, 1999). But children do not seem to use this strategy spontaneously until they are older, if at all (e.g., Klahr et al., 2011; Kuhn et al., 2008). And, as we have discussed, even some middle school students can struggle with constructing meaningful and well-controlled interventions without instruction (e.g., Kuhn & Dean, 2005).

But in some cases, younger children seem to have nascent understanding of the strategy and the ability to recognize its efficacy. One example comes from work by Sodian et al. (1991), who showed that first and second graders could recognize which interventions would be effective in solving a particular problem. Children in this study were presented with a story in which two characters knew there was a mouse in their house but did not know whether the mouse was big or small. The characters designed two mousetraps. One mousetrap had a large opening, which either mouse could get into; the other had a small opening, which only a small mouse could get into. Children were asked which mousetrap they wanted to use to figure out the size of the mouse. Most of the second graders (86%) and about half of the first graders (55%) in this study recognized that the mousetrap with the small opening should be used to figure out the size of the mouse: If the mouse was caught, it was the small mouse; if not, it was the large one. These children also correctly recognized that the mousetrap with the large opening would be useful if the character's goal was to feed the mouse, because either mouse could access it.

Piekny and colleagues (Piekny et al., 2014; Piekny & Maehler, 2013) replicated and extended this work to a larger age range (4- to 12-year-olds), and they also looked at development in a longitudinal sample between the ages of 4 and 6. In general, 4-year-olds were not different from chance responding on this measure. Five-year-olds were better on some of the measures (for example, they were better than chance on choosing which mousetrap to use if the goal was to feed the mouse), but there were clear differences between 5-year-olds and 6-year-olds.

These data suggest that children can choose an intervention that would provide them with the appropriate evidence to make a causal conclusion or to achieve a goal. The ability to evaluate evidence to draw explicit conclusions has a similar developmental trajectory. For example, Ruffman et al. (1993) introduced children between the ages of 4 and 7 to correlational evidence (e.g., that children who ate food of a certain color had fewer teeth than those who ate food of another color). After children reported on the correlation, the experimenter rearranged the data, so that it now presented a "fake" relation—suggesting that food of the other color promoted tooth loss. Over a number of studies, they showed that children by the age of 6 could unambiguously infer that one might come to a particular conclusion from a set of data, even if those data did not represent the actual state of the world.

These studies suggest that children can determine what would be an appropriate test of a hypothesis and can come to appropriate causal conclusions about hypotheses by approximately age 6 or 7. Additionally, evidence reviewed earlier suggests that preschoolers (and possibly even younger children) might be able to use evidence to make causal inferences when those data are presented to them, and they might even perform informative interventions in their exploratory play. But none of these abilities are sufficient for scientific thinking as we have defined it. For instance, hypothesis testing requires more than just causal inference; it requires the kind of strategic thinking involved in appreciating what kinds of data one needs to observe, and what conclusion to draw given observing those data or alternative results that speak against a given hypothesis (Morris et al., 2012). This understanding seems to have a more prolonged developmental trajectory.

But is it the case that young children lack these capacities entirely? We think that the answer is no. For example, van der Graaf et al. (2015) demonstrated that 4- to 6-year-olds could generate unconfounded interventions with ambiguous data using a scientific thinking task that presented only two potential causes, which were specifically explained to the children. While this suggests children might be able to design their own interventions, children in this study were also given much scaffolding and direct instruction. Can children engage in these behaviors more spontaneously?

Recent research suggests that they can. For example, Walker et al. (2019) presented 4- and 5-year-olds with information about a new causal system: A block that had both a dot on its front and an antenna on its top made a machine play music. Children then observed two people try to figure out what made the machine work. One of these people used a control of variables strategy, choosing a block with a dot but no antenna to put on the machine (i.e., changing only one variable while keeping the others fixed). The other person did not use the strategy, choosing a block with neither a dot nor an antenna to put on the machine (i.e., changing two variables at once). In both cases, the machine failed to activate, and both people drew the conclusion that the antenna was causally responsible for the music. When children were asked to endorse one of the experiments, they chose the one that used the control of variables strategy.

Lapidow and Walker (2020) followed up on this work by showing 4- to 6-year-olds a novel causal system: a gear machine in which two gears could be interlocked and made to activate. Across several experiments, they introduced children to the gear machine and different gears that were either "working" (could spin, and possibly cause other gears to spin) or "broken" (would not spin or could only be spun by another gear). In one experiment, children were told that the two gears would spin together if at least one was a "working" gear. They saw two gears spin and were told to determine whether both gears were working or whether only one was working and the other was broken. Children were given the choice of two interventions to try. One was informative and would tell the child which causal structure was accurate. The other was confounded, so observing the results would not allow children to reach the appropriate causal conclusion. Children mostly chose to make the informative intervention. Those who chose this intervention were more likely to draw the correct causal conclusion from the data that they subsequently saw. This indicated that these children could both select an unconfounded intervention designed for them and could learn from the results of this intervention.

A similar study was performed by Moeller and Sodian (2019). They presented 4- to 6-year-olds with cases in which a set of candidate causes produced an effect, but the evidence about these candidate causes' efficacy was confounded. Children were asked whether they knew which of the candidate causes was the actual cause. The majority of children (around 70%) said that they did know, but a significant minority routinely stated that they did not and justified their responses appropriately. The same children also saw trials in which two or three candidate causes produced an effect. Children were asked to choose which experiment they would want to do in order to figure out whether a particular candidate cause was the actual one. Like in Lapidow and Walker's study, children were shown confounded or unconfounded interventions. Around half of the children correctly chose the experiment that was unconfounded. Further, about 20% of the children not only chose that option but also justified their choice by appealing to the control of variables strategy. There were also significant effects of age, with the older children in the sample (6-year-olds[5]) outperforming the younger children (3-year-olds), and of complexity, with all age groups performing better when there were only two candidate causes than when there were three.

We expanded on these results in a follow-up experiment, which produced similar results, but which also considered how well children could determine which of the candidate causes was the actual cause based on the intervention they chose and the data they observed (Moeller, Sobel & Sodian, 2021). Critically, children who chose the unconfounded intervention came to the appropriate causal conclusion about 70% of the time, significantly more often than chance levels and significantly more often than children who chose to make the confounded intervention.

These data suggest that, during the preschool years, children might have some nascent understanding of the control of variables strategy and of what is involved in constructing experiments that will allow one to learn about a system's causal structure. This is quite impressive for children who have not yet had any formal training in science. However, these studies do not show that children can design their own experiments that use the control of variables strategy. In all of these studies, children observed experiments that had been designed by someone else, and they had to choose which intervention was appropriate. This is a far cry from designing a successful experiment oneself (as those of us who mentor students in our labs know well).

Learning from Exploration and Play

The studies we reviewed in the previous section suggest that children can distinguish confounded from unconfounded interventions if those actions are designed by another person. Further, if they observe unconfounded data, they are more likely to learn a system's causal structure. But this work does not demonstrate that young children can design those unconfounded interventions on their own. Indeed, Bullock and Ziegler (1999) showed that children could select interventions consistent with the control of variables strategy in a scientific thinking context, but they struggled to produce such interventions (although see Cook et al., 2011, for a demonstration that they can do so).

A number of studies that examine how children learn from exploration could be interpreted as consistent with the possibility that young children are not always designing unconfounded interventions in their exploration and learning (e.g., McCormack et al., 2016; Meng et al., 2018; Nussenbaum et al., 2020; see also Reiber, 1969, for similar findings[6]). In these studies, children are asked to learn about a causal system through actions that they take on the system. For example, McCormack et al. (2016) showed 5- to 8-year-olds a causal system of three possible events and possible models for how the events were related to one another (for example, event A could cause both events B and C as a common cause, or event A could cause event B, which in turn caused event C, as a chain). They then asked children to learn the causal structure through intervening on the system. Children could learn some structures better than others. For example, common causes seemed to be easier to learn than chains. More interestingly, children in these studies often tried to make events occur. Younger children in particular more often intervened on a "root" node—an event that causes the most outcomes (for example, in the $A \rightarrow B \rightarrow C$ model described above, they mostly intervened on event A). Results like these have led some to describe children not as little scientists but rather as little engineers, whose primary goal is to produce as many effects as possible (following Schauble et al., 1991).[7] Contrary to rational constructivism (which views children as little scientists, as reviewed in chapter 2), these data suggest that children are not naturally motivated to generate data that facilitates learning.

Nussenbaum et al. (2020) proposed a mixed model that integrates aspects of both of these approaches. They suggest that young children have two exploratory strategies, one that involves maximizing information gain in their exploration and one that displays what they call a "positive testing bias" (following work by Coenen et al., 2015, on adults), which motivates children to make as many effects occur as possible. As children move into adolescence, the balance of strategy use shifts away from a positive testing bias and toward more of an optimal information-gain strategy for self-guided learning.

This approach is sensible, but it might tell only part of the story, particularly when contrasted with the exploration and learning that children engage in when they are in more authentic learning environments like a children's museum. As we discussed in chapter 3, when looking at parent-child interaction at a set of gear exhibits, children's systematic exploration of exhibits was related to their causal knowledge of the exhibit components (Callanan, Legare, Sobel et al., 2020). Moreover, numerous studies that we have already discussed suggest that children can learn from their play, both on their own and collaboratively (see Weisberg, Hirsh-Pasek & Golinkoff, 2013, for a review). One of the reasons that play can be such a powerful mechanism for learning is that play involves self-generated actions; children decide what actions to take and then observe the results of those actions.

But self-generated action has some drawbacks, particularly for younger children. As noted in chapter 3, preschoolers can be more influenced by the data they themselves generate than by the data they observe, even if this leads them to make the wrong conclusions. For example, Kushnir and Gopnik (2005) asked 4- to 6-year-old children to make inferences about the efficacy of objects that had probabilistic relations with the blicket detector. In their critical experiment, children observed two objects (A and B) that were each placed on the machine twice. Object A activated the machine both times, and object B did not activate the machine both times. The children were then given both objects and asked to put each on the machine. When the child put A on the machine, it did not activate, but when the child put B on the machine it did. When children were asked which of the two objects had "more special stuff inside" (which they had established was responsible for the machine's activation), children in this condition were more likely to choose object B—the one that worked for them, even though it did not work the majority of the time. In a control condition, children observed A and B being placed on the machine with the same efficacy.

The data were the same, but in this condition, children never touched the objects. Now, they (correctly) chose object A more often. This result suggests that there are drawbacks to self-generated interventions: They might lead children to overvalue what they themselves do as opposed to objectively inferring general patterns of efficacy.

In general, preschoolers overestimate the importance of action for learning (Sobel & Letourneau, 2018) and to some extent discount actions that they generate that are not efficacious (Gweon & Schulz, 2011). This overreliance on self-generated action indicates that how children learn from their own actions might depend on the results of those actions.

We investigated this issue in our lab in two ways. First, we asked parents to play together with their 3- or 4-year-old to learn the rules of how a blicket detector worked. Those rules had previously been shown to be difficult for preschoolers to learn (Walker et al., 2017), but were fairly straightforward for parents. It turned out that the actions that children made during their play did not predict how well they learned. What did predict their learning was the nature of the parent-child interaction. Children whose parents allowed them to play freely with little guidance learned the rules the least, whereas children whose parents guided their play or engaged in more direction by setting goals for their play learned the rules better. However, children in this latter group (the one that was more parent-directed) were the least engaged by the task (Medina & Sobel, 2020). That is, children can learn from both directed and guided play, but are more engaged when parents do not explicitly tell them what to do.

Second, we looked at how 4- to 7-year-olds played with a blicket detector on their own after being shown ambiguous data (Sobel, Benton, Finiasz, Taylor & Weisberg, 2021). In this case, we restricted children to a certain type of intervention: They could not simply test individual potential causes one at a time, but rather could only test pairs of potential causes together. This required them to use Tschirgi's (1980) strategy of varying one thing at a time if they wanted to resolve the ambiguity.

We found that, independent of children's age, their first action and its results influenced how children played. If children's first action varied only one thing relative to the initial demonstration, they were more likely to generate the information necessary to disambiguate the initial data, even if that first action did not itself provide unconfounded information. Moreover, if children's first action was efficacious, they were more likely to continue to activate the machine on subsequent trials. This suggests that children might sometimes be motivated to uncover causal structure in their play, but what happens when children start playing might influence their later goals. That is, young children *can be* systematic in their exploration, even if they are not using the most efficient control of variables strategy and even if they do not engage in such behavior all of the time.

This study, together with the findings reviewed in this section and the last one, shows that children in preschool and in early elementary school seem to understand something about the logic of the control of variables strategy. They can recognize its utility in forced-choice tasks and can sometimes interpret data generated from this strategy. In turn, this suggests that children's difficulties with tasks like Earthquake Forecaster may lie in that system's causal complexity or its contextualization and not necessarily its requirement to generate unconfounded experiments, though more research is needed on these issues.

Diagnosis and Belief Revision

Mature scientific thinking involves a suite of skills, importantly including the ability to diagnose what underlying causal structure was likely responsible for a currently observed set of data (often in the service of making predictions or inferences about the system) and the ability to continually revise and update one's beliefs in light of new evidence. Although the causal graphical model framework discussed in chapter 2 can provide mathematical formalizations for these processes, studies of children's early causal reasoning abilities have not tended to measure whether children can engage in them.

Consider again most experiments using a blicket detector. Children presented with this paradigm are tasked only with saying which objects are blickets (or efficacious in some other way). That is, they need to make predictions about which block or blocks will turn the machine on. These predictions might have come from a process of diagnostic inference, but that process was minimal at best. More importantly, most blicket detector studies do not ask children to revise their articulated beliefs in light of new evidence, so they provide no evidence that children are able to update their hypotheses explicitly. Given that these processes are crucial to scientific thinking, these arguments strongly suggest that the skills that allow children to succeed at young ages in causal reasoning tasks might be different from the skills that they will eventually need for mature scientific thinking.

Indeed, a small number of recent studies suggests that these kinds of reasoning skills continue to mature well after the preschool years. When preschoolers see systems that do not provide them with full information about a system's causal relations, they are less able to construct causal models and to draw appropriate conclusions than are older children (Sobel et al., 2017). In addition, children's diagnostic reasoning capacities interact with some of the variables discussed above, particularly the ability to reason about uncertainty (Fernbach et al. 2012; Kuhn, 2007a), making it even harder to determine when and how these abilities develop.

With respect to this kind of belief revision in young children, the literature is scarce. It is well known that even adults have difficulty considering information that contradicts their existing beliefs, and they will often distort the new information to conform to their beliefs or reject altogether in favor of seeking out confirming evidence (*confirmation bias*; see Nickerson, 1998). While it seems likely that children are implicitly forming and revising hypotheses over the course of a blicket detector experiment, no experiment that we know of has examined this directly. To address this issue, our labs are currently conducting research to examine these processes in children. In one study (Macris & Sobel, 2017), 4- and 5-year-olds were shown that a cube with a particular kind of internal property activated the blicket detector. We asked children whether they thought cubes made the machine go or whether other objects with that same kind of inside made the machine go. Crucially, no matter what children chose, they were provided with evidence that this initial guess was wrong. In some cases, children were shown weak counterevidence—one demonstration that their initial guess was wrong. In other cases, children were shown stronger evidence—three demonstrations that their initial guess was wrong. In still other cases, children were given verbal feedback about that counterevidence. For example, an experimenter told them, "I've seen these objects before, and I know that the cubes make the machine go."

These experiments found that children have some tendencies to change their belief on the basis of counterevidence (see also Bonawitz et al., 2012;

Kimura & Gopnik, 2019; Young et al., 2012). But they are much more likely to succeed when they are directly told that their initial guess was wrong than when they observed counterevidence (although the stronger counterevidence was more effective than the weaker counterevidence). In the cases where they were given verbal testimony, they changed their belief about 80% of the time. In the cases where they observed data, they changed their belief more frequently than chance levels, but not more than half the time. There is definitely room for children to improve.

One facet of these studies is that children's initial belief is based on a guess. A different way of testing children's belief revision is to look at how they generate data to test a more firmly held belief. To do this, Köksal-Tuncer and Sodian (2018) introduced children between the ages of 4 and 6 to a blicket detector and four blocks, two that were heavy and two that were light. The heavy blocks activated the machine and the light blocks did not, suggesting that weight was the underlying factor that caused the machine's activation. Children then observed two new blocks that contradicted that belief (the light block made the machine activate and the heavy one did not). At this point, they showed children four new objects, and revealed that one object but not a second one had a sticker on the bottom, hidden from view. Critically, among these four objects, one heavy and one light object had a sticker, and the objects with a sticker activated the machine, thus invalidating the "weight" hypothesis by offering a hidden, alternate cause. Children were allowed to explore these objects while the researchers measured two aspects of their behavior. First, the researchers considered what objects children placed on the machine and the order of that placement (i.e., what they did first). They also measured what children said was the cause of the machine's activation after exploring these objects.

For the most part, children tested blocks that provided confounded data (heavy with the sticker and light without the sticker) as often as blocks that provided unconfounded data (heavy without the sticker, and light with the sticker). Moreover, children's first actions were equally likely to test a block that provided unconfounded data as confounded data. But when they were asked directly what activated the machine, the majority of children (76%) reported that only the objects with the sticker would do so. Toward the end of the study, a second experimenter articulated the belief that heavy objects made the machine go. Children were given the opportunity to argue against this belief, which they did in many cases by showing the adult disconfirming evidence (i.e., the stickers). So while children in this study could articulate changes to their belief (and contradict others who held what they believed to be an incorrect belief), children were not systematic in the way they went about gathering evidence for their belief changes (see also Ronfard, Chen & Harris, 2018, 2021).

These studies can provide insight into why children struggle with scientific thinking tasks like Earthquake Forecaster. The aspects of scientific thinking that are tested in Earthquake Forecaster require children to predict which variables will be associated with earthquake risk on the basis of their current diagnosis of how the causal system works, which in turn is based on their past observations of the system. Crucially, they must continually form and revise hypotheses about how the system works in order to get to the right answer. Given the findings reviewed above, showing that children struggle to revise their beliefs even when given clear instances of counterevidence, it is not surprising that children struggle with this system, although it does provide a more realistic view of their performance with scientific thinking tasks.

Metacognition

One vitally important process that develops during the elementary-school years is metacognition: an explicit awareness of one's own thought processes. Although this process is used in a wide range of contexts, it is also crucial for fully mature scientific thinking, as we argued in chapter 1. Specifically, metacognition allows one to explicitly recognize that a currently held belief does not accord with a new piece of evidence. This ability is thus closely tied to the process of belief revision described in the previous section. Once one understands that one holds beliefs and that these beliefs could be false, then one is in a better position to understand the need to revise these beliefs on basis of incoming evidence. Similarly, in terms of testing hypotheses, one must also understand how evidence could bear on that hypothesis and that the evidence could demonstrate that the hypothesis is incorrect—that is, that one could have a false belief.

Arguably, one reason that younger children respond differently on measures of belief revision than older children could be because of the development of metacognitive capacities. Understanding that others' behavior can be motivated by false beliefs is standardly said to develop around age 4, while the more advanced abilities characteristic of metacognition (sometimes called "interpretive" or "advanced" theory of mind) develop between the ages of 5 and 8 (Carpendale & Chandler, 1996; Flavell et al., 1995; Lagattuta & Wellman, 2002; Osterhaus et al., 2016; Tang et al., 2007).

Following this work, Kuhn, Cheney, and Weinstock (2000) documented a developmental progression in children's abilities to coordinate multiple people's beliefs. Children start out with an absolutist view of beliefs, in which only one interpretation of a situation can be right; there is no room for subjectivity. Around age 7, they shift to a multiplist view, in which two people can have different interpretations, but any interpretation has equal merit (sometimes called relativism). A later developmental achievement, which might not occur until adulthood, is to attain an evaluativist view: Different interpretations of a situation are not all equally valid. This view recognizes that these interpretations must be evaluated in light of other knowledge and theories to determine which is closest to the truth, and that this process of evaluation can be subjective (see also Barzilai & Weinstock, 2015). While navigating disagreements is not quite the same as aligning one's own beliefs with the world, both processes involve some understanding of how beliefs work and how they are formed, which in turn can allow one to recognize that one is in a situation where one's beliefs could potentially change.

None of the studies of children's causal reasoning or of children's scientific thinking show that children can engage in explicit thinking about their reasoning processes or explicit reflection on how their actions can produce the data necessary for learning. As noted above, the preschoolers recruited in studies that use the blicket detector are generally too young to have this kind of metacognitive understanding and hence are not able to explicitly reflect on how their knowledge can change. Although older children are sometimes asked to reflect on their thought processes as they engage in control of variables tasks like Earthquake Forecaster, it is unclear whether this kind of metacognitive reflection is too complex for children in general. That is, we know little about how children's developing metacognitive abilities might bear on their abilities to think scientifically, particularly in cases that involve hypothesis generation and belief revision.

In chapter 7, we describe a set of studies that begins to address this gap by examining how children's developing understanding of conflicting beliefs might bear on their scientific thinking abilities.

Building Bridges

We have hoped to make clear in this chapter that there are many points of disagreement between the literature on infants' and preschool-age children's precocious causal reasoning abilities and the literature on elementary-age children's scientific thinking abilities. Briefly, studies of children's causal reasoning tend to recruit young children, present them with simple and minimally contextualized causal systems, and ask them direct questions to see whether they can make predictions about those systems' structures. Studies of children's scientific thinking tend to recruit children in elementary school (and beyond), present them with richly contextualized and causally complex systems, and allow them to explore these systems in order to draw conclusions on their own. Small wonder, then, that preschoolers tend to succeed in the former tasks while older children are often not successful on the latter.

While we do not claim to have identified all the points of disagreement that exist between these two bodies of literature, we hope to have made clear that there are many differences in the methods used to test these two groups of children. This is more than just a problem of inconsistent methods, however. The issue is that, because of these inconsistencies and their interrelations, we are not yet in a position to understand how scientific thinking develops. Do older children genuinely lack scientific thinking abilities, or do the complexities of the tasks with which they are presented mask their skills? Do younger children genuinely possess the abilities to reason scientifically, or do the tasks with which they are presented oversimply the situation to the point of being too unrealistic to test those abilities? Do early causal reasoning abilities develop into scientific thinking abilities, and if so, how? These are the questions that we aimed to tackle in the empirical work that we present in the next section of the book. Our approach has been founded on the need to take seriously the results from both sets of studies: Young children really may possess important causal reasoning skills, but they still have much to learn regarding certain facets of scientific thinking. Further, some aspects of scientific thinking may have a more prolonged developmental trajectory, and developmental trajectories may differ in different cultures. Given this, we aimed to find a way to put these results into closer dialogue. In the next few chapters, we review these studies and their results, then discuss how this work can help to paint a fuller picture of the development of scientific thinking.

II Bridging Causal Reasoning to Scientific Thinking

5 A New Blicket Detector Task

The rational constructivist framework that we reviewed in part I of the book takes seriously the idea that children are little scientists (Gopnik et al., 1999). But if this is the case, when and how do they become able to explicitly form hypotheses, revise beliefs, and collect and interpret data? Answers to this question have been elusive. Although we have argued that children's learning is well-described by the causal graphical model framework, which has been incorporated into the rational constructivist view of development, these theories are not yet full descriptions of how learning and development work (as discussed in chapter 3). Most relevantly for our purposes, there is a disconnect between work on children's causal reasoning abilities, which seem to develop early and which arguably underlie the processes involved in scientific thinking, and work on children's later-developing scientific thinking abilities, which documents children's struggles to think about more complex and realistic scenarios (as documented in chapter 4). In this chapter we present an empirical measure that begins to reconcile some of these disparate sets of findings. In the remainder of part II, we employ this measure in several studies investigating how children's causal reasoning abilities provide a basis for their developing abilities for scientific thinking. Importantly, these studies are meant to illustrate how we might bridge the gap between causal reasoning and scientific thinking, and thereby identify some of the important developmental changes that are required to link these two sets of abilities. These studies, however, do not present a definitive or unique way to do so.

As we reviewed in chapter 4, many variables differ between tasks designed to study children's precocious causal reasoning and tasks designed to study children's scientific thinking. And, despite their similarities, there are also many differences between causal reasoning and scientific thinking. Given this, as noted above, our goal with this new empirical measure is simply to provide one potential bridge between these abilities, one way to begin to address the complexity of these issues. To do so, we focus on two of the variables identified in chapter 4: age and causal complexity. First, in terms of age, we use a task that can be understood by younger children, but that is still difficult enough to present a challenge to children at a wide range of ages. We are thus able to administer exactly the same measure to children in preschool and in elementary school, as well as to adults. This consistency allows us to track the emergence of diagnostic reasoning abilities over time.

Second, as is common in work on older children's scientific thinking abilities, our task presents a complex causal system. It includes four potential causes, each of which could be in one of two possible states, and there are three possible effects in this system. This is a major departure from traditional work on causal reasoning, which typically focuses on a smaller number of potential causes (usually two) and outcomes (usually the presence of absence of one). Further, the structure of the causal system is interactive: Two of the causes separately have one effect, but jointly they create a different effect. This type of system resembles much more closely the messiness and complexity of real-life scientific problems and of the tasks designed to capture these problems. Indeed, we based this structure on ones instantiated in investigations of scientific thinking (e.g., Dean & Kuhn, 2007), though with somewhat fewer potential causes and effects, to make it easier for our youngest participants to grasp. In addition, our procedure for demonstrating this system presents combinations of causal factors and does not show the effect of each cause in isolation. Participants thus have to manage some degree of uncertainty and draw inferences from the sets of data that they have observed to determine how the system works.

However, as in standard work on causal reasoning, this system does not present any explicitly scientific content (though it can be used to do so; see chapter 6). Specifically, it uses a procedure with a novel blicket detector, in which colored blocks are placed on a machine in various combinations. Our choice to maintain the minimal contextualized nature of this task, at least for these initial studies, is deliberate: We want to test how children perform with a higher degree of causal complexity before additionally examining their abilities with real-world scientific content.

In addition, rather than asking participants to explore the system for themselves, we present them with sets of data that we constructed. We also ask them to select an answer to our test question rather than to generate one themselves. These design choices are again more in line with work on causal reasoning than with work on scientific thinking, which puts more of an emphasis on children's own exploration of a system. Again, we chose this method of presentation primarily to bring the task within the reach of our preschool-age participants, because it might be easier for children to demonstrate their causal reasoning abilities by recognizing a correct answer that is presented to them. Additionally, as we saw in chapter 4, there will likely be differences in how different children interact with the causal system, which could in turn affect the inferences that different children are able to make; as we've discussed, children's exploration of causal systems is influenced both by the social nature of the interactions around these systems and by the efficacy of their actions on these systems. We want to ensure that all participants observe the same set of data, so that we can describe how children across a range of ages respond to exactly the same task. Of course, this same causal system could be presented as an exploration task; we are aiming to do this in future work. For now, let's meet this new blicket detector task and see how children (and adults) respond to it.

How the Task Works

Our new task begins with an experimenter presenting the blicket detector to a participant. This version is a battery-powered black box with a translucent white pressure-sensitive plate on top (see figure 5.1). The pressure-sensitive plate really works, so that the machine activates when something is placed on top of it. Like the blicket detectors we have already described, though, the actual activation of the machine is a magic trick—any object that pushes down the pressure plate can make the machine go. The pressure plate is controlled by an enabling switch on a remote. The colors of the lights that are displayed through the translucent top are also controlled by the remote, which is operated by an experimenter out of sight of the participant.[1]

The experimenter explains that the box sometimes lights up and plays music when blocks are placed on it. She then introduces a set of four blocks, all identical in size and shape but different in color. To illustrate the task,

Figure 5.1
Blicket detector used in the experiments described in this chapter. This detector could activate two different ways—green or red. Each activation played a unique song.

we will describe one version that we ran, using rectangular prisms that were yellow, black, orange, and blue. The exact shape or color of the blocks do not really matter, though it is important to ensure that there is no connection between the color of the blocks and the color of the lights on the machine. This way, participants will not be able to assume that a green block is responsible for making the machine light up green, or that a yellow block and a blue block together are responsible for making the machine light up green. Further, the blocks are put inside a clear plastic container before being placed on the machine so that they all appear to contact the machine at the same time. Again, this ensures that no causal factors appear to be responsible for making the machine light up other than the combination of colored blocks.

The experimenter demonstrates how the machine works by first placing all four blocks (yellow, black, orange, and blue) into the container and putting the container on the machine. The machine lights up green and plays a song (e.g., a MIDI version of *Fur Elise* or "Twinkle Twinkle Little Star"). This is demonstrated a second time, and the experimenter narrates what is happening: "See, the machine is turning green. When I put all four blocks on the machine, it turned green." The experimenter then places a photograph of the four blocks on the table next to the machine, accompanied by a green dot, explaining, "I'm going to put this green dot here so that we can remember what happened. When I put the yellow block, the black block, the orange block, and the blue block on the machine, it turned green." This ensures that the task places few memory demands on participants; they can always refer back to the reminder cards on the table to tell them about the events that they saw.

Then, the experimenter places three blocks inside the container: the yellow one, the black one, and the blue one. When these blocks are placed onto the machine, it turns red and plays a different song. As before, the experimenter demonstrates this twice and narrates her actions ("Look, it's turning red!"), and places a photograph of these three blocks on the table with a red dot as a reminder.

The third event is similar to the second, except that this time the yellow, black, and orange blocks are placed on the machine together. This combination also makes the machine turn red and play the second song. Another reminder card with these objects and a red dot are placed on the table.

Finally, the experimenter places just the yellow block in the container and puts it on the machine. Nothing happens. This is demonstrated twice, to ensure that the participant knows that it is not an error, and the experimenter verbally reinforces the fact that the yellow block does not activate the machine. The experimenter then places a photograph of just the yellow block on the table with no accompanying colored dot.

This is the end of the data-presentation part of the task; participants have now seen four different combinations of blocks and three different outcomes.[2] We then test children's ability to diagnose the causal structure of this system. To do so, the experimenter puts up a cardboard barrier, blocking the participant's view of the blicket detector and the blocks. The experimenter tells the participant that she is placing two blocks on the machine, and she reports that the machine lit up green. In reality, to avoid providing any visual cues as to which blocks were used, the experimenter merely presses down the pressure-sensitive plate without moving the blocks from their position on the table. Participants can confirm for themselves that the machine turned green because they can hear the music that was associated with the green light. They can also see the green light from the detector illuminate the experimenter's face. The activation stops and the experimenter then removes the barrier to reveal the machine and the four blocks. Then, the participant is asked the test question: "Which two blocks did I place on the machine to make it turn green?" To make this question a bit easier for child participants, we present it as a multiple-choice task rather than as an open response. All participants are asked to choose an answer from among three possible pairs: black and blue, black and orange, and orange and blue. Photographs of each of these options are placed on the table in front of the participant, and the experimenter asks them to choose which pair made the machine turn green.

We considered the correct answer to the test question to be the orange and blue pair. Here's our logic: The yellow block is not efficacious (and also not part of any of the test options), so it can be discarded from consideration. Participants have already seen that the combination of black and blue and the combination of black and orange make the machine turn red. The only time that they see the machine turn green is when the blue and the orange blocks are placed on the machine together (though this was only ever done in combination with the other two blocks). For the purposes of scoring participants' performance on this task, we label the choice of the orange and blue pair as correct and the choice of either of the other two pairs as incorrect.

Importantly, the data that we presented in the demonstration phase involved one, three, or four blocks, but the test question asks participants to reason about a situation they have not seen: the effect of two blocks on the machine. Participants must extrapolate from the events that they have already observed to draw conclusions about a new situation. This task provides them with enough information to do so, but not in a way that makes the conclusion obvious.

Moreover, there is a relation between the structure of this task (multiple potential causes and multiple possible effects) and the interactive causality observed in measures of scientific thinking. Yellow and black are inefficacious. Blue or orange are efficacious and individually make the machine turn red, but together change the nature of the effect (the machine turns green). This is similar to the kind of diagnostic inferences children have to make in some of the scientific reasoning measures we discussed in chapter 4. Fundamentally, this procedure examines whether children have the diagnostic reasoning capacity to engage in the kinds of inferences required in measures of scientific thinking.

So how do participants respond to this task?

Adults' Performance

When we started this project, we initially presented it to our colleagues around our two departments. All of these trained scientists got the answer correct, although only some of them could articulate the logic behind their choice; several admitted they were guessing. But to provide a more formal anchor against which to compare the developmental data, we recruited 40 undergraduate students from a psychology participant pool (28 women; mean age = 20 years and 0.8 months; age range 18–27) and presented them with the task that was just described. Thirty-four of these 40 students, or 85%, answered the test question correctly. This is significantly greater than chance performance (33%, because there are three answer options presented)—but, interestingly, it is not at ceiling.

We also asked these participants to justify their answer choices. Many of them were able to describe the logic of the task correctly. For example, one student said, "I discounted the black because it's present in all three [options], so that's left me with these two [orange and blue]." Another student reasoned similarly: "In the one where it lit up red, the orange was present by itself, and then the blue was present by itself, but then when both the orange and blue were present, it made green." Or, more succinctly, "Because the combination of blue and orange isn't in the ones that make it red."

It is worth noting, however, that even some of these college students who responded correctly to the test question were not able to justify their decision explicitly. In two cases, students answered correctly but then admitted to guessing or not knowing how the machine actually worked. For example, "It was just a guess, because you don't really have much information from the previous stuff about what would happen with two blocks as opposed to three or four."

Conversely, of the six students who answered incorrectly, some of them justified their decision quite logically. For example, one student chose the black and orange combination, saying, "I just kind of figured that each time it lit up with a black block, the black block was necessary for the box to light up. This combination without the orange was red, and for this one [all four blocks], black and orange went green." Some of these students also admitted to guessing, but in four of the six cases, they tried to justify their choice with respect to some combination of the blocks (although sometimes that justification led to a contradiction).

The overall picture, then, is that adults generally reason about this task in the manner we intended and can usually report on the reasons for their choice at test. This implies that adults are explicitly thinking through the task, which allows them to have access to their thought processes when we ask them to justify their choice. Is the same true of children?

Children's Performance

Our initial presentation of this task involved a sample of 72 children (see full report in Sobel, Erb, Tassin & Weisberg, 2017, Study 2). To look at developmental change, we divided our participants into two groups based on age. There were thirty-six 5- and 6-year-olds and thirty-six 7- and 8-year-olds.

Children in the younger group chose the correct response 39% of the time, which was no different from simply guessing (33%). However, children in the older group chose the correct answer 61% of the time, which was significantly better than just guessing. This level of performance was not quite as good as the adults', but it was significantly better than that of the younger children. These data suggest that the ability to reason about complex, but less contextualized, causal systems develops between the ages of 6 and 8.

We found similar levels of performance from a second set of 116 children (see full report in Weisberg, Choi & Sobel, 2020, Experiment 1). These children ranged in age between 4 and 10 years, allowing us to examine a wider spectrum of performance than before. We found that, overall, participants were 45% correct, significantly above chance. There was also a clear developmental trend: 4- and 5-year-olds as a group were 38% correct, 6- and 7-year-olds were 41% correct, and 8- through 10-year-olds were 55% correct. Again, these data show evidence of development between the ages of 6 and 8.

Additionally, we asked a subset of these participants ($n = 39$) to justify their choice following their response to the test question, as we had done with the adults. Many participants were unable to justify their responses coherently or simply said, "I don't know," while about 18% of children referred back to the data that they had seen and gave reasonable explanations for why they chose the answer option that they did. For example, one child said, "Because they [orange and blue] are the only two colors in this [pointed to the picture of all four blocks] when it turned green." Although justifications with this high level of maturity did not appear particularly often, it is noteworthy that children who provided justifications like this one tended to perform better on the task (86% correct) than children who provided other types of justifications (56%), although this trend was not statistically significant, possibly because the sample size was too small. But this pattern could suggest that one part of the developmental trend we are observing is a shift from using a more implicit method of solving this task to using a more explicit one. The latter method would allow children to align their response to the test question with the justification of their choice.

Further Uses of the New Blicket Detector Task

Our goal in constructing this new blicket detector task was to begin to reconcile the findings from work on causal reasoning in young children and work on scientific thinking in older children. This new task combines features from causal reasoning studies, such as the use of systems that lack explicitly scientific content and the role of experimenters in selecting and presenting sets of data, with the causal complexity of scientific thinking studies. Of course, this is just one possible way to more closely align the work from these two fields, and our results using this task should be taken as case studies of ways in which this alignment can happen, not as a definitive method of doing so.

Nevertheless, our findings with this task begin to suggest new ways to think about the development of scientific thinking skills. One thing that seems clear is that reasoning about this kind of interactive causal system in the way that we have described is challenging for the preschoolers we tested; only by about 6 or 7 years of age did children as a group begin to respond to this task at levels greater than expected by chance. This strongly implies that preschoolers' success with tasks using blicket detectors (such as many of the studies described in chapter 2) is due, at least in part, to the type of causal systems that they are asked to diagnose with those tasks. Introducing uncertainty and interactive causality makes our task more difficult. There is thus evidence for a more protracted developmental trajectory for reasoning about these kinds of systems. This conclusion aligns with other work from our labs showing that preschoolers respond at chance levels about causal systems with uncertain causes until about age 6 or 7 (e.g., Erb & Sobel, 2014; Sobel et al., 2017) and that preschoolers' belief revision abilities are present but weak (e.g., Macris & Sobel, 2017).

However, this task in itself does not fully answer the question of when and how scientific thinking skills develop. For one thing, this system is mostly decontextualized; it lacks explicit scientific content. We take up the question of what happens when this type of content is introduced in chapter 6. In addition, it is still not clear exactly how children are solving this task and whether (and when) they are able to reason about it explicitly. Children's justifications indicate that at least some of them (and perhaps even some adults) may not have conscious access to the procedures they use when engaging in the diagnostic reasoning necessary to respond to our test question—even when they answer this question correctly. But, as noted in chapter 4, being able to explicitly reflect on one's own beliefs as one thinks about a causal system is a crucial part of scientific thinking. We address this connection to metacognitive processes in chapter 7.

6 Contextualization in Causal Reasoning and Scientific Thinking

As discussed in chapter 4, one major difference between causal reasoning tasks and scientific thinking tasks is the extent to which these tasks contain real-world scientific content. Although we might be able to conceptualize a kind of "pure" scientific thinking ability that operates independently of any particular content, that is rarely if ever how reasoning proceeds in the real world. Typically, we use our existing knowledge of how the world works to form a hypothesis space in which to reason scientifically (Goodman et al., 2011; Griffiths et al., 2011; Sobel et al., 2004; Ullman et al., 2012; see also Cheng & Novick, 1992, for another version of this hypothesis as applied to adult cognition). We refer to this as the *contextualization* of a problem: the extent to which a problem requires (or seems to require) us to draw on our prior knowledge.

Importantly, contextualization lies on a continuum, such that none of the tasks that we have described sit at either extreme. Nevertheless, the two families of tasks that we have been considering do differ in important ways. Causal reasoning tasks, such as ones involving the blicket detector, are less contextualized—they require minimal prior knowledge about the pieces of the system (i.e., the blocks and the machine) in order to make inferences. The diagnostic reasoning task using the blicket detector that we described in chapter 5 is presented as such. However, this task (and most causal reasoning tasks in general) still relies on some degree of general knowledge about how causal systems in the world work, for example, that causes precede their effects or that the machine works on the basis of contact.

In contrast, scientific thinking tasks are more contextualized; they present information about particular physical or biological systems in the world, like earthquakes, gravity, friction, or disease transmission. Although they do contain specific scientific information, such tasks do not capture the full richness and complexity of real-world causal systems, for which it is virtually impossible to construct a fully controlled experiment. Indeed, when we teach experimental design to undergraduate students, they often get lost when they begin to consider just how many variables they might have to control.

These examples illustrate that causal reasoning tasks and scientific thinking tasks differ starkly in their degree of contextualization. In this chapter, we review a set of studies designed to determine the extent to which contextualization matters to children's reasoning. One possibility is that the inclusion of real-world content helps children's reasoning abilities by instantiating an otherwise abstract system in a familiar context. Being able to tap into their existing knowledge could jump-start children's thinking about the causal system. But a different possibility is that the inclusion of real-world content hurts children's reasoning abilities by distracting them away from the underlying causal structure and toward irrelevant features of the scientific information. To explore these two possibilities, we first describe some prior work on the effects of contextualization on adults' and children's reasoning. Then, we review a series of studies designed to investigate directly how different levels and types of contextualization affect performance on the blicket detector task we described in chapter 5 (previously published in Weisberg, Choi & Sobel, 2020).

The Role of Contextualization in Adult Reasoning

Work in adult cognitive psychology finds that contextualization often matters for reasoning, particularly for causal or logical inferences. The trouble is that a problem's context sometimes facilitates people's reasoning and sometimes interferes with it (see e.g., Stanovich, 2004, for a review).

One example of how the inclusion of more real-world content can help adults' reasoning is the Wason card selection task (e.g., Wason, 1960, 1966, 1968). Adult participants in this task are shown a set of cards. They are told that each card has a letter on one side and a number on the other. For example, they can see four cards: one with the letter E, one with the letter F, one with the number 4 and one with the number 7. Participants are then told a rule: If there is a vowel on one side, then there is an even number on the other side. Participants are asked to turn over all and only the cards that are necessary for proving this rule true or false.

This is a problem of logical reasoning. Participants are given an "if-then" statement (called a *conditional statement*), and they need to determine whether that statement is true. Importantly, conditional statements are only false in one situation—when the antecedent (the part of the statement that starts with "if") is true, but the consequent (the part of the statement that starts with "then") is false. So this rule is only false if there's a card with a vowel on one side and an odd number on the other.

That is a big hint to figuring out which card or cards one has to turn over to verify the truth of the rule (although participants are not given that hint). To be sure that the conditional statement is true, one has to do two things. First, one has to turn over the vowel (E) to make sure that there is an even number on the other side. Most participants do this, and many then stop there, assuming that they have verified the rule. Those that do not stop tend to additionally turn over the even number (4), to make sure that there is a vowel on the other side.

But this is a logical error. Why? Because the rule only goes in one direction: If there is a vowel, then there's an even number. The rule never says that even numbers have to have vowels on the other side; this is the difference between a conditional and a biconditional, or "if and only if." In simple conditional reasoning, it does not matter whether there is a consonant on the back of the even number because the rule only specifies the relations for vowels.

The error that tends to occur on this task is that the majority of participants do not check the other part of the conditional; they do not turn over the card with the odd number (7). Turning this card over is necessary because one has to make sure that there is not a vowel on the other side. If there is a vowel on the other side, then the rule is obviously wrong, because the rule states that a vowel on one side means that there is an even number on the other, not the odd number 7.

This abstract—relatively decontextualized—version of the problem might seem counterintuitive and difficult to follow. But there is a much easier version of this problem, which presents the same logical inference to participants. Let's substitute the letters and numbers on the cards for ages on one side and beverages on the other (following Griggs & Cox, 1982, Experiment

3). For example, say the cards show two ages (17 and 22) and two drinks (beer and soda). Now, consider the legal drinking age in the United States: If you are under the age of 21, then you cannot (legally) drink alcohol. Which cards should you turn over to verify that the four people (represented by the cards) are following the rule? Griggs and Cox gave participants a version of the task in this context, further emphasizing the rule by having participants imagine that they were police officers trying to enforce the law. The answer in this case should be obvious. You turn over the card with the 17 on it, because you want to know what the underage person is drinking. And you also turn over the beer card to make sure that there is someone 21 or older on the other side.

Note that the structure of the problem is the same as the one described above. Participants are presented with conditional statements that they have to validate, and four possible values for the antecedent and the consequent that represent the possible worlds that could exist in this conditional. In the less-contextualized case—the original Wason selection task with letters and numbers—participants were not great at answering these questions. In fact, none of the participants in Griggs and Cox's experiment who were given this version of the problem got it right (i.e., turned over the E and the 7). But in the more contextualized case—the one with the rule about the legal drinking age—participants were pretty good; 71% of Griggs and Cox's participants succeeded at turning over both cards in this version. There are a number of follow-ups to this task, all of which point to the idea that adult reasoning obeys certain pragmatic constraints or has certain modes of thinking based on our prior knowledge (e.g., Cheng & Holyoak, 1985; Johnson-Laird et al., 1972). Moreover, adults tend to search for explanations of less contextualized problems (or strangely contextualized ones) in order to make the conditional (or other causal construct) make sense to them (e.g., Korman & Malle, 2016).

These findings suggest that, if we can place a reasoning problem into a familiar context, then reasoning about that problem becomes easier. However, as noted above, there are also cases in which the presence of a familiar context makes a problem harder. For example, Kuhn (2007a) described an experiment where she presented a set of adults with the task of determining which combination of entertainment options (e.g., door prizes, comedian) would be most effective at boosting the success of a fundraiser. Although clear evidence demonstrated that some factors were causal (because they were always related to the desired outcome of high fundraising) and some factors were not (because they occurred in all scenarios, regardless of outcome), participants had trouble judging which factors were related to the outcome. Further, participants used inconsistent logic in justifying their responses, seeming to be swayed by their preexisting knowledge of door prizes and comedians and ignoring the underlying logical structure of the task. She writes, "[the participants'] responses revealed that their judgments were in fact influenced by their own ideas about how effective these features ought to be" (p. 47).

Another example comes from the logical reasoning literature. Consider a simple (hypothetical) example of syllogistic reasoning. One can be told that all things that are smoked are good for one's health, and then that cigarettes are smoked. In this case, adults are less likely to draw the correct logical conclusion (in this example, not in real life, of course) that cigarettes are good for one's health. Because this conclusion is false in the real world, participants have a difficult time saying that it is correct, even though it is supported by the premises (Evans et al., 1983; see also Markovits & Nantel, 1989).

These lines of research on adults' logical reasoning show that context matters for adults' inferences. Sometimes adding more context facilitates reasoning, but in other cases it can impair adults' ability to come to a logical conclusion. What remains unclear is whether including a real-world context in a reasoning problem changes how children respond to reasoning tasks.

The Role of Contextualization in Development (or, What's Fantasy Got to Do with It?)

Research in developmental psychology has also looked at the question of how contextualization impacts learning. Interestingly, much of this work has compared children's reasoning within typical, everyday contexts to their performance when presented with fantastical situations. As a whole, these studies tend to find that taking a problem out of its everyday context and placing it into a fantasy context improves children's performance.

One example comes from work by Dias and Harris (1988, 1990). These researchers presented logical syllogisms problems to 4- to 6-year-olds that were contrary to fact, such as, "All cats bark. Rex is a cat. Does Rex bark?" Preschoolers were not that good at reasoning about these kinds of problems, although the older children were a bit better than the younger children.

In these studies, children tended to answer on the basis of their real-world knowledge, saying that Rex would meow and not bark. (Note that this response tendency is similar to what adults did in the study about smoking, described above.)

Rather than concluding that young children could not reason about such syllogisms, Dias and Harris had the insight to put these reasoning problems into a pretense or a fantasy context. For instance, in one case, they introduced children to a game that was about "another planet" where everything is different. They then gave the children the same types of syllogisms, but with respect to this other planet: "On this planet, all cats bark. Rex is a cat. Does Rex bark?" Children's performance in this study improved markedly under these circumstances; they were able to draw the correct (logical) conclusion even though it was contrary to their real-world knowledge (see also similar findings by Hawkins et al., 1984). This work suggests that children are capable of syllogistic reasoning but get tripped up by their real-world knowledge. They reason differently where there is less real-world content or where they are better able to understand that they need to suppress their real-world knowledge.

Research on children's self-control points to a similar conclusion. White and Carlson (2016) showed that 5-year-olds' performance on tasks that require a high degree of self-control improved when they adopted a third-person perspective on themselves or when they pretended to be an exemplary fictional character, like Batman (see also White et al., 2017). Arguably, because children in these conditions were able to feel more distant from themselves, or because they were able to take on the positive characteristics of the fictional character, they felt more able to control their impulses on the tasks. Four-year-olds also had an easier time understanding the role of mental states in pretending (e.g., someone has to know what X is in order to pretend to be X) if the pretend scenario involved fictional characters or fantastical actions (see Lillard & Sobel, 1999; Sobel & Lillard, 2001). We consider this particular case in more detail in chapter 10.

The same "fantasy advantage" effect can be seen in some cases when children learn from fictional stories. The vast majority of media aimed at children is fantastical, with a recent content analysis finding that 78.8% of popular children's books, 97.8% of popular children's movies, and 100% of popular children's television shows contain at least one fantasy element (e.g., magic spells, anthropomorphized animals; Goldstein & Alperson, 2020; see also

Chlebuch et al., 2021; Taggart et al., 2019). Many of these pieces of media are also explicitly educational, designed to teach children new languages or cultural practices (e.g., *Dora the Explorer*), new scientific information (e.g., *Sid the Science Kid*), or new vocabulary (e.g., *Word Girl*). Given the findings reviewed above, it is possible that the fantastical contexts of these stories may facilitate children's learning.

There is some evidence for this possibility. One of our studies (Weisberg, Ilgaz, Hirsh-Pasek, Golinkoff, Nicolopoulou & Dickinson, 2015) taught new vocabulary words to preschoolers in Head Start programs using some books that had realistic themes (like farming) and some that had fantastical themes (like dragons). Children in this study tended to learn the words better when they were embedded in the books with the fantastical themes; they specifically performed better on a measure asking them to provide definitions for these words. Other work from our labs corroborates this finding, showing that preschoolers are more likely to transfer some types of new science content from fantastical stories than from realistic stories (Hopkins & Weisberg, 2021; Weisberg & Hopkins, 2020).

Despite cases where children learn and reason better in unrealistic contexts, it is important to stress that the vast majority of research on children's learning from fantasy has found better learning from contexts that are more realistic. The logic of this conclusion is simple: Learning involves transferring information from the original context where one first encountered it to different contexts where it might apply. This transfer is made vastly easier if a learner can see the similarities between the learning context and the application context (e.g., Daehler & Chen, 1993; Gentner, 1983; Holyoak et al., 1984). Realistic contexts, because they are more similar to reality, provide children with more support for this transfer process.

To take one example, preschoolers who heard about a new causal relation ("popple flowers cause hiccups") within the context of a realistic story were more likely to believe that this relation also held true in reality than children who heard about this causal relation within the context of a fantastical story (Walker et al., 2015). Similarly, a series of studies on children's transfer of problem solutions out of fictional stories shows a clear advantage for realistic stories (Richert et al., 2009; Richert & Smith, 2011). In these studies, children heard a story in which a character comes up with a solution for a problem, like hiding behind a robot (fantastical story) or a babysitter (realistic story) to avoid being seen. Children were more likely to transfer this solution to an analogous real-world problem (figuring out where a doll should go to avoid being in a second doll's photograph of their room) if they had heard the realistic story. In these cases, where children are being asked to apply the rules of the story's world to reality, their transfer is facilitated by story contexts that more closely resemble the real world.

In general, the literature is mixed about whether (or under what circumstances) fantastical contexts can aid children's learning or other aspects of their cognitive functioning (see Hopkins & Weisberg, 2017, for a review). The majority of the evidence suggests that realistic contexts are better, particularly for learning, and possibly for various kinds of reasoning abilities as well. But studies like the ones reviewed here suggest that it can sometimes be beneficial to decrease the similarity between the reasoning problem and the real world because fantasy contexts—or other contexts that are more removed from children's everyday experiences—seem to benefit children's thinking under certain circumstances.

However, our primary goal in the current work is not to investigate children's thinking about fantasy, but about science (although, to misquote Arthur C. Clarke, fantasy is just science we don't understand yet[1]).

Context in Causal Reasoning and Scientific Thinking

The previous two sections illustrate the difficulty of drawing general conclusions about the role of contextualization in reasoning, especially in development. We aimed to investigate this issue directly by examining how scientific contexts might help or hinder children's reasoning.

In these studies, we conceptualized scientific contexts as being those aspects of a reasoning problem that reflect some aspect of the real world. Although these aspects could take many forms, for the studies on young children that are our primary concern in the current work, they tend look like the kind of content that would standardly be found in science classrooms or in museum exhibits aimed at children, such as earthquakes, volcanos, sinking and floating, or dinosaurs. This means that the children engaging in these tasks should understand that they are being asked to think about some aspect of the natural world.

In terms of whether these contexts help or hinder children's reasoning, as noted above, past research provides little guidance. In the case of fantastical contexts, there is some suggestion such situations help children's thinking in certain cases, although more realistic situations have also been shown to be beneficial in others. Further, although some of those studies have tested aspects of children's reasoning, none have presented a complex diagnostic reasoning task like the one that we introduced in chapter 5. For such cases, it is possible that including some type of real-world science content could jump-start children's thinking about the causal system by tapping into their existing knowledge and providing a supportive context for their thinking. It might also lead children to explicitly realize that they are in a task that requires scientific thinking, which could itself boost their performance, as we argued in chapter 1 (and will investigate in more detail in chapter 8).

But it is also possible that such content could hinder children's thinking by introducing additional task demands, such as having to remember a set of unfamiliar or complex labels for categories or to access prior category knowledge, which could lead them to draw an erroneous conclusion. For example, to understand that different kinds of rocks could be potential causes for earthquake risk, one has to understand that the different labels (e.g., igneous, sedimentary) represent different categories of rocks, and to remember those categories as potentially having different properties, which in turn could lead to different causal relations. This might introduce task demands that are not inherent in less contextualized causal systems.

Our approach to addressing this issue began with the diagnostic reasoning measure described in chapter 5, which uses a blicket detector. We then began by taking the smallest possible step toward greater contextualization, keeping as much of the original task intact as possible.

Blickets to Butterflies

As described in chapter 5, the diagnostic reasoning measure we used involved putting combinations of four colored blocks (for example, yellow, black, blue, and orange) onto a machine. There were three possible outcomes: the machine turned green and played music, or it turned red and played a different piece of music, or it did not activate. This version of the task was relatively decontextualized, in that it presented no explicitly scientific content. Further, participants did not need to know anything about the system before starting the task in order to solve it correctly. Apart from a few general assumptions about how causes and effects operate, as noted above, all the information required to answer the test question was provided.

As we shifted toward a greater degree of contextualization, we wanted to maintain as many aspects of this structure as possible, to test whether contextualization alone could affect children's performance. This meant that we had to choose a real-world causal system in which colors could plausibly be considered causal factors, to give the task some degree of realism. We also wanted to be sure that any real-world domain knowledge that children possessed would not automatically put them at a disadvantage in terms of solving the task (as the snake activity in the Earthquake Forecaster procedure might do).

We thus initially decided to focus our more contextualized version of this task on butterflies and flowers. Specifically, instead of discovering which combination of colored blocks would make the machine light up green, we asked children to discover which combination of colored flowers would bring a certain kind of butterfly to a field. This allowed us to introduce some real-world scientific content, specifically about a biological system in which color could plausibly play a role. Many of the children tested in this procedure were recruited at the Academy of Natural Sciences in Philadelphia, which houses a butterfly exhibit, making this context even more relevant to them. Otherwise, however, the task itself was exactly the same as the diagnostic reasoning task with the blicket detector, described in chapter 5.

To create the butterfly version of the task, the colored blocks were replaced by a set of four identical silk flowers: white, black, orange, and blue. We also constructed a "flower pot," which was made out of a rectangular block of Styrofoam. It had four holes punched in it, into which we could "plant" the flowers; this was the equivalent of the plastic container into which we placed the blocks in the blicket version of the task (see figure 6.1).

To illustrate the effects of flowers in the system, we used red and green plastic butterflies. We glued each set of butterflies onto wooden sticks, which were painted sky blue. To display these stimuli, we constructed a box out of foam board, which was rectangular with no top and with one side shorter than the other so that participants could see the butterflies peek out over the shorter side (see figure 6.2). We also recorded a musical cue to accompany the appearance of the green butterflies, just like the blicket detector played music when it lit up green. Finally, as in the blicket version

Figure 6.1
Flowers and field block used in the butterfly version of the diagnostic reasoning task.

of the task, we made laminated pictures of the combinations of flowers displayed in the task, along with green and red dots.

This apparatus allowed us to present the same causal structure as the blicket task: Two of the potential causes on their own led to one effect (the orange flower and the blue flower separately brought red butterflies), while their combination led to a different effect (the orange and the blue flower together brought green butterflies).

We presented this butterfly version of the task to 126 children in the same age range as we had previously tested with the blicket detector task: 4- through 10-year-olds (see full report in Weisberg, Choi & Sobel, 2020, Experiment 1). We found that, overall, participants who saw the butterfly task were 41% correct, which is marginally better than chance performance. Additionally, we observed roughly the same developmental progression as we had for the blicket version of the task, reported in chapter 5. Performance improved with age, but only by about 9 years old did they choose the correct combination at significantly above-chance levels (see table 6.1).

In addition to asking our main test question, we asked children to justify their responses. We were interested primarily in whether children's justifications referred to some aspect of the system's causal structure that they had observed in the demonstration phase, implying that children understood that their task was to use that information to determine how this system worked. For example, "because in this picture [with all four

Figure 6.2
Butterfly apparatus used in the butterfly version of the diagnostic reasoning task.

Table 6.1
Children's average percent correct on the blicket and butterfly tasks across age groups, as reported in Weisberg et al., 2020 (standard deviation in parentheses)

	Youngest third	Middle third	Oldest third
Blicket task	38.46	41.03	55.29
	(49.29)	(49.83)	(50.39)
	$n = 39$	$n = 39$	$n = 38$
	mean age = 58.84 months	mean age = 84.65 months	mean age = 108.17 months
Butterfly task	26.19	38.10	59.52
	(44.50)	(49.15)	(49.68)
	$n = 42$	$n = 42$	$n = 42$
	mean age = 58.76 months	mean age = 80.44 months	mean age = 108.44 months

flowers] the green butterflies came out and it has the orange and blue." We categorized these justifications as "data-based," because they referenced the data about the system that the child had seen. All other justifications were categorized as "other." These tended to refer to some irrelevant aspect of the task or to its surface features, for example, "because they're brighter" or "because these two are next to each other." We found that about 24% of children provided a data-based justification. Although we had initially thought that this kind of justification would indicate that children were thinking about the task at a more explicit, mature level, providing a data-based justification was not related to better performance on the task. Interestingly, this implies that children might be able to successfully solve the task without thinking it through explicitly, and only later come to an explicit understanding of their thought processes about the data they had observed.

Comparing Contexts

More important than responses to this task, though, is the question of how this performance compares to the blicket version: Do children perform better with the butterflies than the blickets, or worse, or about the same? To answer this question, we compared the children who had engaged with our butterfly task to a group of 116 children who engaged with the blicket version of the same task (we previously reported these data in chapter 5).

To make this comparison as close as possible, both groups of children were recruited from and tested at the same science museum. Interestingly, our comparisons across these two groups of subjects reveal no differences, either overall or when considering performance within each age group (see table 6.1 above; full analyses available in Weisberg et al., 2020, Experiment 1). This suggests that this contextualization, minimal though it may be, does not help or hinder children's reasoning abilities.

Interestingly, although there was no difference in performance, there was a difference in justifications: Children who saw the butterfly version of the task were somewhat more likely to justify their response with respect to the data that they had observed (24%) than children who saw the blicket version of the task (18%). This difference was not statistically significant. There was, however, an error in our data collection—only about a third of the children in the blicket version of the task were asked to provide justifications.

A follow-up study corrected this error. We replicated the entire procedure in a second sample of 103 children who each saw both the blicket version of the task and the butterfly version of the task, rather than just one or the other (order counterbalanced; full report in Weisberg et al., 2020, Experiment 2). This within-subjects design allowed us to test connections between the more and less contextualized versions of the task even more strongly. Here again, performance improved with age for both versions of the task, with only the oldest group (8- to 10-year-olds) performing significantly above chance: 71% for the blicket version and 59% for the butterfly version. Importantly, we again found no differences between children's performance on the two versions of the task. We also found a similar difference in the percentage of children providing data-based justifications in this version of the study (14% for the blicket version and 25% for the butterfly version).[2] This suggests that something about the more real-world scientific context presented in the butterfly condition may help children focus more on the causal structure of the task, or that the more abstract version of the task may hinder children's abilities to use the relevance of the evidence we provided in the demonstration phase of the task in their explanation.

In general, these results confirm the conclusion from the between-subjects version of the task: At least this minimal way of including real-world science content in this causal structure neither benefited nor detracted from children's reasoning abilities. However, the real-world context may have helped children justify their responses in a more mature way.

Blicket-saurus

The butterfly version of the task presents one way to provide more real-world context to the same causal structure presented with the blicket detector. Although we found no differences in reasoning between these two versions of the task, it remains to be seen whether presenting children with a task that integrates even more real-world content would help or hinder their reasoning. To address that issue, we constructed a third version of the same task, this time focused on dinosaurs. Again, this context was chosen because of our partnership with the Academy of Natural Sciences in Philadelphia, which has an extensive dinosaur exhibit. This choice of context allowed us to take advantage of the fact that many children are interested in dinosaurs and even know quite a bit about them, which would test more strongly the role of prior knowledge in solving this task. Like the blicket and butterfly versions, though, no actual knowledge of dinosaurs was necessary for reaching the correct answer; thinking about the system's structure alone was enough.

This version of the task told children about different features that were characteristic of certain kinds of dinosaurs. Specifically, we selected dinosaurs from the family of ceratopsians (horned-headed dinosaurs). We told children about four different features that these dinosaurs could have or lack, in parallel to blocks being on or off the machine and to flowers being planted in the field or not: having a beak-shaped mouth (or not), being larger (or smaller) than a human when fully grown, having a large (or small) a head crest, and having backward-facing (or forward-facing) horns (see figure 6.3).

We first[3] introduced Einiosaurus, which has all four features. It has a beak-shaped mouth, is larger than a human when fully grown, has a large head crest, and has backward-facing horns. This dinosaur is thus the parallel of placing all four blocks on the machine or planting all four flowers in the field. The blocks or flowers are now dinosaur features, and the effect of combining all four of these variables (features) is that the dinosaur is an Einiosaurus.

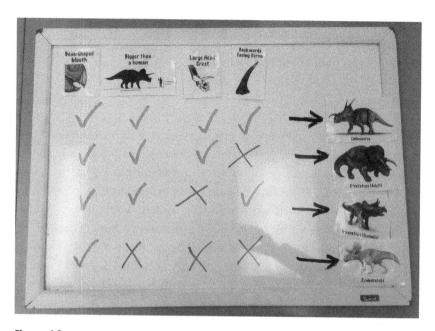

Figure 6.3
Whiteboard chart of the data presented to participants in the dinosaur version of the diagnostic reasoning task.

The next dinosaur has three of these four features. It has a beak-shaped mouth, is larger than a human when fully grown, and has a large head crest, but it does not have backward-facing horns; its horns face forward. This combination of features is an adult Triceratops. The next dinosaur has a different combination of three features: It has a beak-shaped mouth, is larger than a human when fully grown, and has backward-facing horns, but it has a small head crest. This combination of features is also a Triceratops, just a baby one. We chose Triceratops for this study because this species of dinosaur only grows a head crest in maturity, and its horns shift direction as it ages, from backward-facing to forward-facing (Horner & Goodwin, 2006). This allowed us to have two different combinations of causes (i.e., features) lead to the same effect (i.e., the same species of dinosaur), just as two different combinations of three blocks (or three flowers) both led to the effect of a red light (or red butterflies) in the other versions of this task.

Finally, we showed a dinosaur that had only one of the four features: It has a beak-shaped mouth, but is smaller than a human when fully grown, lacks a head crest, and has forward-facing horns. This dinosaur is a Zuniceratops.

We presented these four dinosaurs and their combinations of features using a chart that we drew on a whiteboard (figure 6.3). As we described each feature, we stuck a picture of this feature on the title row of the chart. As we described each dinosaur, we put a green check mark or a red X in each column of the chart and stuck a picture of that dinosaur next to this row, so that children could see the full set of feature combinations by the end of our presentation.

Participants were then asked the test question: Which combination of two features defines an Einiosaurus? Specifically, we told them that paleontologists have found a new set of fossils that they know is an Einiosaurus. But they used only two clues to figure that out. Children were asked to choose which combination of the two clues the paleontologists had used: (a) being bigger than a human and having backward-facing horns, (b) being bigger than a human and having a large head crest, or (c) having backward-facing horns and having a large head crest. In parallel to the other two versions of this task, we consider the correct answer to be (c); only the Einiosaurus has this combination of two features.

We presented this version of the task to a new set of 110 participants between the ages of 4 and 10, as in the previous version of the task (Weisberg et al., 2020, Experiment 3). We found that children responded correctly on this version of the task only 27% of the time, which was not significantly different than chance. Unlike the blicket and butterfly tasks, there was no relation with age; even the oldest children did not respond differently from chance levels.

This comparison gives us a better picture of how introducing one kind of scientific contextualization affects children's thinking in this diagnostic reasoning task. The minimal contextualization of the butterfly task did not disrupt children's performance, while the richer and more realistic context presented in the blicket-saurus task did—even though the underlying causal structure of all three tasks was identical.

Granted, the dinosaur version of the task differed from the blicket and butterfly versions in a number of ways: It presented different features (e.g., size and shape) as relevant for causal structure, as opposed to different levels of a single feature (color). It included names of different and potentially novel dinosaurs. And it couched the task in terms of categorization rather than in terms of causal relations. These differences were introduced deliberately in order to present the same system in a way that more closely resembled other scientific thinking tasks like Earthquake Forecaster. This combination of changes does not allow us to draw firm conclusions about which aspect of this contextualization may have affected children's performance, but it does clearly illustrate that the amount of prior knowledge that children must bring to bear (or feel that they must bring to bear) on constructing a hypothesis space for a problem impacts their reasoning abilities.

At least in the case of scientific contexts, then, the more real-world knowledge a task requires (or appears to require), the less children might be able to demonstrate their reasoning capacities. It thus does not necessarily make sense to draw general conclusions about the impact of "contextualized" and "decontextualized" systems. Contextualization occurs on a continuum, and the amount and type of contextualization appears to be critical. Further, the amount and type of contextualization might have different impacts in different contexts; there may be some domains where more contextualization helps children's reasoning and others where it hurts.

In turn, the results we have discussed in this chapter should encourage us to take a closer look at past work with older children and adults. It may be the case that some aspects of the realistic scientific contexts used in those tasks are indeed interfering with these participants' abilities to demonstrate their reasoning abilities. Or, put differently, it may be the case that both children and adults are capable of scientific thinking only in cases that bear little resemblance to real-world scientific contexts. If participants possess those reasoning skills, but are unable to demonstrate them in some tasks that have real-world contexts, then we must reconsider the extent to which we wish to conclude that young children possess genuine and sophisticated scientific thinking abilities.

7 Causal Reasoning and the Development of Metacognitive Thinking: Cross-Sectional and Longitudinal Investigations

Here in part II, we have charted the development of reasoning capacities related to scientific thinking, using a new diagnostic reasoning measure to begin to bridge the gap between children's early success at engaging in causal reasoning and their later difficulties with scientific thinking. We have already seen that causal complexity is a factor in this developmental trajectory. In the studies we have presented, children begin to make inferences about our reasoning task in a similar manner to adults at ages 7 to 8 years, but not earlier (chapter 5). Further, we saw that making this reasoning problem more contextualized by including more real-world science content makes it more difficult for children to solve (chapter 6). Although more work needs to be done on both of these topics to really settle how different types of causal structures and different aspects of a particular contextualization interact with children's scientific thinking abilities, here we turn to an exploration of a different variable that could affect this process: children's ability to understand and resolve disagreements.

This ability is part of children's developing metacognitive capacities, a suite of abilities that involve explicit reflection on the nature of beliefs and how they function. In general, children by about the age of 4 can understand that a particular belief can be false and that a single agent can hold a belief that does not match the truth of current reality (e.g., Flavell et al., 1992; Gopnik & Astington, 1988; Perner et al., 1987; Wellman & Liu, 2004). But there is much more to understanding beliefs than understanding that others' beliefs can be false.

A good example of the continued trajectory of children's understanding of beliefs can be seen in their performance on a battery of tasks aimed at probing their *advanced theory of mind* abilities. Advanced theory of mind includes abilities such as recognizing the complexity of others' mental states (e.g., understanding the recursive nature of belief and its relation to other mental states; e.g., Eisbach, 2004; Lagattuta & Wellman, 2001; Perner & Wimmer, 1985), broad perspective-taking capacities (e.g., Carpendale & Chandler, 1996), and understanding and recognizing others' emotions (e.g., Baron-Cohen et al., 2001). This kind of mental-state knowledge continues to develop during the elementary school years (e.g., Osterhaus et al., 2016). Further, such understanding is composed of multiple facets, meaning that there is not a single developmental trajectory for advanced theory of mind.

One aspect of children's metacognitive understanding that we believe is critical for scientific reasoning is the ability to negotiate situations in which agents hold different beliefs, or situations in which agents' own beliefs need to change in light of new evidence—that is, cases of disagreement or belief conflict (Beck, Robinson et al., 2011; Heiphetz, Spelke et al., 2013, 2014; Walker et al., 2012). Many of the studies cited above, such as studies on *interpretive* theory of mind (e.g., Carpendale & Chandler, 1996) demonstrate that children show broad improvement in their understanding of disagreement between the ages of 5 and 8.

The ability to understand disagreements is a major part of mature scientific thinking. One's currently held beliefs might disagree with newly observed data, or two individuals might disagree about the correct interpretation for a set of observations (Barzilai & Eshet-Alkalai, 2015; Barzilai & Weinstock, 2015). Because of this, metacognition may underpin various aspects of scientific cognition (Kuhn, 1989, 2002), an argument that we outlined in chapter 1. The diagnostic reasoning measure we introduced in chapter 5 provides children with several opportunities to navigate among conflicting beliefs. For example, when shown the initial data, children might think that only the combination of all four blocks makes the machine turn green. But then they are told that there is a combination of two blocks that can make the machine turn green. Hearing this information might encourage children to change their beliefs about which blocks or combination of blocks activate the machine.

More broadly, children must learn to integrate the information they hear from others with the data they observe. In one of our investigations of this ability (McLoughlin, Finiasz, Sobel & Corriveau, 2021), we found that 5-year-olds tended to learn from an informant whose level of certainty matched the evidence they observed (though 4-year-olds were less likely to do so). Specifically, when children observed deterministic data, the 5-year-olds we studied learned about the causal efficacy of this system better from an informant who was certain that particular events were or were not efficacious as opposed to an informant who was uncertain. When children observed probabilistic data, 5-year-olds showed the opposite pattern and learned better from informants who were uncertain than from informants who were certain. These data are part of a broader research program on children's ability to calibrate the relation between social information that indicates epistemic competence on the part of a speaker and the truth value of the speaker's claims (e.g., Birch et al., 2020; Fitneva et al., 2013; Tenney et al., 2011). All these findings show development in children's abilities to appropriately calibrate their expectations about others' testimony after the preschool years. Children develop these further metacognitive capacities to aid their scientific thinking as they enter formal schooling, and this development involves integrating together distinct kinds of data (e.g., information from verbal testimony and observed data) to reach a causal conclusion.

We wanted to investigate how developing an understanding of conflicting beliefs and how they are resolved might link to children's performance on our diagnostic reasoning measure. To do so, we presented a task investigating children's understanding of disagreement and our blicket detector task using both a cross-sectional and a longitudinal design. This allowed us to investigate the development of children's understanding of disagreement in general and to probe any possible relations between this understanding and the kind of causal reasoning necessary to solve this blicket detector task.

The Disagreement Task

In this study (Haber, Sobel & Weisberg, 2019), we wanted to investigate cases in which children had to reason about two characters who held conflicting beliefs. Specifically, we asked whether children believed that there could be legitimate disagreements, in which two characters could both have some degree of correctness, or if children take a more absolute view of beliefs. The development of this ability is interesting in its own right, but we focus on it here because it is a common situation in mature scientific thinking. Different sets of evidence might lead two researchers to draw different conclusions, or incoming evidence that does not match with one's expectations might lead one to question or revise their previous beliefs, even if one does not reject these previous beliefs entirely.

Prior work in this area has outlined a developmental progression through different levels of understanding of disagreement, as noted in chapter 4 (Kuhn et al., 2000). Children might first believe that knowledge is just based on reality, hence beliefs must be the same for everyone (Wellman, 1990). Then they progress to an understanding that different people can have different preferences and appreciate that others' beliefs can be false, but they remain *absolutist* about other types of information: Knowledge is objective and true beliefs must be shared. On this view, when there are disagreements, one person is right and the other is wrong. Following this stage, children become *relativists* (or *multiplists*), claiming that everyone can be right about whatever belief they may hold, because all beliefs are subjective. That is, they have gained the understanding that different people can believe different things, but they still misunderstand that some of these beliefs can be more well-grounded than others. This understanding comes at the final stage of the developmental progression, called *evaluativism*, in which two people who disagree can both be right, but there is also a sense in which one can be "more right" than the other (see also Barzilai & Weinstock, 2015; Weisberg et al., 2021). The disagreement task that we used with our participants presents several situations in which characters disagree in order to probe how this developmental progression unfolds and how children's thinking about disagreements might impact their performance on the diagnostic reasoning task we described in chapter 5.

Although this progression is framed in terms of disagreements between two characters, the same idea applies to disagreements between one's own beliefs and the world or between one's current belief and a past belief. In order to think scientifically, children must understand how to reconcile disagreements between their beliefs and the data that they observe from the world, as well as between an idea that they hold currently and an idea that they held in the past (see Gopnik & Slaughter, 1991). An absolutist framework would lead them to completely accept one and completely reject the other, which is often not warranted in a true scientific investigation. Relativism is equally problematic, as there are many cases in which conflicts between different hypotheses cannot be reconciled, or in which two conflicting ideas do not have equal merit. The evaluativist framework allows children to be more successful at both understanding and conducting scientific investigations, as they understand that older ideas can give way to newer and more correct ideas, even if the older idea still has some merit and even if the newer idea is not perfect. In this way, children's understanding of disagreements between characters can be informative as to their approach to changing their own beliefs over time.

Our measure of children's understanding of disagreement showed children two characters, attributed a belief to each character, and asked children to judge whether each character can be correct in their belief (based on Heiphetz et al., 2013; Walker et al., 2012). Using this framework, we ask about disagreements involving three different types of beliefs: beliefs about facts, beliefs about interpretations, and beliefs about preferences.

In the Fact trial, children were told about an action: An experimenter hid a penny in a certain location. They were introduced to two characters, one who said that the penny was hidden where the experimenter said that she hid it, and the other who said that the penny was hidden somewhere else. Thus, one is objectively right and the other is objectively wrong (at least if you believe that the experimenter is a reliable source of knowledge). The experimenter then asked children whether each character could be right and to justify their responses. We considered whether children responded correctly and also whether children justified their responses by referring back to what the experimenter said (*testimony-based* justifications) or by referring to the state of the world itself (*world-based* justifications).

The Interpretation trial was similar to the Fact trial, except that the experimenter's statement was ambiguous, so the penny could be hidden in one of two possible locations. One character stated that the penny was in one of the locations, while the other character stated it was in the other possible location. The penny, not being Schrödinger's cat, can be in only one of those locations. So although only one of the characters is right, we cannot know which. We again asked children whether each character could be right. We also asked them to justify their responses, which were coded into the same two categories (testimony-based and world-based) as for the Fact trial.

Finally, in the Preference trial, children were told about two characters who liked different things. Children were asked whether each character could be right about what they liked and to justify their responses. Justifications for this trial were coded as referring to the characters themselves or their preferences (*character-based*) or to the nature of preferences themselves (*preference-based*, e.g., "Because it's an opinion. There's no right or wrong").

School Partnership and Longitudinal Sample

For this line of work, we recruited students from a single school district, instead of through our labs or from museums. Working in partnership with a school district allowed us to look at how individual children's responses to our measures might change over time, as well as to address some of the issues with our previous samples. Importantly, recruiting children from museums and other venues does not allow us to control for the educational background of these participants. In contrast, for the studies described in this chapter, we know what children have been exposed to in school. We were also able to obtain these students' scores on standardized assessments in literacy, math, and science, which allowed us to determine whether there were any relations between our measures and performance on these tests.

Our school district partner for these studies was in Springfield, Pennsylvania, located in the western suburbs of Philadelphia. This district is primarily white and primarily mid- to high socioeconomic status, but it does incorporate a small degree of racial and economic diversity. In our sample specifically, 82.2% of participants were white, 6.6% were Black, 5.4% were Asian, 0.3% were Native American, and 2.0% were mixed race (3.4% of the sample did not report their race). Twelve percent qualified for free or reduced lunch.

We first began working with a cohort of 120 first graders from this district in the spring of 2014–2015 school year; we call this the 2015 group. We were able to return to the school again in the spring of 2016–2017 school year (2017 group) to retest some of these children and to recruit new participants. The 2017 sample included 78 third graders we had previously tested in 2015, providing us with a longitudinal sample. We also tested 39 third graders who had not been tested before and a new group of 112 first graders. See table 7.1 for details about these samples.

Interestingly, as part of this study, we were afforded the opportunity to link children's performance on our tasks to a change in school's curriculum. In the year between our two testing sessions (2015–2016), the school district chose to change its curriculum for the elementary grades. During our first year of testing, the curriculum was fairly traditional, emphasizing text-based content learning. This curriculum asked children to rely mainly on teachers to acquire knowledge, and students had few opportunities to actively engage in investigations, ask questions, or develop higher-level thinking or cognitive processing skills (e.g., analyzing data, constructing arguments based on

Table 7.1
Participants from the Springfield School District

	2014–2015 school year	2016–2017 school year
First graders	$n = 120$ 60 female, 60 male mean age = 86.9 months age range = 73.7 – 97.4 months	$n = 112$ 58 female, 54 male mean age = 87.3 months age range = 79.9 – 103.9 months
Third graders		$n = 117$ (includes 78 children previously tested in 2015) 59 female, 58 male mean age = 111.0 months age range = 100.6 – 120.5 months

evidence, reflecting on knowledge). Furthermore, little time was dedicated to learning social studies or science.

Over the 2015–2016 year, the school chose to incorporate more inquiry-based learning, which emphasizes the idea that children learn and actively construct knowledge through exploration, question-asking, and experimentation (e.g., Edson, 2013). This new curriculum integrated content knowledge with process skills and focused heavily on bringing more science content into the classroom, especially in the primary grades. It was structured around essential questions that focused on fostering students' critical thinking, reasoning, and analytical skills. These essential questions were designed to be open-ended, thought-provoking questions that require high-order thinking (e.g., making predictions, analyzing findings, reflecting on knowledge) in order to lead students to ask additional questions (McTighe & Wiggins, 2013; see Haber et al., 2019, for more details about the differences between these curricula as implemented in this school district).

The structure of this sample allows us to make two key comparisons. First, comparing the children who were tested both in 2015 and in 2017 gives us a longitudinal view on the development of children's understanding of disagreements and on the development of children's diagnostic reasoning. Second, comparing the first graders tested in 2015 to the first graders tested in 2017 potentially gives us insight into whether different curricula can affect children's understanding of disagreements.

Although students exposed to either a direct-instruction curriculum or an inquiry-based curriculum can learn the same content, the process by which this knowledge is acquired looks different. Instead of solely relying on the teacher for information, children who receive an inquiry-based curriculum are obtaining their own information through direct interactions with the world. They are also asked to evaluate evidence, develop arguments, and reflect on their own knowledge as they prepare to explain their decisions to their classmates. This process of acquiring knowledge, in contrast with direct instruction, places students at the center of their own learning process. Children's experiences with these different methods of learning may thus affect their understanding of the objectivity of knowledge and hence of the ways in which different individuals may disagree.

At both time points (2015 and 2017), all participants received three tasks: the test of understanding disagreements, described above, the diagnostic reasoning measure that we introduced in chapter 5, and an interview about the meaning of the word "science." Motivation for and results from the science interview are discussed in chapter 8. In the remainder of this chapter, we focus on charting children's developing understanding of disagreements (as a proxy for metacognitive development) and their developing causal reasoning skills (as measured with the blicket detector task), and in testing relations between these two tasks.

Performance on the Disagreement Task

Figure 7.1 shows how children from all three groups responded to the three types of trials in the disagreement task. For the Fact trial, we scored responses as correct if children said that the character who agreed with the experimenter's assertion was right and that the other character was wrong. For the Interpretation trial, we scored children's responses as correct if they said that both characters could be right. For the Preference trial, we scored children's responses as correct if they said either that both characters could be right or that both characters could be wrong in having their different preferences; we only considered a response incorrect here if children said that one character was right while the other was wrong.

An in-depth discussion of the difference between the 2015 and 2017 first graders is presented in Haber et al. (2019). Here, we highlight the main finding from this comparison, which is the difference between performance on the Fact trial in these two groups. The 2017 first graders, who received the more inquiry-based curriculum, had a better understanding of situations of disagreement about a known, objective truth than the 2015 first graders.

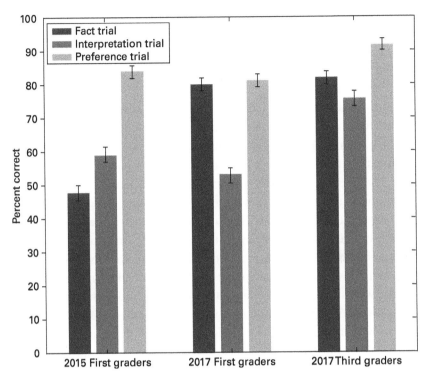

Figure 7.1
Performance of children from the Springfield School District on the disagreement task, by grade and year tested.

Further, children in these two cohorts justified their responses differently. While the proportion of testimony-based justifications did not differ between the first graders tested in 2015 and 2017, the proportion of world-based justification did. First graders tested in 2017 appealed to the actual state of the world significantly more often (39% of the time) than those tested in 2015 (16% of the time). And even though their correct responding on the Interpretation trial did not differ, a similar difference was found in their world-based justifications for this trial (32% in 2017 vs. 24% in 2015). This suggests that the curriculum difference might have focused children more on the objective truth of the situation.

We also want to discuss the contrast between the first graders and the third graders tested in 2017 (i.e., data not published in Haber et al., 2019). We found that the third graders' performance on all three trial types was above chance.[1] Comparing these children to the first graders tested in the same year shows that third graders were better only on the Interpretation and Preference trials.[2] On the Fact trial, the first graders tested in 2017 performed in the same way as the third graders.[3] This provides evidence for the positive effect of the inquiry-based curriculum on first graders' understanding of fact-based disagreements; their performance as first graders was no difference than the performance of children who were two years older.

The pattern of justifications for the Fact trial for the third graders also did not differ from the pattern generated by the first graders tested the same year: 37% of the third graders' justifications were testimony-based and 39% were world-based (compared to 40% and 32% for the first graders).[4] However, for the Interpretation trial, the third graders generated significantly more testimony-based justifications (19%) than the first graders (8%). The third graders did not provide more world-based justifications for Interpretation trial (32%) than the first graders (32%).[5] For the Preference trial, first graders were more likely to refer to the characters and their likes and dislikes (56% of first graders, contrasted with 31% of third graders), while third graders were more likely to refer to the idea that opinions cannot be right or wrong (17% of first graders, contrasted with 59% of third graders).[6]

Longitudinal Analyses

We also considered the performance of the 78 children who participated in this task both in first grade and in third grade. The patterns we observed here generally aligned with the patterns in the data as a whole: Children improved significantly in their performance between first and third grade on the Fact and Interpretation trials, but improved only marginally in their performance on the Preference trial, likely because performance was already quite high on this trial in first grade.[7]

In terms of justifications, on the Fact trial, children in this longitudinal sample gave significantly more world-based justifications in third grade than in first grade.[8] This aligns with the findings from the larger data set, in which we also saw a relation between a greater proportion of correct responses being associated with more references to the state of the world. There were no developmental differences in children's justifications for the Interpretation trial.[9] For the Preference trial, children were more likely to justify their responses with reference to the subjective nature of opinions in general and less likely to refer to an individual character's preferences in third grade than in first grade.[10]

Summary

The overall goal of this task was to assess how children think about disagreement across a range of different types of situations. One of the main lessons to draw from this work is the overall developmental pattern. Even first graders understand that two people can hold differing beliefs with respect to preferences. This replicates several earlier results in this area, which also show that children respect the subjective nature of preferences by the age of 7 or so (e.g., Heiphetz et al., 2014). Although the pattern of correct responding stayed relatively stable over these two years, we did find developmental changes in how children tended to justify their responses to this question. Younger children were generally more likely to think about the immediate situation, saying that a character could be right because of his or her individual preferences (e.g., "she really likes yellow"). In contrast, older children were more likely to think about the more abstract issue of how preferences work, saying that either character could be right because opinions cannot be right or wrong. This indicates that, with age, children develop a more abstract and nuanced understanding of the ways in which preferences function.

While understanding differences in two characters' preferences seems to be a relatively straightforward development achievement, understanding how to coordinate cases of conflicting beliefs about facts or interpretations is more difficult. We did see maturation here, with third graders tending to perform better than first graders. Within this general developmental context, though, the educational setting can make a significant difference. Students who were used to learning through more direct instruction were more likely to refer to the experimenter's testimony as a reason for why a character could be right. That is, the fact that these children had received a curriculum that emphasizes learning through direct testimony possibly led them to be more likely to rely on the experimenter's testimony as a source of information. A curriculum that encouraged children to engage in discovery on their own behalf seems to have made them more prone to examining the state of the world for themselves. Although there are other possible reasons for this difference, because both curricula were administered by the same teachers in the same district, we feel reasonably confident in concluding that the curriculum played a role in this difference.

Finally, disagreements that were open to different interpretations were more difficult for children to understand than disagreements about objective matters of fact, and more difficult for children to understand than disagreements about preferences. This makes sense in light of other work on interpretive theory of mind, finding that children throughout the early elementary years have trouble understanding that two people can interpret an ambiguous situation differently (e.g., a duck-rabbit figure; Carpendale & Chandler, 1996; Mitroff et al., 2006). The ability to coordinate multiple interpretations thus may be a later developmental achievement. Indeed, this is a domain in which even adults struggle (e.g., Barzilai & Weinstock, 2015), and where development is ongoing.

Performance on Causal Reasoning Tasks

In addition to the disagreement task, we also presented some of these children with the diagnostic reasoning measure introduced in chapter 5 as a test of their causal reasoning abilities. As a reminder, in this task, children are presented with a blicket detector that lights up red or green.[11] Children observe what happens when the experimenter places different combinations of blocks on the machine. They are then asked to figure out which pair of blocks (out of three possibilities) make the machine turn green and play music, based on the data that they observed. We also asked children to justify their choices.

First Graders Tested in 2015

Half of the first graders we recruited in 2015 ($n=60$) were given this task. We found that 45% of them provided the correct answer to this task, which is marginally better than chance performance (33%).[12] This is generally in line with the findings from our other samples (described in chapters 5 and 6), suggesting that the performance of children in those samples was not due to where or how they were recruited.

The distribution of these children's justifications also looked similar to those from the other samples. About 18% of these children made some mention of the data that they had observed. But referring to the data from the demonstration phase did not relate to their performance; about half of the children who gave data-based justifications passed the task (55%), while children who did not give such justifications were not more successful (45%).[13] This pattern does not match what we found earlier, where data-based justifications were associated (though not significantly) with better performance. However, this sample includes only first graders, whereas the sample we discussed in chapter 5 spanned a much wider age range. It might be that these links between performance and explicit ability to justify one's own choices only appear later in development.

The other half of our first graders in this year ($n = 60$) received a different scientific thinking task: the mouse task (based on Sodian et al., 1991, which we described in chapter 4). We chose to vary which task children received because we wanted to be able to compare the performance of our participants to a published task that was already established as being a test of some aspect of scientific thinking, specifically of distinguishing hypotheses from evidence.

Like the blicket detector measure of diagnostic reasoning, the mouse task assesses children's abilities to reason backward from effects to causes. But rather than using a machine, it uses a short story that presents a reasoning problem. Specifically, children are told about two brothers who both know there is a mouse in their house. The characters want to find out whether the mouse is big or small. To do this, they make two different mouse houses. One house has a small door that only a small mouse could fit through. The second house has a large door that both a small and large mouse could fit through. The key test question asks which house the characters should choose to find out the size of the mouse. We were interested in whether children would understand that the brothers have to use the box with the small door to answer this question. Only the small mouse can get into this box, so only this box could help them to tell whether the mouse is big or small. This task assesses whether children can recognize what actions they would need to take to construct a diagnostic test.

We found that 40% of the children correctly chose the box with the small door, no different from chance (50%).[14] First graders' performance on this task is thus comparable to their overall performance than their performance on the blicket task (45% correct), although it is difficult to make direct comparisons because the mouse task presents two answer options while the blicket task presents three, making children more likely to respond correctly on the mouse task by chance.

Further, we found that children in this task tended to provide informative justifications for their choices (e.g., "because the big one can't get into this one") about 30% of the time. Unlike for the blicket task, children's justifications related to their performance on the test question, with children who responded correctly to the test question being more likely to provide informative justifications (67%) than children who responded incorrectly (6%).[15]

Similar to Sodian et al. (1991), in addition to this main test question, we also asked which box the brothers should choose to allow the mouse to get food, regardless of its size. The correct answer to this question is the box with the large door, because both mice can get into it. We found that 88% of first graders were able to answer this question correctly, showing significantly better performance on this question than on the main test question.[16] This demonstrates that children understood the parameters of the task and the relations between the sizes of the mice and the sizes of the boxes' doors. In turn, this suggests that their worse performance on the main test question was due to them not understanding how to construct a controlled experiment in this system.

In Sodian et al.'s (1991) original report of this task, children were categorized based on how they had responded to both of these questions. Those researchers found that 55% of their first graders responded correctly to both the main test question and the checking question described in the last paragraph. Thirty-five percent of their first graders responded correctly to the question about how to feed both mice, but not to the main test question of finding out whether the mouse is big or small. When we analyze our data in the same way, we see a reversal: 33% of our participants responded correctly to both questions, while 55% of our participants correctly answered the "feed" question but not the "find out" question. This might suggest that our participants are lagging somewhat behind participants in Sodian et al. (1991) in their understanding of this aspect of diagnostic reasoning. However, these differences are not statistically significant, perhaps because the sample sizes are so small (20 first-graders in the Sodian et al., 1991, study and 60 in ours).

In general, then, there are some interesting differences between the blicket task (which measures children's diagnostic reasoning abilities) and the mouse task (which measures children's abilities to evaluate a diagnostic test). Specifically, the first graders we tested who received the blicket task were able to answer correctly at above-chance levels, but generally were unable to justify their responses in an adult-like way, suggesting that their intuitive sense of the correct answer was not yet accessible to their explicit cognition. Conversely, the first graders we tested who received the mouse task had more difficulty responding correctly overall, but those who did so were generally able to justify their responses accurately. This contrast underscores our arguments about the important role that individual tasks play in our conclusions about whether children can reason scientifically. The fact that children do so for some tasks but not others, and the fact that they can think explicitly about their decisions for some tasks but not others, highlights the need to be cautious about drawing broad conclusions about the development of scientific thinking from any single task.

First Graders Tested in 2017

In 2017, we tested a new cohort of 112 first graders on the blicket task.[17] We found that 43% of these first graders provided the correct answer, significantly better than chance.[18] As noted above, the school district adopted a new curriculum between 2015 and 2017, but the proportion of correct answers from the 2017 first graders was no different than those of the 2015 first graders.[19] This suggests some stability in this measure.

In terms of their justifications for their responses, only 15% of these children referred back to the pattern of blocks that had been placed on the machine in the demonstration phase. Unlike the 2015 sample of first graders, children in the 2017 sample who gave such justifications were significantly more likely to respond correctly to the main test question (12%) than those who did not give such a justification (4%).[20] This aligns with our findings from the studies reported in chapter 5, in which an attention to the previously observed data was associated with better performance. It is also possible that the inquiry-based curriculum that these children received might have encouraged greater attention to the data that they observed about the blocks and the machine; however, this is speculative.

Third Graders Tested in 2017

Overall, 56% of the third graders we tested in 2017 provided the correct answer, significantly better than chance.[21] This level of performance aligns with our previously reported results (chapters 5 and 6) on children of this age recruited from museum settings. The performance of the 2017 third graders was also (perhaps unsurprisingly) significantly better than performance of the 2017 first graders (43%).[22] Again, then, we have good reason to believe that our earlier results with this diagnostic reasoning measure were not due to children being recruited from or tested in a museum setting or in a lab setting; the developmental trajectory we outlined in chapters 5 and 6 seems to be fairly robust, at least based on these data.

In terms of justifications, 26% of the third graders justified their responses with reference to the data they had previously observed, significantly more than the first graders tested in the same year (15%).[23] However, we found no relation between providing this type of justification and their performance on the blicket task.[24]

As noted above, 78 of the third-grade participants had participated in our study two years earlier, when they were in first grade. However, because some of the first graders we had tested in 2015 received the mouse task, only 36 participants received the blicket task at both time points. We found that 14 of these participants responded correctly as first graders (39%) and 17 responded correctly as third graders (47%). However, there was no relation between children's performance at the two time points: 7 children responded correctly both times and 12 children responded incorrectly both times. There were 10 children whose performance improved, responding correctly in third grade but not in first grade, but there were also 7 children whose performance declined, responding correctly in first grade but not in third grade. These numbers are small, though, so we hesitate to draw strong conclusions about these trends.

To summarize, the results from these children's performance in the blicket task align nicely with our findings from other populations. First graders (mostly 6- to 7-year-olds) were marginally above chance at choosing the correct response, while third graders (mostly 8- to 9-year-olds) were significantly above chance. We thus replicate the general developmental pattern that reasoning about this kind of complex but minimally contextualized causal system emerges starting around 7 to 8 years of age.

Relations between the Causal Reasoning Task and the Disagreement Task

As noted earlier, we included the disagreement task in this study because we anticipated that it might relate to performance on the blicket task, which served as a proxy for children's scientific thinking abilities. Understanding disagreements is a metacognitive skill that involves children's thinking about what beliefs are and how they work—crucial processes that contribute to scientific thinking in general. We thus examined relations between children's performance with the three different types of disagreement and their responses to the blicket task.

For the first graders, we found no relations between their performance on the blicket task and their performance on the disagreement task, either when considering the three trial types individually[25] or when combining them into a single scale of overall correct performance.[26] For the third graders, by contrast, we did find a relation: Children who answered correctly on the blicket task scored significantly higher on the scale of overall correct performance in the disagreement task (mean=2.64 out of a possible 3, SD=0.52) than children who answered incorrectly on the blicket task (mean=2.31 out of a possible 3, SD=0.73).[27] We thus find some support for our hypothesis that metacognitive reasoning is related to children's scientific thinking skills. Interestingly, we found this relation only in the third graders, who were the only group to consistently score above chance on the blicket task. This further suggests that there are some developmental links between metacognitive thinking skills and scientific thinking skills.

Relations to Standardized Metrics of Academic Achievement

Although we hypothesized that children's scientific thinking might depend on their developing metacognitive capacities, it is also possible that children who answered correctly on both tasks may simply be smarter or more verbally able. Luckily, we can investigate this possibility directly. Because this project involved working in partnership with the school district, we were able to obtain our participants' scores on standardized measures of literacy and math. The main test that was used in this case was the Measures of Academic Progress (MAP), an assessment that is linked to the Pennsylvania Common Core standards. This test was administered to both the first graders and the third graders in our sample.[28] We were also able to obtain scores for a different set of standardized assessments for our group of third graders. One year after their participation in our study (i.e., when they were in fourth grade), these students were tested with a state-level assessment known as the PSSA (Pennsylvania System of School Assessment), which examines performance in English language arts, math, and science and technology.[29]

Relations with the Disagreement Task

We found some relations between performance on the disagreement task and students' scores on these standardized assessments. For first graders, higher scores on both the reading assessment and the math assessment were associated with better overall performance on the disagreement task.[30] The same was true for third graders.[31]

More specifically, first graders who responded correctly on the Fact trial earned significantly higher scores on both the reading MAP[32] and the math MAP.[33] Similarly, both first and third graders who responded correctly on the Interpretation trial earned higher scores on both the reading MAP[34] and the math MAP.[35] Third graders who responded correctly on the Preference trial achieved significantly higher scores on the reading assessment.[36]

For the PSSA test, we found positive relations between all three subtests (language, math, and science) with the third graders' overall performance on the disagreement task.[37] This latter result seemed to be driven by these students' performance on the Interpretation trial, as this was always significantly associated with higher test scores, while performance on the other two trial types was more inconsistent.

Relations with the Causal Reasoning Task

Looking at the first graders' scores on the standardized assessments in relation to the blicket task revealed no difference in performance on the blicket task based on their MAP reading scores.[38] But we did find that the first graders who scored higher on the math portion of this test tended to respond *incorrectly* to the blicket task.[39] We do not have a ready explanation for this pattern, but at the least, it suggests that skill in math is not related to the kind of diagnostic thinking abilities that are necessary to solve our blicket task. For the third graders, correct responding on the blicket task was associated with higher scores on the reading assessment[40] but not on the math assessment.[41]

With respect to the PSSA, we found, as before, that students who answered correctly on the blicket task had significantly higher language scores,[42] but there was no relation with math scores.[43] The important finding is that there was a relation with science scores: Students who answered correctly on the blicket task performed significantly better on the science section of this test.[44] This suggests that something about our blicket task is tapping into an underlying process in scientific thinking, which was our original intention for developing this task.

What Do These Data Tell Us about the Relation between Causal Reasoning and Science Education?

We began our partnership with the Springfield School District primarily for reasons of experimental control. Recruiting participants from museums and other community sites did not allow us to examine the effect of schooling, as those participants came from a wide variety of schools and educational backgrounds. By testing children within a single school district, we were able to examine the effect of curriculum in more detail, as well as to control for prior science experience at school. What we found with the blicket task in this population was in line with our findings from museums: The ability to solve this diagnostic reasoning task emerges between the ages of 6 and 9, with our first graders performing roughly at chance and our third graders performing significantly above chance. We are thus narrowing in on this age range as a critical time for the development of these sets of skills.

This is also a critical time for the development of children's advanced theory of mind skills and their metacognitive understanding, as demonstrated in prior work (e.g., Heiphetz et al., 2013, 2014; Kuhn et al., 2000; Osterhaus et al., 2016). Here we tested one facet of metacognitive thinking: the ability to understand disagreements about different types of beliefs. We found that even first graders understood that two people can reasonably disagree about matters of preference, but they did not demonstrate an understanding of disagreements about matters of fact or matters of interpretation until third grade. The one exception to this pattern was the first graders who had received a more inquiry-based curriculum. Their performance on questions about fact-based disagreement mirrored that of the third graders, illustrating that experience with exploration and investigation in the classroom can help to bolster the development of this important skill.

A more general conclusion to draw from the body of work reported in this chapter is that the relation between basic causal reasoning skills and more mature scientific thinking skills is complicated. Our diagnostic reasoning measure goes some way toward bridging this gap, bringing in a more complex set of causal structures without necessarily contextualizing these structures with particular scientific content. That performance on this measure relates to performance on a standardized measure of scientific thinking in a statewide assessment suggests that this task does relate to some aspects of scientific thinking, which is promising. But it requires much more investigation to understand the nature of the relation and to translate these basic findings into potential interventions in the classroom. One relevant finding from this work is that there is development between the ages of 7 and 8, possibly coinciding with the development of more advanced metacognitive capacities. This suggests that we should be looking at this time point to discover other developmental shifts that may underlie the further development of scientific thinking.

Our work in this area is far from over; we have barely begun to scratch the surface of the variables that define scientific thinking and how these develop. As noted at the start of part II, we intended this set of investigations using our new blicket detector task to serve as a series of case studies for how work in cognitive development can be brought into better dialogue with work on scientific thinking abilities. We have by no means exhausted the ways in which fruitful connections can be made across these areas.

Luckily, many other researchers are starting to take up the challenge of bridging this gap and have contributed additional insights into how this process works. As reviewed in chapter 4, one major area of research within this framework is examining children's abilities to engage in belief revision: the process of weighing incoming evidence against one's existing beliefs and deciding to change one's beliefs on this basis (e.g., Bonawitz et al., 2012; Macris & Sobel, 2017; Young et al., 2012). This work generally demonstrates that young children can successfully do so under some circumstances, especially when they receive direct testimony that their initial belief was incorrect, though again there is growth in the ability to use other evidence to do so starting around age 7.

A second major area of research examines whether children understand the control of variables strategy, even if they cannot produce this strategy themselves. These studies tend to show children a confounded causal system and ask them to choose which test will allow them to draw a definitive conclusion about how the system works. This work generally finds that children can choose or recognize unconfounded experiments for simple causal systems, although they may not yet be able to produce such experiments themselves (e.g., Lapidow & Walker, 2020; see also Sobel, Benton et al., 2021, described in chapter 4). Another body of research is examining how children's nascent scientific thinking skills can be nurtured in school-based settings and transferred to real-world settings (see Sandoval et al., 2014).

Moving forward from here will involve assembling these different lines of work into more unified sets of studies that can address multiple aspects of the differences between young children's and older children's reasoning abilities. It should also involve examining in more detail the gaps between the kind of implicit causal reasoning seen in young children and the kind of explicit scientific thinking expected of older children and adults. We have begun this investigation by asking children to justify their responses, which has illustrated that children's understanding of their own actions is still quite nascent. Even at ages where children (as a group) tend to answer correctly on the test question itself, their justifications for their choices do not necessarily reflect an explicit understanding of the system's causal structure. As a first step toward a fuller investigation of this issue, part III of this book begins to address children's explicit conceptions of science, learning, and other related concepts. Our goal in that body of work is to chart the development of these explicit definitions and to examine the relation between this developmental trajectory and children's abilities to engage in causal reasoning and scientific thinking.

III   Children's Explicit Definitions of Abstract Concepts

Intensions are creatures of darkness, and I shall rejoice with the reader when they are exorcised.

—W. V. O. Quine, 1956

8   Children's Definitions of "Science"

Part II (chapters 5 to 7) of this book described a series of case studies aimed at charting how children's causal reasoning abilities develop into more mature scientific thinking abilities, with particular focus on some of the variables that might affect this transition. Here in part III (chapters 8 to 10) we turn our attention to children's explicit understanding of concepts related to scientific thinking. Specifically, we ask how children's thinking about science, learning, and pretending (among other concepts) develops across the preschool and early elementary school years. Our goal in this work is to try to understand more about the gap between children's early abilities with causal reasoning, which tend to be implicit, and their later abilities with scientific thinking, which involve some degree of explicit reflection.

Intensions and Extensions

There is a long philosophical literature on the role of *intensions* in the field of philosophy of language (e.g., Kripke, 1972; Putnam, 1975). This literature mostly examines the relation between propositions described by linguistic utterances and their truth values. Roughly speaking, the intension of a word is the function that maps its usage to its meaning. Putnam (1975) called it the "something else" that represents the meaning of a word inherent to the concept itself, as opposed to the entities that are members of the category represented by the concept.

If that previous paragraph is confusing to you, don't worry; it is to us as well. Talking about intensions is confusing, and we suspect this is why Quine (quoted at the beginning of this part of the book) wanted them exorcised. From the point of view of the psychological literature, this term has mostly been used to describe the difference between our internal representations of the meanings of words (our concepts) and the external objects or actions or events that are denoted by those words (their referents). The idea, as we understand it, is that linguistic utterances have meaning beyond simply the objects or actions or events in the world that they refer to, because they are embedded in our psychological representations of these objects, actions, or events.

Understanding concepts—particularly in the philosophical literature—has often been characterized as constructing definitions for concepts. This approach determines a set of necessary and sufficient features for a particular thing to be an instance of a particular concept. To take a classic example, the necessary and sufficient features of being a bachelor are being unmarried and being a man. Although this seems sensible at first glance, this approach has been widely criticized in both the philosophical and psychological literatures, primarily because so many of our concepts are not able to be adequately defined in this way (see, e.g., Keil, 1989; Lakoff, 1987; Murphy & Medin, 1985; Wittgenstein, 1958).

Putting aside these philosophical foundations, within the field of cognitive development, researchers have tended to use "intension" to just mean "definition." Work on children's category-based intensions has generally taken two forms. First, there's a long literature in which children are asked to provide definitions for specific concepts. For example, Piaget (1929) asked children questions such as, "Do you know what it means to think of something?" This probed children's understanding of what he called "the notion of thought" (p. 37). From children's definitions and their reflection on their thinking, he suggested that children do not initially understand the differences between mental and physical entities, and that this understanding must be acquired (see also Johnson & Wellman, 1982). These conclusions were part of his theory of egocentricity, in which younger children had difficulty differentiating between the self and the other. The development of that differentiation marked an important transition in development, specifically forming one facet of the emergence of concrete operations.

Second, there is another long literature relating children's intensions (their definitions of categories) to their *extensions* (which in cognitive development mostly refers to their judgments about category membership). In this research, children are asked whether particular objects or entities are members of a category given descriptions of the objects or entities that emphasize different features. These studies tend to show that, between the ages of 4 and

10, children's conceptual representations transition from being based more on superficial, perceptual, or nonessential features of objects (e.g., taxis are yellow) to ones that are more central to their meaning (e.g., taxis give people rides for money). This is known as the "characteristic-to-defining shift" (Keil, 1989; Keil & Batterman, 1984) or the shift between concrete and abstract definitions (Anglin, 1977).

Despite these interesting findings, Keil (1989) warned researchers against relying too heavily on the method of asking children to define words explicitly (i.e., asking about intentions). His concern was that, "because of the wide range of definition types children give" (p. 69), responses from children would be difficult to code systematically. He adds, "The task of giving definitions places unusual demands on children, and the younger ones may have little experience with it. . . . The definition task is at least partly a metalinguistic skill" (p. 69). His concern was that asking children for category-based intensions is an artificial task, which might capture children's understanding about pragmatics or about social interactions and not their understanding of the target concepts or categories. More generally, the concern is that asking children to define categories probes their linguistic capacities, not their cognitive capacities.

We want to point out three reasons why investigating children's category-based intensions might be more informative than Keil suggests. First, there are several instances in which even preschoolers' intensions and extensions show a great deal of coherence (Caplan & Barr, 1989; Gentner, 2003; Maguire et al., 2008). For example, Caplan and Barr (1989) showed that the ways in which children defined categories related to the ways in which they made judgments about category membership. This implies that asking children to articulate their definitions of a concept might tell us something about their understanding of that specific concept, particularly what should and should not be categorized as a member of the category.

Second, the majority of this research has investigated children's definitions for artifacts or natural kinds. Asking children to define "taxi" or "island" may provide insight into how they represent those specific concepts, as well as others that have salient characteristic features. But work in this area has rarely investigated how children define concepts that are more abstract or concepts that are related to an activity (as opposed to an object or entity). Our approach in this part of the book is to query precisely these kinds of concepts (specifically, science, learning, teaching, play, and pretending) in an effort to understand more about how those concepts develop and about how children's understanding of those concepts develop. This approach fits with one of the major tenets of rational constructivism, namely that children are actively responsible for constructing theories about the world based on input they receive, whether via testimony from others or based on the results of their own exploration. As part of this process, children not only develop intuitive theories about how the world works, they also develop the ability to reflect explicitly on the content of these theories. The conceptual structures that children construct should be coherent, and the relation between children's intensions and extensions of their conceptual structure is an example of this coherence. It is even possible that children's reflection on this coherence could promote further conceptual change (see also Sobel, 2004b).

Third, as noted in chapter 1, we suspect that there may be relations between children's understanding of what science is and their abilities to think scientifically. To review briefly, children who understand that science is a set of abilities or a way of learning about the world may approach scientific thinking problems with a more productive mindset for solving such problems than children who believe that science is merely a set of content areas. This may be another case in which children's metacognitive abilities to reflect on their own thought processes or to understand their own actions can positively influence their reasoning abilities in scientific contexts (see Kuhn, 1989, 2002).

Children's Conceptions of "Science"

Although asking young children what they think the word "science" means is a fairly simple matter, few researchers have actually done so. One related body of prior work examines children's responses to the Draw a Scientist task (Chambers, 1983), in which (as the name suggests) children are asked to draw a scientist. Participants are sometimes additionally asked to label parts of their drawing or to provide the scientist with a name. Their drawings are scored on various dimensions, including how many stereotypical elements are included (e.g., being male, wearing a lab coat, having glasses) and whether the child exhibits self-efficacy as a scientist by including elements that resemble themselves or by naming the scientist after themselves.

In a large-scale study of this test, Chambers (1983) asked more than 4,800 children from Canada, the United States, and Australia to draw a scientist.

His qualitative analysis of aspects of the drawings showed that drawings by children in kindergarten and first grade had few or none of the elements of our culture's stereotypical image (i.e., a white male chemist working alone in a lab, wearing glasses and a lab coat). These elements emerged starting in the second grade, and their inclusion continued to increase through the elementary years. Students from lower-socioeconomic status (SES) backgrounds in this study showed an even later emergence of this stereotypical image, and their drawings were also less likely to include varied equipment. Chambers took these findings to indicate that these students had a less well-developed understanding of science instruments than students of higher-SES backgrounds. Further, it was extremely rare for any child to draw a woman in response to the task prompt, and all female scientists in the sample were drawn by girls. In terms of other elements, almost no child drew naturalists or explorers, tending instead to focus on labs and chemistry. Chambers also found some rare but consistent images of a stereotypical mad scientist, including underground labs, high secrecy, explosions, and so on. More recent treatments of this task tend to find similar results, with the stereotypical view of scientists as male chemists dominating children's drawings across age, gender, and national boundaries (Finson, 2002).[1]

Although the Draw a Scientist measure primarily focuses on when and to what extent children internalize a culture's stereotypical view of a scientist, it also provides insight into how children conceptualize science in general. To the extent that children internalize a view of scientists as male chemists working alone in labs, this internalization shapes the kinds of activities they consider to be scientific, as we will see from their responses to our questions (described below).

Aside from this drawing task, most of the work done on children's explicit conceptions of science and scientists has been done with children in late elementary school or later, and these measures usually do not ask directly about these children's definitions of "science" (e.g., Aikenhead & Ryan, 1992; Bourdeau & Arnold, 2009; Halloun, 2001; Kahle et al., 2000; Klopfer & Cooley, 1961; Lederman et al., 2002; Osterhaus et al., 2015; Rubba & Anderson, 1978; Schwartz et al., 2008; Tsai & Liu, 2005). Conversely, while some explicit measures of scientific understanding are aimed at younger children, these tend to examine their conception of themselves as able to learn science (e.g., Mantzicopoulos et al., 2008), which may not be directly related to their understanding of what science itself is.

The one instrument that we know of that does ask younger children to define the word "science," as we do, is the VNOS-E (Views of Nature of Science Questionnaire—Elementary School; Lederman et al., 2002; Lederman & Lederman, 2004). This question is typically administered as part of a larger interview probing students' understanding of the nature of science. For example, Walls (2012) presented a version of this instrument to a group of 23 African American third graders. These children said that science was for learning about the natural world, making inventions and other creative products, and making discoveries. Most students mentioned experiments as being important for science, and they explained experiments in terms of finding something out or testing something or answering a question one might have. Specifically, most of the students in this study mentioned topics within the natural world (91%) as well as experimentation (74%) in their responses. Additionally, and in line with results from the Draw a Scientist task, a large minority (39%) mentioned potions or mixing substances together.

Studies using the VNOS-E provide a good baseline for understanding how children view the practice of science. However, the approach taken with this instrument differs from the approach we describe below in two main ways. First, children's responses to this question are usually analyzed qualitatively, rather than the more quantitative analysis we present below. Second, the children who are interviewed tend to be in third grade or higher. We believe that this is past the point when they would have begun to form conceptions about what science is.

Our sample thus included children between the ages of 3 and 11, to track children's early experiences with science and to probe how their conceptualizations of science develop over time. In support of this approach, we note that most of young children's knowledge about science is learned from exposure to formal and informal environments and from natural experiences, such as parent-child interaction, peer-to-peer interaction, or their own observations (Callanan & Jipson, 2001; Callanan & Oakes, 1992; Evans, 2000; Evans et al., 2016; Fender & Crowley, 2007; Jant et al., 2014; National Research Council, 2009; Szechter & Carey, 2009), all of which are available to children well before third grade.

Documenting what children understand about science has potentially important implications for early science education, which aims to impart a sense of what science is and how it works in addition to knowledge of science content (e.g., National Research Council, 2012; NGSS Lead States, 2013). While we agree that we should remain skeptical about the extent to which young children can reflect on the content of their own concepts, querying children's definitions of "science" can serve to illustrate the processes that connect an implicit conceptual understanding to a more explicit awareness of one's beliefs. Understanding more about children's definitions of "science" can allow us to trace the connections between the development of children's abilities to engage in explicit reflection and the development of their scientific thinking in general.

Asking children about the definition of science can also help elucidate what Osterhaus et al. (2017) refer to as an "epistemological understanding of science"—the understanding of how data relate to theories (following Kuhn, 2002). Osterhaus and colleagues demonstrated that 8- to 10-year-olds showed relations between their performance on a battery of advanced theory of mind measures (described in chapter 7) and their epistemological understanding. This epistemological understanding of science was also related to a third capacity, children's understanding of experimental design. These researchers measured this understanding by stepping children through confounded or unconfounded experimental designs and their resulting conclusions to see whether children registered that unconfounded procedures would result in legitimate causal conclusions while confounded procedures would not (see also Osterhaus et al., 2015).

That study indicates that there is a relation between children's understanding of the scientific process and their ability to determine whether certain experimental designs would result in causal conclusions. Our goal in having children define science, which we discuss in the next section, was to investigate a similar relation. We wanted to explore what children believe science is and how these definitions relate to their understanding of scientific methods and conclusions, as well as to their reasoning and exploration more generally.

"What Is Science?"

Our approach in this study was simple. We asked our participants an open-ended question: "What is science?" If they said that they did not know or if they refused to answer, the question was rephrased: "What do you think the word 'science' means?" If they still said that they did not know, the interview ended. If they did provide a response, they were given further open-ended prompts to elaborate on their response, for example, "Can you tell me anything more about that?" or "What else can you tell me about science?" These prompts were repeated until they indicated that they had provided all the information that they could, following a procedure for eliciting children's understanding of the meaning of a word from Blewitt et al. (2009). It usually took about a minute for each child to complete the interview.

We administered this interview to many of the participants that we tested in the other investigations reported in this book. Although their experiences in our studies differed in other ways, here we report just their responses to this interview. This complete data set includes 940 children ranging between 3 and 11 years of age (mean age = 84.7 months; age range 37.3–143.6 months), with 481 girls and 457 boys (the gender of two children was not reported). Some of these children were tested in the lab, some at preschools or elementary schools, and some at museums or playgrounds.

We also tested a sample of 113 adults (85 female), some of whom were undergraduate students recruited from psychology department participant pools and some of whom were the parents of children that we tested. Our analyses here focus primarily on children's responses, but these adults' responses provide a useful developmental comparison.

Unsurprisingly, children's responses to our question were wide-ranging (see figure 8.1). Many participants talked about a specific domain of science or a particular science-related activity (e.g., "nature, like a frog or outside"; "mix stuff together"). Other participants focused more on science as a general process of inquiry (e.g., "an experiment; when you test something," "discover something, like what the moon is made of; like about learning"). Still others responded that they did not know what the word meant or provided responses that were off-topic or repetitive (e.g., "artwork," "science museum"). As in previous work, a significant minority of participants gave responses that conformed with social stereotypes of scientists, referencing potions,[2] mad scientists, or science-fiction events (e.g., "It means a lot like math, except you do potions and stuff"; "It's really dangerous because most scientists want to make dinosaurs come back alive, meat eaters too").

Given the wide variety of responses to this question and the fact that many of the answers included multiple themes, we adopted a quantitative

Figure 8.1
Word cloud representation of children's responses to the "what is science" interview. The larger the word in the cloud, the more frequently it appeared in children's responses.

coding scheme that allowed us to apply multiple nonexclusive codes to each answer. There were five coding categories: Specificity, Science Word, Learning, Other Process, and Personal Experience.

A response received a Specificity code if it mentioned a specific area or aspect of science or related topic, such as "bones," "math," "earth," or "fossils." The point of this code was to consider whether children's definition of "science" referenced a particular activity or topic.

A response received a Science Word code if it contained a science-oriented keyword such as "experiment," "observe," "hypothesis," or "research." The goal here was to determine whether children's definitions were just key words that they might have been exposed to, whether in school or through media. For example, children's television shows often use words like "hypothesis" when teaching science; not all of them, however, use it correctly.

A response received a Learning code if it referred to a process of knowledge change. These responses often contained the word "learning" but also could refer to other processes, such as figuring things out or making discoveries. The goal of this code was to consider whether children recognized the relation between science and knowledge change and recognized that scientific investigations could be used for the purpose of acquiring knowledge or changing beliefs. In contrast, a response received an Other Process code if it referred to an activity other than learning, such as "mixing things together," "inventing," or "making stuff." Unlike the other categories, the Learning and Other Process categories were treated as mutually exclusive. If a child mentioned an active process that involved learning, or if they mentioned a definition with both a learning process and another process, the Learning code took precedence.

The final category, Personal Experience, was applied if the response either described an experience that the child had with science or expressed the child's personal opinion. Responses such as "I go to science camp" or "I like science" received this code. We included this category because some prior work suggests that talk about personal connections might be related to children's engagement, as well as their understanding of parents' causal explanations (Callanan et al., 2017).

Six independent coders, blind to participants' ages and genders and performance on other measures, scored the responses. Reliability was established by asking each coder to categorize 25 responses. Agreement was measured and the coding scheme was iteratively refined twice until agreement on each code was 85% or higher. At that point, each coder independently coded a section of the remainder of the sample.

We employed this coding scheme to try to rein in the complexity of this data set, though we acknowledge that it glosses over much of the richness of children's and adults' responses. It is also important to note that we did not conduct these analyses with the goal of determining whether children or adults provided a correct answer to the question. The issue of how science should be defined is a complex one, and there likely is not a simple, single correct answer (see Godfrey-Smith, 2003). Rather, we aimed to discover developmental trends in children's responses and to examine how and when they used the different coding categories in their responses.

Distribution of Coding Categories

We first examined the overall usage of each of the five coding categories across all children and adults (table 8.1). For the children, we found that about 30% to 40% of the sample used the Specificity, Science Word, Learning, and Other Process categories. Science Word was significantly less common than the other three categories,[3] which did not differ from each other.[4] The Personal Experience category was significantly less likely to be used than the other four.[5] Adults, in contrast, used the Science Word and Learning categories most, followed by the Specificity and Other Process categories, followed by Personal Experience.[6]

Age differences. We compared the use of each of the five codes by age to test for developmental differences. Within the child sample, we found no significant relations with age in the rates of referring to specific topics,[7] non-learning processes,[8] or personal experiences.[9] Comparing children to adults, we found no differences in children's overall appeal to specific topics[10] and non-learning processes.[11] These findings are somewhat surprising; they indicate that even young children share adults' views about what science is with respect to these features. However, children were significantly more likely to refer to personal experiences than adults.[12]

In contrast, the use of science words increased significantly with age across the child sample,[13] and adults were even more likely than children to use these words (figure 8.2).[14] It is somewhat unsurprising that older children and adults were significantly more likely than younger children to use

Table 8.1
Use of the five coding categories for the "what is science" interview by age

Coding category	Children aged 3–11 ($n=940$)	Adults ($n=113$)	Total ($n=1053$)
Specificity	377 (40.5%)	38 (33.6%)	415 (39.8%)
Science Word	250 (26.9%)	69 (61.1%)	319 (30.6%)
Learning	349 (37.5%)	72 (63.7%)	421 (40.3%)
Other Process	354 (38.0%)	36 (31.9%)	390 (37.4%)
Personal Experience	76 (8.2%)	2 (1.8%)	78 (7.5%)

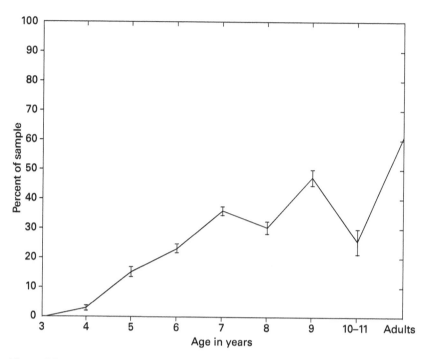

Figure 8.2
Proportion of sample using a science word, by age.

science-specific vocabulary. This indicates that science education already includes these terms even in early elementary school or in informal educational materials aimed at this age group.

We find it more intriguing that older children were significantly more likely than younger children to refer to science as a process of learning,[15] and adults were yet more likely to do so[16] (figure 8.3). This more general and flexible way of understanding science could be seen as a hallmark of developing a more mature understanding of science. For this reason, we had initially expected references to specific topics to decline with age while references to general learning processes rose. But this was not the case. Rather, it seems that both adults and children retain an idea of science as being about particular topics or involving particular kinds of activities. What develops in parallel to this understanding is the idea that science is a general way of learning about the world.

Gender differences. In addition to age, we examined whether there were any gender differences in children's definitions (table 8.2). Looking at the

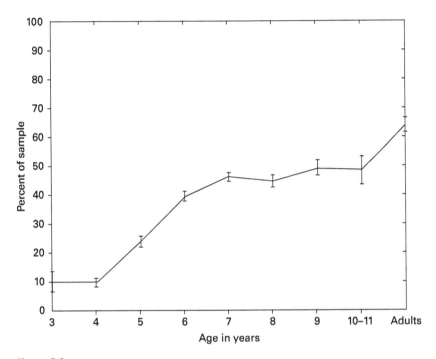

Figure 8.3
Proportion of sample referring to learning, by age.

entire child sample, we found that girls were marginally less likely to name a specific topic than boys[17] and significantly less likely to refer to learning.[18] Girls, however, were marginally more likely to use a science word than boys,[19] and girls were also significantly more likely to refer to a personal experience than boys.[20] The pattern for these four coding categories remained consistent across children of all ages in our sample.

For the Other Process category, we found no gender differences in the overall sample.[21] However, when we examined the effect of both gender and age, we found an interaction effect: As children got older, girls became more likely to refer to an active process while boys became less likely to do so.[22] No other coding category showed this pattern.

Overall, then, we see some evidence of gender's effects on how children talk about science, with girls and boys choosing to refer to different aspects of science in their definitions. These differences might be related to the persistent gender bias in science, whereby male scientists are disproportionately more common than female scientists (see e.g. Huang et al., 2020; Larivière et al., 2013). Additionally, in some studies, adults perceive men as being more competent in science (Nosek et al., 2009), and men often receive preferential treatment in science labs (Moss-Racusin et al., 2012). Even children are aware of these gender stereotypes: By the age of 6, they report that men are more likely to be "really, really good" at science than women (Bian et al., 2017), and their drawings of scientists are disproportionately male, as reviewed above (e.g., Finson, 2002). Girls' persistence in a challenging science task are also disproportionately negatively affected by instructions to "be a scientist" rather than to "do science," as described in chapter 1 (Rhodes et al., 2019). These findings suggest that early exposure to science and early socialization of children with respect to these gender biases could have influenced children's responses to our questions.

What is not clear from our data, however, is whether these differences in children's definitions have consequences, either for children's performance on related tasks or for their attitudes toward science more generally. Looking just at the subset of children from this sample who participated in our blicket task in one form or another ($n = 590$), we found no relation between gender and performance: 42.9% of girls and 44.6% of boys responded to the task in a manner consistent with understanding the interactive causality. For the moment, then, we cannot say with any certainty whether the gender differences we discovered in response to our interview might be symptomatic of girls' attitudes toward science or diagnostic of their performance on science, technology, engineering, and mathematics (STEM) tasks. Indeed, the fact that girls in our sample were proportionally more likely to refer to a personal experience when asked about science could reveal that they feel particularly connected to science or competent at doing science, contrary to the research cited above.

Table 8.2
Use of the five coding categories for the "what is science" interview by children's gender

Coding category	Girls ($n = 481$)	Boys ($n = 457$)
Specificity	178 (37.4%)	198 (43.7%)
Science Word	141 (29.6%)	109 (24.1%)
Learning	163 (34.2%)	186 (41.1%)
Other Process	188 (39.5%)	165 (36.4%)
Personal Experience	48 (10.1%)	27 (6.0%)

There were no gender differences in the adults' uses of these categories in our data. This suggests that the patterns we observed in children may not be maintained throughout their lives. Later science instruction, caregiver-child or peer-to-peer interaction, or other experiences in high school or in college may negate any earlier gender differences in our sample.

This conclusion was corroborated by data from a different sample: parents who participated in a set of studies about parent-child interactions, particularly in museum settings. These data were taken from several samples we collected (Callanan, Legare, Sobel et al., 2020; Letourneau et al., 2021; Sobel et al., 2021), which include 600 parents of children participating in research at a set of museums around the United States. The adults completed the Attitudes Towards Science questionnaire (Szetcher & Carey, 2009), which consists of fifteen questions that look at how adults think about science and scientists. Responses to these questions are scored on a scale from 1 to 7, with higher scores indicating that participants had more positive attitudes toward science. The overall questionnaire is divided into three subscales. The first subscale is Personal Interest in Science, which measures how much participants are interested in science and scientific topics. This subscale includes items like, "I would enjoy being a scientist." The second subscale is Views of Science and Scientists, which measures how much participants think science is an important topic and scientists are engaged in important work. This subscale includes items like, "Scientists are among the most successful people." The third subscale is Utility of Science, which measures the extent to which participants think science is useful for society. This subscale includes items like, "Thinking like a scientist is only useful when taking a test in science class."

The goal of administering this measure in this parallel project was to see whether children's causal reasoning, learning, and engagement with STEM activities at museum settings were related to parents' attitudes toward science. The short answer was that they were not, at least in this set of studies (although other facets of parents' background in science did relate to the ways in which children explored the museum exhibits). However, we can use this set of data to examine whether there are any gender differences in parents' attitudes to try to probe more deeply into the patterns that we found in the children's responses to our "what is science" interview.

Of the 600 parents, 157 self-identified as male and 436 self-identified as female (the other 7 left this question blank). Overall, the group of parents who identified as male had significantly higher scores on this questionnaire (mean=5.41 out of 7, SD=0.77) than the group who identified as female (mean=5.12 out of 7, SD=0.73).[23] Although that result might suggest that there are gender differences with respect to parents' attitudes toward science, the picture becomes more interesting when we examine the subscales. The entirety of the difference between these two groups was in the Personal Interest subscale, where males scored higher than females (5.43 vs. 4.94).[24] Otherwise, males and females in the adult sample did not significantly differ in their views about science and scientists (scores of 5.07 vs. 4.81)[25] or their beliefs about the utility of science (scores of 5.59 vs. 5.52).[26] Contrary to typical findings about gender biases in science, but in support of our results with children, these results suggest that adults may not fully endorse the stereotype of men being more competent at science. However, this sample may be somewhat biased because these parents chose to participate in a scientific study.

Relations with Language

Unsurprisingly, adults' definitions of "science" (mean word count=28.14, SD=22.68) tended to be longer than children's (mean word count=17.47, SD=17.12).[27] The length of children's definitions increased linearly with age.[28] Further, longer definitions tended to be coded into more categories for both children and adults.[29]

Although it makes sense that longer responses received more codes, we were particularly interested in examining whether longer responses were also more likely to be coded as indicating a learning process, which seems to be a hallmark of more mature responses. This code was more common in older children and adults, so children who gave longer responses (i.e., children who might be more linguistically advanced) might also have been more likely to mention learning regardless of their age. This was indeed the case: Older children, as well as those who give longer responses, were more likely to receive a Learning code.[30] Interestingly, though, these two effects were independent from each other; there was no interaction between age and response length here.[31]

Because we administered the "what is science" interview to all the participants in our Springfield school sample (see chapter 7), we could also examine relations between children's responses to this question and their scores on standardized tests of language ability. For the first graders who were tested in 2017 ($n = 112$), students who used a science word were marginally more likely to score higher on the MAP test of reading.[32] This makes sense, because both the Science Word code and the reading test capture aspects of children's vocabulary. The relations for the third graders ($n = 116$) were more interesting: These students received significantly higher scores on the MAP test of reading when their definitions of "science" mentioned learning,[33] but significantly *lower* scores when their definitions mentioned another active process.[34]

For the statewide PSSA test that these students went on to take in fourth grade, again the students who received a Learning code had higher scores on the language section of the test and the students who received an Other Process code had lower scores, though these relations were only marginally significant.[35] Even more interestingly, there were significant positive relations between these students' use of the Learning code and these students' scores on the math[36] and science[37] sections of the PSSA test. Children who have more mature language, math, and science skills, as measured by these standardized tests, were thus more likely to view science as a way of learning about the world.

Relations among Coding Categories

Our five coding categories were designed to capture orthogonal concepts in our participants' definitions. However, with the exception of the Learning and Other Process categories, they were not mutually exclusive. We were thus interested to determine whether there were relations among these codes. When examining responses from all children together, we found that their use of the five categories was generally not related to each other; children whose definitions were coded into one category (e.g., Specificity) were not necessarily more or less likely to have their definitions also coded into a second category (e.g., Learning).[38] This indicates that these codes captured independent features of children's definitions of "science," as intended.

There were two exceptions to this general pattern. First, children whose definitions used a science word were significantly less likely to also talk about a non-learning process.[39] Potentially, children who chose to use scientific vocabulary in response to our prompt may not have felt the need to also talk about a particular scientific activity. Second, children whose definitions *did not* mention an active process were significantly less likely to also talk about a personal experience.[40] Given that so few children referred to personal experiences, though, we do not believe that this finding reflects a particularly important pattern in children's thinking about science.

In terms of adults, as with the children, there were generally no relations among the coding categories. Here, however, the exception to this general conclusion was with their use of science-specific vocabulary: When adults used a science word, they were also significantly more likely to refer to learning and to mention a different active process.[41] That is, adults who used science-specific vocabulary in their responses also tended to conceptualize science as a process of learning, potentially reflecting a more complete understanding of science.

Summary of Findings from the "What Is Science" Interview

This investigation was designed to uncover how children and adults talk about science. Our question here was intentionally broad and open-ended, as we wanted to find out what subjects would say about science with little guidance. We applied a five-category coding scheme to narrow down the scope of these responses and to investigate how other variables affected what children and adults said. In general, both children and adults saw science as a type of activity involving specific content. Older participants, as well as children in our sample with more advanced linguistic and mathematical skills, were additionally more likely to define "science" as a process of learning.

Testing the Relation between Definitions of "Science" and Measures of Scientific Thinking

Diagnostic Reasoning Task

Having obtained a view of how children define "science," we wanted to test whether there were any relations between these definitions and children's scientific thinking. We began this investigation with the first sample of children we tested using our diagnostic reasoning task with the blicket detector (introduced in chapter 5). This sample included thirty-six 5- and 6-year-olds and thirty-six 7- and 8-year-olds (previously reported in Sobel et al., 2017). We measured whether children responded correctly to the diagnostic reasoning task and whether they defined science as an active process rather than as a particular type of content. In this sample, as predicted, we found a significant relation between responding correctly on the blicket task and children's definitions of "science."[42] Children who mentioned some kind of active process in their definition of "science" were significantly more likely to succeed at the blicket task (choosing the correct response 83% of the time) than children who did not (choosing the correct response only 39% of the time). Even more importantly, this correlation remained significant when age was factored out.[43] That is, children who had a conception of science as a process also performed well on our diagnostic reasoning task, whereas those who provided other types of definitions did not—regardless of their age.

These results imply that children who believe science to be a learning activity or another active process might be better able to engage in diagnostic reasoning, as we hypothesized. We also have some support for this conclusion from our study in the Springfield School District, described in chapter 7, in which students who correctly solved our blicket task in third grade performed significantly better on a standardized assessment of their science knowledge in fourth grade.

We next attempted to replicate these findings in a larger sample of children, which included 590 children who both completed our blicket task (or the butterfly version of this task) and provided a definition of "science" (303 girls, 285 boys, 2 unreported gender; mean age = 85.9 months; age range 46.2–132.9 months). These children were drawn from our studies of the role of contextualization (previously reported in chapter 6) and from our partnership with the Springfield School District (previously reported in chapter 7). In this sample, children who defined science as a way of learning about the world were marginally more likely to succeed on the diagnostic reasoning task that they were administered (blicket or butterfly).[44] Although these findings were not as strong as those from the earlier sample, they still support our argument that something about understanding science as a learning process may benefit children's reasoning abilities.

We also found a relation between these definitions and children's justifications for their responses on the diagnostic reasoning task. Children whose definitions of "science" referred to learning were more likely to justify their responses by referring back to the data they had observed in the blicket task (24%), relative to children whose definitions did not refer to learning (15%).[45] This could suggest that understanding science as a process of learning may help children explicitly understand the structure of our diagnostic reasoning task, which does require an examination of the data from the demonstration phase to solve. But we are hesitant to draw strong conclusions from this relation, primarily because children's justifications of their choices in the blicket (or butterfly) task did not generally relate to their performance in this task.

We also attempted a more direct test of the hypothesized connection between definitions and task performance in a subset of 89 participants (48 girls, 41 boys; mean age = 102.0 months; age range 84.2–129.7 months).[46] These children were all administered the blicket detector task. Half of these children heard the task described as "a science game" and the other half heard this task described as "a puzzle game." Given the arguments above, we had anticipated that participants who heard the task described as "a science game" would perform better, because this framing would encourage them to bring their scientific thinking abilities to bear on answering our questions. However, there was no difference in performance between these two groups.[47]

Children's Exploratory Behavior

We also looked at the relation between children's definitions of science and their exploratory behavior at museum exhibits. One such project is reported in Callanan, Legare, Sobel et al. (2020), previously described in chapter 3. In this study, we asked 325 children between the ages of 3 and 6 to define "science" after they had played with a parent at a gear exhibit in a children's museum. These children also played with an experimenter in a set of tasks that were designed to test their understanding of gears. As reported in that monograph, children's definitions of science did not relate to their causal knowledge about gears.[48]

Here, we consider a different facet of these data: how children's definitions of science related to the way they played at the exhibit with their parent. One aspect of children's play at the exhibit was what we called *systematic exploration*, in which they constructed a gear machine by connecting one gear to a set of gears and then tested how that machine functioned by spinning the machine. As reported in chapter 3, the proportion of children's systematic exploration in their free play related to their understanding of the causal system on a set of follow-up measures. The proportion of systematic exploration also significantly positively correlated with whether children used a science word in their definition and whether they defined science in terms of learning. This proportion also correlated with whether they defined science in terms of a specific activity or topic, although this correlation was negative,[49] indicating that the more children thought science was a specific activity, the *less* systematic exploration they engaged in during their play. The relations between systematic exploration and generating a specific definition[50] or one that involved learning[51] were also still significant after controlling for children's age. So although responses to the "what is science" interview did not have much predictive power on the way children responded to questions about gears when tested in a laboratory setting, there were direct relations between how children defined science and their exploratory behaviors. How children defined "science" thus seems to have some bearing on how they chose to explore a STEM activity in a museum setting.

In a more recent study (Sobel, Letourneau, Legare & Callanan, 2021), we investigated parent-child interaction at a different museum exhibit, this time about electric circuits. Parents and 4- to 7-year-olds played together at the exhibit. We then presented children with a set of progressively more difficult circuit-construction challenges. After each challenge, children were asked whether they wanted to continue, and we coded the number of challenges they participated in as a measure of their engagement with the activity. We also coded the proportion of challenges that they solved as a measure of their understanding of the causal complexity of the circuits.

These same children were also given the "what is science" interview. Both of our measures—the number of challenges that children engaged in and the proportion of those challenges they could solve on their own—correlated negatively with whether they described science as a specific activity.[52] These correlations, however, did not remain significant when age was factored out.[53] Older children in this sample were more engaged by the challenges and performed better on them, and older children were also less likely to generate a definition of science that appealed to a specific activity.[54]

In this study, we also looked at the circuits that children and parents built during free play and coded when particular circuits were completed. We examined the 30 seconds prior to the completion of each circuit and counted the average number of actions that parents and children engaged in before completing the circuit. This provides a measure of who is completing the circuits. As reported in Sobel et al. (2021), parents' actions during this time frame negatively related to children's performance on the challenges: In general, the more that parents involved themselves in their children's play, the less children learned about the circuits when tested on their own. Interestingly, one facet of children's responses to the "what is science" interview did relate to the number of actions that both parents and children engaged in: If children generated a personal connection to science in their definition, they were more likely to engage in more actions in the 30 seconds prior to completing a circuit during their free play, and their parents were less likely to act.[55]

Do Definitions of "Science" Relate to Scientific Thinking?

Taken together, these data point to potential relations between children's definitions of "science" and their scientific thinking, including solving diagnostic reasoning tasks and engaging in exploratory actions. Some of our studies find strong relations, some find weak relations, and some find no relations. The museum studies with the gears and circuits are illuminating here, as children's definitions of "science" sometimes relate to the way they actively explore an environment and to whether this exploration leads to learning. Interestingly, we sometimes saw negative relations between children's performance in these museum tasks and their tendencies to refer to specific topics in their definitions, suggesting that this way of thinking about science may be somewhat less mature or less helpful to their exploration. And that makes sense: Thinking that science is just mixing things together, for example, ignores the process of scientific thinking that is the focus of these investigations.

Although we have some evidence for our hypothesized relation between explicit definitions and children's scientific thinking, we cannot conclude definitively that we have found strong relations because of the inconsistencies in our results. This might mean that our argument for the connection between explicit definitions of "science" and scientific thinking might be flawed. It is also possible that the relation is just weak from the perspective of statistical power, but is still applicable in certain circumstances. We take up these arguments in more detail in chapter 11, but for right now, what is important to take away from this discussion is that there are some relations between children's definitions of science and their reasoning behaviors, although the relations are not clear and require more investigation.

Finally, in agreement with the arguments from Keil that we reviewed at the beginning of this chapter, it is worth noting that our definition task is a bit odd. We are committed to the value of exploratory measures like this one to give us a sense of children's thinking, but the broad range of possible answers might be both a feature and a bug. Children may never before have had to explicitly reflect on how they conceptualize science, and their responses to this question may not have captured the full range of their thinking. Given this, the fact that any of our tasks found positive relations with performance suggests that there may still be some reason to believe that our hypothesis about this connection holds. A more focused definition task or a more highly structured interview could uncover more precisely how children think about science, and these responses might show stronger relations with their abilities to engage in scientific thinking on some tasks. These studies take the first step in this direction, and future work should tease out these potential connections in more detail.

"Is That Science?"

One of the major strengths of our open-ended interview is that it allows children to talk about anything that they believe about science, providing a rich view of the development of this concept. But this open-endedness is also one of its weaknesses: Children generated many different kinds of responses, and the conversation sometimes ran off course, particularly when the children we tested had little understanding about what the word means. We thus conducted two additional investigations, which asked closed-ended questions to probe children's beliefs about what science is in more detail.

The first investigation provided children with a list of actions and asked them to report whether each one was an example of science. These questions were asked to a subset of 88 children from the main sample, ranging from 3 to 10 years of age (44 female, 44 male; mean age=81.5 months; age range 37–131 months), as well as to 24 adult participants (22 female; 2 male). Each participant was presented with a list of 12 specific actions and asked whether each action was science (shown in table 8.3). Some of these actions were general processes used in scientific endeavors but that lacked specific scientific content (e.g., asking questions), some were specific scientific actions (e.g., finding out what happened to the dinosaurs), and some were not canonically scientific (e.g., doing jigsaw puzzles). We chose these items to reflect topics studied by several scientific disciplines (e.g., biology, chemistry, psychology) and other nonscience disciplines covered in elementary school (e.g., history, mathematics). These items were presented in a random order for each subject. For each item, participants were asked, "Is that science?" and were asked to respond "yes" or "no."

In looking at patterns of agreement for individual items, we uncovered a few interesting patterns. First, children and adults did not differ in their rates of agreeing that the following activities are science: trying to figure out what happened to the dinosaurs, mixing things together, and learning about the world.[56] These activities, from the earliest age that we tested, seem to fall squarely into the category of science for all participants. It is not particularly surprising that something involving dinosaurs and a chemistry-related action were seen as scientific. It is reassuring, however, that the general activity of finding out about the world was also seen as scientific at all ages. Similarly, children and adults did not differ in their rate of agreeing that doing jigsaw puzzles is not science.[57]

Second, there were several items on which children and adults differed because adults or older children were more likely to claim that an activity is scientific than younger children. This was the case for trying to figure out why babies cry, asking questions, playing in the dirt, and finding out the names of different birds.[58] Figuring out why babies cry and playing in the dirt were not seen as scientific until adulthood. This latter item may have gained a high level of agreement from the adults because this study often recruited the parents of the children we were testing, who may have been particularly likely to see open-ended exploratory activities like this one as scientific. Only the 3- and 4-year-olds were ambivalent about whether finding out the names of different birds is scientific; children at all other ages, as well as adults, agreed that this is a scientific activity.

The item about asking questions showed an unexpected U-shaped pattern, with 3- and 4-year-olds as well as 9- and 10-year-olds and adults claiming that this is a scientific activity. Children in the middle of our age range, from 5 to 8 years old, tended to deny that this is science. We are not sure of the source of this U-shaped curve, but we suggest three speculative explanations. First, participants at the ends of our age spectrum may genuinely have had a different understanding of science, such that they agree that asking questions is science while children in the middle do not. Second, it is possible that participants across our entire age spectrum believed that asking questions is science, but the children in the middle of the age range might have been more influenced by their educational experiences, which tend to be highly didactic at these ages and which hence might have led

Table 8.3
Closed-ended categorization questions and percent agreement by age group

"Is that science?"	3- and 4-year-olds (n=22)	5- and 6-year-olds (n=24)	7- and 8-year-olds (n=23)	9- and 10-year-olds (n=19)	All children (n=88)	Adults (n=24)
Trying to figure out what happened to the dinosaurs[a]	90.9%	91.7%	91.3%	50%	93.2%	100%
Trying to figure out why babies cry[a,b]	40.9%	33.3%	52.2%	73.7%	48.9%	100%
Mixing things together[a]	77.3%	83.3%	91.3%	89.5%	85.2%	95.8%
Learning about the world[a]	86.4%	75.0%	91.3%	94.7%	86.4%	95.8%
Asking questions[a,b]	63.6%	58.3%	39.1%	68.4%	56.8%	87.5%
Playing in the dirt[a,b]	45.5%	29.2%	26.1%	31.6%	33.0%	83.3%
Finding out the names of different birds[a,b]	50.0%	83.3%	73.9%	89.5%	73.9%	79.2%
Studying what happened a really long time ago[a,c]	81.8%	79.2%	87.0%	52.6%	76.1%	70.8%
Doing jigsaw puzzles	50.0%	45.8%	26.1%	26.3%	37.5%	54.2%
Trying to add numbers together[b]	86.4%	66.7%	17.4%	31.6%	51.1%	37.5%
Reading stories[b]	50.0%	62.5%	17.4%	31.6%	40.9%	37.5%
Learning to write[b]	77.3%	75.0%	17.4%	10.5%	46.6%	29.2%

Note: [a]Indicates that more than 70% of the adult sample categorized the item as being science. [b]Indicates a significant difference among the age groups in rates of categorizing the item as being science. [c]Indicates a marginally significant difference in the age groups in rates of categorizing the item as being science.

them to respond that asking questions is not a scientific activity. Finally, and relatedly, children in the middle age range may have seen question-asking as an activity that applies across multiple disciplines and is not specifically characteristic of science, so they may have been reluctant to say that it was scientific because it is not limited to science.

A third interesting pattern that we uncovered in these data is that there were several items on which children and adults differ because children were more likely to claim that an activity is scientific than adults. This was the case for studying what happened a long time ago (marginally), trying to add numbers together, reading stories, and learning to write.[59] For studying what happened a long time ago, 9- and 10-year-olds were marginally less likely to claim that this is science, possibly because they are beginning to study history and social studies in a more in-depth way or in classes that are separate from their science classes, leading them to see this subject as different from science. For the other three activities, children younger than 7 tended to claim that these activities are scientific, while older children and adults tend to claim that they are not. This may reflect a general trend on the part of these children to see any kind of school-like activity as science.

Composite Scores

To analyze the overall development of children's understanding of scientific activities, we coded their responses to this set of closed-ended questions into a composite score. For all the items in which more than 70% of the adults categorized that activity as science, we gave children a score of 1 if they said "yes" for that item and a score of 0 if they said "no." If less than 70% of the adults categorized the item as science, we reversed this scoring, giving children a score of 1 for the item if they said "no" and a score of 0 if they said "yes." We then summed these scores into a scale from 0 to 12 (see table 8.4). Importantly, these scores reflected only the extent to which children's responses agreed with adults' modal responses; we remain neutral about whether these are the correct answers to the question of what science is.

These composite scores significantly correlated with age.[60] That is, older children were more likely to provide an adult-like pattern of responses about which activities were scientific.[61] More specifically, the 3- and 4-year-olds and the 5- and 6-year-olds did not differ from one another,[62] and the 7- and 8-year-olds and the 9- and 10-year-olds also did not differ from one another.[63]

Table 8.4

Children's average composite scores

	3- and 4-year-olds ($n=22$)	5- and 6-year-olds ($n=24$)	7- and 8-year-olds ($n=23$)	9- and 10-year-olds ($n=19$)
Mean score (standard deviation)	6.73 (2.31)	6.83 (1.93)	8.74 (1.19)	9.00 (1.15)

However, scores for the 7- and 8-year-olds were significantly higher than scores for the 5- and 6-year-olds.[64] This age range might thus be an important time for the development of children's ideas about science. That conclusion aligns nicely with the work reported in part II of this book, which also finds that this time period is important for the growth of children's scientific thinking, such as the diagnostic reasoning tested by our blicket detector task, and for many of the age-related changes that we documented on children's metacognitive capacities. We return to the question of whether these two developmental trends might relate to each other in chapter 11.

Relations between Closed- and Open-Ended Questions

We also considered the relation between children's categorization of items in this closed-ended measure and their responses to our open-ended question asking them to define "science." We found that children who used a science word in their definitions had higher composite scores ($M=8.81$ out of a possible 12, $SD=1.52$) than children who did not ($M=7.53$, $SD=2.22$).[65] Additionally, children who referred to learning in their definitions had higher composite scores ($M=8.86$, $SD=1.42$) than children who did not ($M=7.39$, $SD=2.24$).[66] This confirms the trends noted above, because higher composite scores and definitions of "science" that mentioned learning both reflect more adult-like patterns.

These results imply that, before the age of about 7, children's understanding of science is somewhat undifferentiated, and it is around this age that children begin to appreciate science as a process of learning and discovery. This developing understanding is reflected not only in their explicit definitions, but also in their judgments of what activities are science, and possibly also in how they are able to demonstrate their capacity for scientific thinking.

Children's Understanding of What Makes an Investigation Scientific

In our second closed-ended investigation, we probed children's understanding of science by telling them brief stories about characters who were conducting different kinds of investigations. Our general approach was to introduce children to a character who had a question in a particular area of science (chemistry, biology, or psychology). We varied what answer the character received to their question and how the character went about obtaining that answer.

Specifically, in one set of conditions, characters obtained answers to their question that were either accurate or inaccurate, but no information was provided about how these answers were obtained. In another set of conditions, instead of being told the answer that the character found, children were told only the method by which the character went about learning the answer. These methods were either consistent or inconsistent with "systematic empiricism" (Stanovich, 2012, p. 9): experimental interventions designed to answer scientific questions. In all cases, children were asked whether the character had done science and to justify their answers.

In general, judgments about whether investigations into certain questions are scientific should be independent of topic and independent of the exact answer that one obtains. Chemistry is no more or less scientific than biology or psychology; thinking that certain disciplines are more scientific than others places undue emphasis on the content of science, rather than on its process (see chapter 1 of Stanovich, 2012; see also our discussion in chapter 1). If children understand this, they should judge all cases with appropriate methods as science, regardless of the topic that the character is investigating or the answer that they receive.

Recent work, however, suggests that both adults' and children's folk thinking about science is influenced by the topic under investigation (Fernandez-Duque et al., 2015; Hopkins et al., 2016; Keil et al., 2010). This research suggests that individuals believe the natural sciences are more scientific than the life sciences, which in turn are more scientific than the social sciences. Results from our own work are consistent with these ideas. In the closed-ended study described earlier in this chapter, children readily judged certain topics (dinosaurs, chemistry) as science, while they were less willing to judge psychological investigations (such as finding out why babies cry) as science. These results suggest that children might be biased by topic when making judgments in our tasks, focusing less on an investigation's method or its outcome and more on the nature of the question itself.

General Materials and Methods

To parallel the work on children's judgments of activities reported above, we chose three phenomena for these experiments, one to represent each of three scientific disciplines: chemistry (why salt disappears in water), biology (what happened to the dinosaurs), and psychology (why babies cry). For each phenomenon, we constructed four vignettes that described a character who wanted to find out about that phenomenon. There were 12 vignettes in total.

Half of these vignettes focused on what the character found out—that is, the outcome of the investigation. These six vignettes followed a 3 (Discipline: psychology, biology, chemistry) × 2 (Outcome: appropriate/inappropriate) design; both variables were within-subject such that each child received six vignettes. For these stories, children received no information about the methods used to obtain this result. For example, the vignette that presented the appropriate outcome for biology said, "Joanne/Robert [gender matched to the child participant] wants to know what happened to the dinosaurs. She/He finds out that they all died because a meteor crashed into the Earth." The matched vignette with the inappropriate outcome said, "Joanne/Robert wants to know what happened to the dinosaurs. She/He finds out that they all died because they ate too much cake at a party." We tested 76 children in this condition (38 girls, 38 boys; mean age 88.03 months; age range 61–136 months).

The other half of the vignettes focused on what the character did, that is, the methods of the investigation; no information was provided about the investigation's outcome. These 6 vignettes also used a 3 (Discipline: psychology, biology, chemistry) × 2 (Method: appropriate/inappropriate) design. For example, our biology/appropriate method vignette said, "Joanne/Robert [again, gender matched to the child participant] wants to know what happened to the dinosaurs. So s/he looks carefully at fossils." The paired inappropriate method vignette said, "Joanne/Robert wants to know what happened to the dinosaurs. So s/he takes a picture of his/her dog." No information was ever provided about the result of these investigations. We tested 101 children in this condition (55 girls, 46 boys; mean age = 77.33 months; age range 37–143 months).

All the vignettes were accompanied by two pictures. The first picture was the same across conditions and always showed the character thinking about the question. The second picture varied by condition and depicted either the character's idea about the answer or the character engaging in an investigation (figures 8.4 and 8.5). Following each vignette, children were asked whether the main character of the vignette was doing science (e.g., "Is Joanne/Robert doing science?"). After answering this yes/no question, children were asked to justify their response: "Why/why not?"

Figure 8.4
Example stimulus item from the study on outcomes. The character wants to find out what happens to the dinosaurs and discovers that a meteor hit the Earth and killed them.

Figure 8.5
Example stimulus item from the study on methods. The character wants to find out about the dinosaurs and does so by looking at fossils.

What Kinds of Outcomes Are Scientific?

To test children's understanding of the role of outcomes, we counted the number of times that children accepted an appropriate outcome as being science or rejected an inappropriate outcome as not being science. The proportion of vignettes that children judged in this way are shown in table 8.5, with data separated by age group (via a median split). The average age of the younger group was about 6 years, and the average age of the older group was about 8 years.

Children responded similarly to all three of the vignettes with appropriate outcomes, correctly judging all of these as scientific.[67] But there were

Table 8.5
Proportions of correct responses across vignette types and age in the study on outcomes (standard deviations in parentheses)

	Chemistry Appropriate	Chemistry Inappropriate	Biology Appropriate	Biology Inappropriate	Psychology Appropriate	Psychology Inappropriate
All Children (n=76)	.82 (.39)	.53 (.50)	.70 (.46)	.68 (.47)	.74 (.44)	.61 (.49)
Younger Children (roughly 5–7-year-olds, n=38)	.76 (.43)	.34 (.48)	.68 (.47)	.50 (.51)	.84 (.37)	.34 (.48)
Older Children (roughly 7–11-year-olds, n=38)	.87 (.34)	.71 (.45)	.71 (.46)	.87 (.34)	.63 (.49)	.87 (.34)

differences in children's responses to the vignettes with inappropriate outcomes. Most importantly, only for the chemistry vignettes did responses differ significantly between the appropriate and inappropriate vignette.[68] That is, children only reliably distinguished between investigations with appropriate and inappropriate results for chemistry.

When we examined these results separately for the two age groups, however, different patterns emerged. The younger children tended to accept vignettes with appropriate outcomes as scientific,[69] but did not generally reject vignettes with inappropriate outcomes as not being science.[70] While this response pattern could reflect a simple bias to say "yes" to all items, a closer look at the data does not support this possibility. These children's responses did not differ between the two biology vignettes, which were judged as equally scientific,[71] but their responses did differ between the two psychology and two chemistry vignettes.[72] For these latter two topics, younger children were more likely to correctly distinguish between appropriate and inappropriate outcomes.

For the older children in the sample, responses were all significantly above chance,[73] except on the appropriate psychology vignette, which did not differ from chance.[74] Additionally, they were significantly more likely to reject the inappropriate psychology vignette as not being science than they were to accept the appropriate psychology vignette as being science.[75] This difference was not seen in the other two topics.[76]

Overall, then, there were marked differences between the younger and older children in our sample with respect to which features they considered in their judgments. Between the ages of 3 and 7, children correctly accepted investigations with appropriate results as scientific. But these younger children were much less likely than the older children to reject investigations with inappropriate results. For children between the ages of 7 and 11, topic played more of a role in their judgments: Older children were less likely to accept investigations with appropriate outcomes in psychology than for other two topics. In general, rejecting each of the inappropriate vignettes as not being science increased with age. However, accepting the appropriate vignettes as being science did not improve with age for the biology and chemistry vignettes; these were seen as scientific at all ages. Further, accepting the appropriate psychology vignette as being science actually decreased with age.

These results suggest that the outcome of an investigation is of primary importance in children's decisions about whether that investigation is scientific, but the investigation's topic (especially for psychology) and the age of the child also play some role. Again, we see important differences emerging around age 7.

Justifications of Responses. We categorized children's justifications for their responses into three main categories. *Metacognitive* responses referenced mental states or character's intentions (e.g., "because he's trying to figure out how dinosaurs died in the past"). *Topic* responses referred exclusively or primarily to the topic of investigation (e.g., "because babies is not a science"). *Outcome* responses referred exclusively or primarily to the result of the investigation (e.g., "because that's what [salt] does, the water melts it"). All remaining responses were coded as *Irrelevant*. Distributions of these justifications by age are shown in table 8.6.

For the most part, the distribution of justifications did not differ between the age groups (see last row of table 8.6). This suggests that both older and younger children in this sample understood the criteria they were using to make their judgments. What did differ is the number of Metacognitive and Outcome justifications between the appropriate and inappropriate vignettes for each topic. For each topic, children generated more justifications that referred to outcomes for the inappropriate vignette than for the appropriate one.[77] Conversely, children generated more justifications that referred to metacognition for the appropriate than for the inappropriate vignettes, though this pattern was only statistically significant for the psychology and chemistry vignettes.[78] This means that, when justifying their acceptance of an investigation with an appropriate outcome, children tended to refer to the character's beliefs and intentions. But when justifying their rejection of an investigation with an inappropriate outcome, children tended to refer to the outcome itself.

We also examined justifications in relation to their responses to the main test question. We first examined the number of children who justified a correct response to the test question for each vignette in terms of the outcome of the study. For the three appropriate vignettes, participants who judged that the character was doing science were more likely to justify that response in terms of the investigation's outcome than were participants who judged that the character was not doing science.[79] We found the same

Table 8.6

Distribution of justifications by age group (# of children) for the study on outcomes

Younger Group ($n=38$)

	Chemistry Appropriate	Chemistry Inappropriate	Biology Appropriate	Biology Inappropriate	Psychology Appropriate	Psychology Inappropriate
Metacognitive	9	7	10	5	10	4
Outcome	12	20	13	20	13	22
Topic	2	1	2	2	2	2
Irrelevant	15	10	13	11	13	10

Older Group ($n=38$)

	Chemistry Appropriate	Chemistry Inappropriate	Biology Appropriate	Biology Inappropriate	Psychology Appropriate	Psychology Inappropriate
Metacognitive	14	5	6	4	13	4
Outcome	16	31	21	28	13	26
Topic	2	1	3	2	4	2
Irrelevant	6	1	8	4	8	6
Difference in age groups	$\chi^2(3)=5.52$, $p=.14$	$\chi^2(3)=10.07$, $p=.02$	$\chi^2(3)=4.27$, $p=.23$	$\chi^2(3)=4.71$, $p=.19$	$\chi^2(3)=2.25$, $p=.52$	$\chi^2(3)=1.33$, $p=.72$

relation for the inappropriate vignettes, in which participants were more likely to justify their response that the character was not doing science by referring to outcomes.[80] This suggests that these children explicitly believe that appropriate outcomes are of primary importance in making an investigation scientific.

We also examined children's metacognitive justifications in this way and found generally the same pattern: Children were more likely to use metacognitive justifications when they accepted appropriate outcomes or rejected inappropriate outcomes.[81] The one exception to this pattern was for the psychology vignettes that had inappropriate outcomes. In this case, children's justifications did not differ depending on how they had responded to the main test question.[82]

Summary of Study on Scientific Outcomes. This study was designed to investigate the impact of an investigation's topic and its outcomes on children's explicit judgments of whether that investigation was scientific. Our results showed that children have a rudimentary understanding of scientific outcomes. They seemed to have a good understanding of what kinds of outcomes were scientific at the early end of the age range that we studied, and they seemed to be developing an understanding of what kinds of outcomes are not scientific over the early elementary years. However, they were also beginning to believe that different kinds of questions can be more or less scientific—in line with adults' intuitions that psychology is less scientific than fields like biology or chemistry (e.g., Keil et al., 2010).

What Kinds of Methods Are Scientific?

The second condition in this study considered whether children correctly accepted or rejected the use of a scientific method of investigation, and whether these judgments varied by topic (biology, chemistry, and psychology), by children's age, and by whether the method was appropriate or inappropriate. The proportion of correct responses are shown in table 8.7.

In general, children considered both an investigation's topic and its method in their judgments of whether that investigation is scientific. Specifically, when the method was appropriate, children responded correctly at above-chance levels to both the chemistry[83] and the biology[84] vignettes, but not to the psychology vignette.[85] When the method was inappropriate, children performed at chance on the psychology[86] and biology[87] vignettes and, unexpectedly, performed below chance on the chemistry vignette.[88] That

Table 8.7

Proportions of correct responses across vignette types and age in the study on methods (standard deviations in parentheses)

	Chemistry Appropriate	Chemistry Inappropriate	Biology Appropriate	Biology Inappropriate	Psychology Appropriate	Psychology Inappropriate
All Children (n =101)	.92 (.27)	.36 (.48)	.94 (.24)	.54 (.50)	.44 (.50)	.56 (.50)
Younger Children (roughly 3–7-year-olds, n=50)	.88 (.33)	.28 (.46)	.88 (.33)	.51 (.50)	.48 (.50)	.46 (.50)
Older Children (roughly 8–11-year-olds, n=51)	.96 (.20)	.44 (.50)	1.00 (0)	.58 (.51)	.39 (.49)	.67 (.48)

is, children often incorrectly said that the chemistry vignettes were science, even when the method used to investigate the question was inappropriate.

As can be seen in table 8.7, these same patterns were found independently in both age groups; there were few differences between the younger and the older children in this study. But when age was treated continuously rather than categorically, several relations emerged. For the chemistry vignettes, correct responses increased with age on both the appropriate and inappropriate cases.[89] For the biology vignettes, correct responses increased with age when the method was appropriate but not when the method was inappropriate.[90] For the psychology vignettes, correct responses increased with age when the method was *inappropriate* but not when the method was appropriate.[91] That is, older children were more likely to say that inappropriate psychological methods were not science, but they did not get better at saying that appropriate psychological methods were science.

These results show that children do not simply ignore the method of investigation; overall, they correctly judged investigations that used scientific methods as scientific more often than investigations that used nonscientific methods. However, this tendency varied significantly by topic. Children were much more likely to accept investigations using scientific methods as scientific and to reject investigations using nonscientific methods as not scientific when asked about a biology vignette. In contrast, children responded at chance for both types of psychology vignettes, regardless of method. And both chemistry vignettes were judged to be scientific, regardless of whether the method was appropriate. As children got older, they were more likely overall to reject nonscientific methods as not scientific. This was especially true for of the biological and psychological vignettes we presented.

Justifications of Responses. As in the previous study, we also coded children's open-ended justifications of their responses. We coded responses as *Metacognitive* if they contained some reference to mental states or intentions (e.g., "because he's learning about why babies cry"). *Exclusive* justifications involved children using a particle like "just" or "only," which indicated that the character's actions were inadequate to be scientific (e.g., "because he's just taking pictures").[92] All remaining responses were coded as *Irrelevant*. The distributions of these justifications by age group are shown in table 8.8.

Older and younger children generated different types of justifications for all six vignettes (see last row of table 8.8). Specifically, older children tended to generate more Metacognitive justifications while younger children tended

Table 8.8

Distribution of justifications by age group (# of children) for the study on methods

Younger Group (n=50)

	Chemistry Appropriate	Chemistry Inappropriate	Biology Appropriate	Biology Inappropriate	Psychology Appropriate	Psychology Inappropriate
Metacognitive	12	5	14	5	4	6
Exclusive	0	2	0	7	9	7
Irrelevant	38	42	36	37	36	36

Older Group (n=51)

	Chemistry Appropriate	Chemistry Inappropriate	Biology Appropriate	Biology Inappropriate	Psychology Appropriate	Psychology Inappropriate
Metacognitive	25	22	27	16	19	17
Exclusive	1	2	1	3	4	3
Irrelevant	24	26	22	32	26	30
Difference in age groups	$\chi^2(2)=8.83$, $p=.02$	$\chi^2(2)=15.02$, $p=.002$	$\chi^2(2)=10.07$, $p=.02$	$\chi^2(2)=8.81$, $p=.03$	$\chi^2(2)=16.69$, $p<.001$	$\chi^2(2)=7.62$, $p=.06$

to generate more Irrelevant justifications. We did not find differences in the pattern of justification types for the appropriate as opposed to the inappropriate methods for the chemistry[93] or psychology vignettes,[94] but there was a significant difference for the biology vignettes.[95] Specifically, participants tended to give more Exclusive justifications for the inappropriate methods and more Metacognitive justifications for the appropriate methods.

We also looked at the relation between responses to the test question and justifications. Here, we looked just at Metacognitive justifications, primarily because there were so few Exclusive justifications, but also because we believe that Metacognitive justifications are more likely to reflect a mature view of how science works.

We found that children were more likely to justify a correct acceptance of an appropriate method by referring to a metacognitive process, and less likely to justify an incorrect answer to these questions in this way.[96] However, correct rejections of inappropriate methods were only associated with metacognitive justifications for the chemistry vignettes; justifications for biology and psychology vignettes with inappropriate methods did not follow this pattern.[97]

General Summary of the Outcomes and Methods Studies

The results of these two studies reveal different patterns in how an investigation's topic, results, and methods affect children's judgments of whether that investigation is categorized as science. For vignettes that gave information only about the investigation's results, the topic of the investigation mattered little. Across the ages we investigated, most children were likely to accept investigations that came to appropriate outcomes as scientific and reject investigations that came to inappropriate outcomes. One exception to this general pattern was a negative correlation with age for psychology investigations that came to appropriate conclusions, suggesting that psychology was seen as increasingly *less* scientific as children grew older. This is in line with adults' views about psychology as being less scientific than other topics like physics, chemistry, or biology (e.g., Keil et al., 2010). In contrast, all three topics showed strong positive relations with age when the result was inappropriate: As children got older, they were more likely to reject inappropriate outcomes as not science. Moreover, on the vignettes where the character came to an inappropriate conclusion, children also provided more sophisticated justifications for those correct rejections as they got older. Children thus seem to be learning what are appropriate outcomes for each topic, leading them to judge that coming to such erroneous conclusions should not be counted as science.

Note that we do not necessarily endorse this view of science. Finding a correct answer may loom large in children's views, and it is likely to be emphasized in school. But true scientific investigations do not necessarily depend on their outcomes. Science aims at finding the truth, but finding an incorrect answer or getting a null result does not on its own make a particular investigation any less scientific.

In contrast, for vignettes that gave information only about the investigation's method, both the topic of investigation and the method influenced children's judgments. Children judged an investigation of a biological or chemical phenomenon as scientific more often than an investigation of a psychological phenomenon, regardless of whether the method was appropriate or inappropriate. At least during the preschool years, chemical phenomena are seen as so scientific that even investigations using nonscientific methods are judged as scientific the majority of the time.

But children in our studies did not simply ignore the method of investigation; they were generally more likely to judge investigations that used appropriate methods as scientific than they were to judge investigations that used inappropriate methods as scientific, though this tendency did not exceed chance levels. Here, what may be developing is an understanding that appropriate methods are scientifically valid, as the tendency to justify a correct answer with reference to the character's intentions and goals increased with age.

Combined with the less-nuanced pattern of responses found for vignettes that described only outcomes, these results indicate that, during the preschool and elementary-school years, children have an outcome-focused view of what science is, and they believe that investigations need to get the right answer in order to be scientific. This is potentially part of the more undifferentiated concept of science that children might have before the age of 7.

Interestingly, this focus on outcomes parallels developments within the domain of moral reasoning, in which young children of roughly the same age tend to initially privilege information about the outcome of an action, rather than an agent's intention, when deciding on issues of blame and punishment (see Cushman, 2008). The fact that children might be more influenced by the outcome of an investigation as opposed to its method suggests that science educators should focus on incorporating discussions of what makes a good scientific method into their pedagogy, and that such instruction should also take topic into account. These data can thus help point the way toward better science instruction in early elementary school, suggesting that rudimentary lessons about the philosophy of science—ones that focus on imparting an understanding of what methods are scientific and of how appropriate methods relate to doing science—might help even the youngest learners.

Developing an Understanding of What Science Is

In closing this chapter, we want to say two things. First, all the research we have described here has examined a relatively wide age range (3-year-olds to 11-year-olds). Given this, it is not particularly surprising that we found developmental changes in how children answered our questions. It is thus more important to consider what specifically is changing and why. We found that children at the younger end of this range were not particularly adept at understanding science as an abstract method of inquiry; their explicit definitions were more tied to specific topics or actions. The tendency to describe "science" as a series of domain-specific facts was maintained with age, but older children and adults were additionally more likely to define "science" as a process of learning than younger children were. Also, responses to our closed-ended questions about which activities are scientific reveal an important period in development between about 6 and 9, roughly corresponding to the ages in which we saw rapid development in children's scientific thinking abilities in the studies described in part II of the book. Further, we have some preliminary evidence in favor of our proposed connection between children's explicit understanding of science as a process of learning or knowledge change and their abilities to engage in various facets of scientific thinking.

Second, we want to repeat the point that our research did not aim to discover when children come to know the correct answer to the question about what science is. There may be no single answer to this question, and indeed there seems to be a tension even in adults' definitions between thinking of science as a set of topic areas (as captured by our Specificity code) and thinking of science as a process of inquiry or experimentation (as captured by our Learning and Other Process codes). This tension is reflected in the research in this area, with some studies examining how children come to acquire specific pieces of scientific knowledge, and others examining how children come to demonstrate particular kinds of scientific skills, as reviewed in chapter 1. Both the process of science and its specific disciplinary content are important for a full understanding of what science is; what we wish to highlight is that an explicit conception of what science is should also be part of this understanding.

9   Children's Definitions of "Learning" and "Teaching"

Chapter 8 introduced our focus on children's definitions of abstract concepts to provide insight into their development and looked at children's definitions of "science." We found rapid growth in these definitions over the late preschool and early elementary-school period, particularly between the ages of 6 and 8, similar to our investigations of children's diagnostic reasoning abilities (chapter 5) and their understanding of disagreement (see chapter 7). We also found that the younger children we investigated (3- to 6-year-olds) tended to see only certain kinds of questions or activities as scientific and to provide specific topics or actions in response to our open-ended definition question. By contrast, older children (7- to 11-year-olds) and adults accepted a wider range of questions and activities as being scientific, and their definitions included more references to learning and knowledge change.

This chapter focuses on a related set of questions: how children define the concepts of "learning" and "teaching." Analyzing how children talk about these concepts might help us understand how they learn and teach. Moreover, because many children (and most adults) used the concept of learning in their definitions of "science," we can use these investigations to gain greater insight into how they think about the kind of learning activities that might be involved in doing science. Much like our studies on how children conceptualize "science," here we examine the relation between children's appreciation of the intensions of lexical items and their extensions. Specifically, we examine how children define "learning" and "teaching," and we relate those definitions to the inferences that children make about whether someone is learning or teaching.

"What Is Learning?"

Much research in developmental and educational psychology is devoted to describing not only what children learn, but also, more importantly, *how* children learn. Given the importance of this topic, it is interesting that only a handful of studies have examined children's explicit understanding of learning. Most of this work comes from research in formal learning environments, where it has long been thought that children's understanding of learning might affect their engagement with learning and relate to their academic achievement (Dweck, 2006; Dweck & Leggett, 1988; Eccles et al., 1998; Li, 2004; Skinner, 1995; Stipek & Mac Iver, 1989). But this work has tended to focus on adolescents' developing identities as learners (e.g., Burden, 1998; Randi, 2009; Rubin, 2007), and does not often consider how preschool-age children or even early elementary-school children understand this concept.

To examine this issue, our work (Sobel & Letourneau, 2015) ran a parallel study to the one on children's understanding of science described in chapter 8: We asked 4- to 10-year-olds, "What do you think learning means?" We coded children's responses to this question into three mutually exclusive categories. In Identity responses, children simply defined "learning" as learning. In Content responses, children defined "learning" based on subjects or topics (e.g., "like reading and math"). In Process responses, children defined "learning" as related either to a source (e.g., "when your teacher tells you something") or to a strategy (e.g., "when you practice again and again until you know it") that would result in knowledge change (see figure 9.1).

One of the clearest trends in this study is that children's process-based definitions increased with age. This provides an intriguing parallel with the "what is science" interview described in chapter 8, whereby children became increasingly more likely to define "science" as an active process, specifically a process of learning or knowledge change, between the ages of 6 and 8.

This parallel could be taken to show that the ability to generate process-based definitions of abstract concepts is domain-general. Children's ability to talk about "science" and about "learning" as an active process could reflect a broader aspect of their language development or of their ability to use metacognitive language. However, this is not necessarily the case; as we will see later, children's talk about other abstract concepts like "teaching" and "play" does not follow this same specific developmental trend.

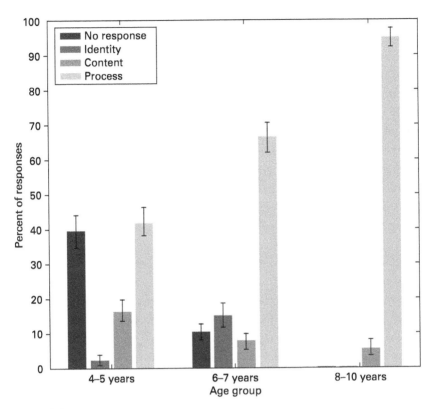

Figure 9.1
How children defined "learning" across age groups.

In addition to asking children to define "learning," in this study, we also asked children to generate examples of what they had learned (e.g., "Can you think of something that you have learned?") and how they had learned in each example (e.g., "How did you learn that?"). We also asked children whether they could think of other ways of learning (i.e., "How else could you learn?"). These questions were repeated several times so that children could generate multiple examples. We compared children's answers to these questions to their definitions in order to gain a deeper understanding of how children conceptualized learning.

We found some connections between children's definitions and the examples they provided. If children generated a process-based definition of "learning," they were also more likely to give an example in which they learned a skill (e.g., "how to tie my shoes") or a fact (e.g., "ants have six legs"). In contrast, children who did not generate process-based definitions often simply did not respond to this question. Children who generated a process-based definition of "learning" were additionally more likely to describe a source (e.g., "I learned it from my teacher") or a strategy (e.g., "I read it in a book") through which they acquired knowledge. These children were also more likely to generate multiple different strategies for learning. Critically, all of the relations just described held regardless of age. That is, children's definitions of "learning" independently predicted how well children were able to reflect on their own learning abilities. Even some 4-year-olds conceptualized learning as a process, and those children seemed to have better access to process-based details about their own learning than children who defined "learning" as a type of content.

Children's Understanding of When Learning Happens

In general, children develop a concept of learning based on their understanding of other mental states, like knowledge and ignorance. Because learning involves changes to knowledge states (e.g., replacing ignorance with knowledge or updating one belief with another), children's understanding of knowledge, ignorance, and beliefs can all influence their concepts of learning. This claim draws on past literature on the development of theory of mind. Research in theory of mind is rarely described as being about learning, but it can be understood as investigating children's developing abilities to understand knowledge and knowledge change. For example, 3-year-olds use the word "know" in conversations to reflect their own epistemic states (Bartsch & Wellman, 1995; Shatz et al., 1983; Taumoepeau & Ruffman, 2008) and spontaneously use different phrases to distinguish between their own knowledge and ignorance (Harris et al., 2017).

Between the ages of 3 and 5, children come to recognize the distinction between others' knowledge and ignorance (e.g., Hogrefe et al., 1986). They also begin to explicitly understand that beliefs represent one's ideas about the world, which are not necessarily the same as reality (e.g., Gopnik & Astington, 1988; Perner, 1991; Perner et al., 1987; Wellman et al., 2001; Wimmer & Perner, 1983). Children at this age also appreciate some information about the sources of false beliefs, such as that they can arise from different sources of information (Flavell et al., 1992). Although not typically thought of in this way, research on children's theory of mind addresses their understanding

of learning in its focus on the distinction between having and not having knowledge about particular situations. The preschool years are a time
of significant development in the ability to track and reason about others'
epistemic states (Wellman & Liu, 2004). The preschool years are also when
children's use of words like "learn" and "teach" begins to emerge, as shown
by an examination of a corpus of children's natural conversations (Bartsch
et al., 2003). In a replication of this analysis, which focused only on utterances
that were spontaneously generated by the child, we found that references
to learning processes increase between the ages of 3 and 5 (Sobel, Li & Corriveau, 2007).

Although these data suggest that children's understanding of learning in
general emerges during the preschool years, this understanding demonstrates
an interesting asymmetry: Children seem better able to understand others'
learning than their own. That is, preschool-age children have not yet developed the metacognitive capacities to reflect accurately on their own thought
processes, and they tend to misunderstand how thinking works (e.g., Flavell
et al., 1993, 1995), which might prevent them from fully understanding their
own learning (e.g., Klahr & Dunbar, 1988; Kuhn, 1989; Kuhn & Dean, 2004).
For instance, preschoolers think that they knew pieces of novel information
all along, even when they just learned them (e.g., Esbensen et al., 1997; Taylor
et al., 1994). Similarly, 6-year-olds overestimate others' ability to learn (Miller
et al., 2003). So while 4-year-olds might understand *that* their knowledge
changes, it is not until around age 5 or 6 that they can track and articulate
how they know their knowledge changes (Gopnik & Graf, 1988; Gopnik &
Slaughter, 1991). These results suggest that the metacognitive awareness that
is required for understanding one's own learning might develop later in preschool and during the early elementary-school years.

However, children do have some understanding of their own learning at these ages, specifically with respect to *source memory*: the ability to
recall from where one learned information. Bemis et al. (2011) asked 4- to
9-year-olds questions they were likely to be able to answer. These children
were then asked to describe how they had learned that piece of information. Even the youngest children in their sample could generate some
information about how they had learned, although there was significant
age-related change (i.e., older children could generate more information).

Bemis et al. (2013) followed up on this finding by teaching 4- and
5-year-olds novel facts and then examining whether those children could

articulate how they had learned that knowledge. Again, even the youngest children they tested were able to state how they had learned the new facts. Tang and Bartsch (2012; see also Tang et al., 2007) similarly showed 4- to 5-year-olds information either using a visual demonstration or through direct instruction. One week later, these children could accurately report whether they had been shown or told the information, although they could not report that this was done a week prior.

To add to these studies of children's understanding of learning, we investigated the connections among learning and different mental states (Sobel, Li & Corriveau, 2007). We found that preschoolers judge whether learning takes place primarily based on whether an individual wants to learn something, regardless of that individual's other mental states, such as their attention to necessary information. We presented 4- to 6-year-olds with two child characters who were in a similar learning environment (e.g., a teacher was teaching the characters to sing a song at school). The mental states of these two characters were potentially in conflict with their learning goals. One character wanted to learn but did not pay attention to the teacher. The other character did not want to learn but did pay attention to the teacher. When asked whether each character learned, 4-year-olds mostly responded based on the character's desires; they judged that characters who wanted to learn did so while characters who did not want to learn did not. Six-year-olds, in contrast, reported that paying attention or practice was also necessary for learning. Children's understanding of the relations among mental states involved in learning thus continues to develop past the preschool years.

In support of that conclusion, in another study, we asked whether children were sensitive to the type of the knowledge being acquired. Specifically, we wanted to investigate whether children understood the difference between learning a fact and learning a skill, especially with respect to the role of intentional action in the learning process. Facts can have deterministic truth-values and are either known or not known at a given point in time. This means that learning a fact involves changing one's mental state from ignorance to knowledge (or from belief A to belief B). Skills, in contrast, are more scalar. One gets better at a certain skill with practice, but performance may vary and can be subject to chance. This means that learning a skill involves a continual process of improving, and knowledge of a skill is rarely deterministic. Similarly, most facts are insensitive to the agency of the learner or teacher; whether you hear a fact on purpose or by accident, you can still learn that fact. In contrast, skill learning is usually intentional; accidental actions that result in a successful demonstration do not show that one has truly learned a skill.

To test whether children understand these contrasts, we (Lai et al., unpublished data, described in Sobel et al., 2016) told a group of sixty-four 4- to 7-year-olds (31 girls, 33 boys; mean age = 71.90 months; age range 50–96 months) stories about a character who wanted to learn either a fact (the location of a teddy bear) or a skill (how to throw a basketball through a hoop). The character then engaged in an action with the intention to either learn (i.e., intentionally looked in a closet or aimed at the basketball hoop) or not (i.e., opened the closet accidentally or accidentally threw the ball in the air). These actions resulted in either a successful or an unsuccessful outcome (i.e., the bear was in the closet or not; the ball landed in the basket or not). After hearing each story, children were asked whether the character had learned the fact (where the bear was) or the skill (how to throw the ball into the basket).

Table 9.1 shows children's responses to the fact and skill questions, depending on the intention of the character and on whether the character's actions resulted in a successful outcome. What is clear from the table is that children made judgments based mostly on the outcome of the action. When the outcome was positive, children tended to say that the character learned, regardless of the other manipulations (91% vs. 18% overall). But there is an

Table 9.1
Children's responses to whether the character had learned a fact or a skill based on whether they intended to learn and they observed the outcome, taken from Lai et al. (unpublished data)

Type of knowledge	Intention	Outcome	Percent of children who said character learned
Fact	Positive	Positive	88 (33)
Fact	Positive	Negative	20 (41)
Fact	Negative	Positive	86 (35)
Fact	Negative	Negative	19 (39)
Skill	Positive	Positive	97 (18)
Skill	Positive	Negative	20 (41)
Skill	Negative	Positive	94 (24)
Skill	Negative	Negative	11 (31)

additional intuition that we wanted to capture with these data. In the fact condition, reasoning on the basis of the outcome is unsurprising. Characters who find the bear have learned where it is, regardless of how they found it or whether they wanted to find it. In the skill condition, however, the intention is relevant; happy accidents do not necessarily indicate learning. This would predict a significant three-way interaction among the type of question (fact vs. skill), the intention of the character, and the outcome of the action, which is what we found.[1] This can be seen by looking at the last row in table 9.1, where the intention and outcome are both negative for the skill question. Children's rates of saying that the character learned in this condition (11%) are lower than the cases where the outcome is negative for the fact question as well as the case where the intention is positive but the outcome is negative for the skill question (an average of 20%).

A general focus on outcomes in learning might be warranted, because acting on the world is an important mechanism for learning in early childhood. In support of this argument, as reviewed in chapter 4, children are more systematic in their exploration when it has the potential to reveal new information (Cook et al., 2001; Schulz & Bonawitz, 2007; Sobel & Sommerville, 2010). And when we translated research on causal reasoning to informal learning environments, we saw that children who generated more systematic exploratory behaviors during their play at an exhibit were more likely to score higher on measures of causal reasoning about the exhibit (Callanan, Legare, Sobel, et al., 2020, as described in chapter 3).

Because actions are so important for learning in early childhood, children might struggle to understand the nuances of the relation between learning and action. To investigate this issue further, we asked whether 3- to 5-year-olds have an explicit understanding of how actions can lead to subsequent learning (Sobel & Letourneau, 2018). We told preschoolers stories about characters learning about novel toys. In one set of stories, the character learned how the toy worked by acting on it. In the other set, the character learned how the toy worked by being told. Children were asked to recall how the characters had learned through a series of open-ended responses. Before the age of 4, children overemphasized the role of action, stating that the character learned by playing with the toy, regardless of condition. That is, the children in this study overweighted the character's actions as being important for learning, much as they overweight their own actions when they themselves learn.

We further examined children's understanding of the relation between action and learning by testing how they think about the difference between claims about what others have learned and whether those individuals actually learned (Sobel, 2015; previously described in chapter 1). In this study, we showed preschoolers vignettes about children in a school who were playing with puzzles. Some of the children claimed they knew how to solve the puzzles, while others did not make this claim. A teacher then asked the children to solve the puzzles. Some of these children succeeded and some failed. Children were then asked whether each character had learned to solve the puzzle. When the character's claims were in accord, as when a character who claimed to know how to solve the puzzle actually did so, children had no trouble stating that the character had learned (or had not learned) how to solve the puzzles. But when the claim and the demonstrative ability were in conflict, children struggled with answering the questions; they said that the character had learned about half of the time. Critically, what seemed to predict children's understanding was their performance on a standard false belief task. If children passed the false belief task, they were more likely to use the character's demonstrative actions as the basis for their response. These children seemed to understand that claims about knowledge could be false, and that the judgment of whether someone had learned something was based on whether they could demonstrate that they had the knowledge.

These results, however, are more about the development of children's understanding of the role of belief in learning; children's understanding of the interrelation between learning and other mental states continues to develop past the preschool years. As noted above, 6-year-old children judge that characters who are paying attention or who have the intention to learn will be more likely to learn than characters who are not or who do not, while 4-year-olds misunderstand these relations. This holds true in children's own learning as well: When the results of an action are identical, understanding that another person generated the action for the purpose of getting others to learn leads to better learning than not having such a rationale (Sobel & Sommerville, 2009). Nevertheless, it appears that children come to this more mature understanding of what learning is and how it works in early elementary school, not in preschool.

Taken together, the results described in this section begin to suggest a developmental trajectory regarding children's understanding of learning. Early on, learning is embedded in action and intention. If you want to learn something, you learn it; if you do it, you learn it. Learning as a mental state might be conflated with action (perhaps specifically goal-directed action). However, over the course of the preschool years, children begin to appreciate the difference between learning and other mental states. This developmental trajectory has strong parallels to children's understanding of science, described in chapter 8. For both concepts, there seems to be a marked shift around age 6 where children move past an outcomes-focused or actions-focused conception and toward a more sophisticated understanding of the role of mental states and knowledge change for both learning and science.

As an example, consider the difference between knowledge and beliefs about one's knowledge. One either knows something or not (particularly for factual knowledge), but one's belief about one's knowledge is not the same as possessing the knowledge itself. One can know something and know that one knows it, or similarly, not know something and know that one is ignorant. These concordant cases seem trivial; the more interesting ones are when knowledge and belief about that knowledge are in conflict. One can believe one knows something (particularly about how to do something or about how something works), but be revealed to be ignorant or to have overestimated one's knowledge (such as in the case of the "illusion of explanatory depth"; see Rozenblit & Keil, 2002). But one can also believe oneself ignorant, but actually possess the knowledge all along, such as in the plot of many mystery stories. More relevant to children's lives, there are many situations where young children might initially think themselves incapable of performing an action that adults know they can actually do, and encouragement and guidance ("scaffolding"; see Mermelshtine, 2017) can lead them to realize their abilities.

Children's Understanding of the Relations between Learning and Play

So far, our discussion has centered on children's understanding of learning generally construed. In this section, we want to expand this discussion to one aspect of learning that we have not considered yet: learning through play.

For many years, play has been thought to be critical to children's healthy social and cognitive development (e.g., Pellegrini & Boyd, 1993; Rubin et al., 1983; Saracho & Spodek, 1998; Smith & Vollstedt, 1985). Play provides children with the opportunity to develop social skills, emotional regulation abilities, prosocial behavior, and empathy (e.g., Coplan & Arbeau, 2009; Ginsburg et al., 2007; Lester & Russell, 2010). Play also supports learning, particularly collaborative or guided play with adults (Hirsh-Pasek et al., 2009; Mayer, 2004; Weisberg et al., 2016). Play is a way that children learn, and this has been a focus for early childhood education.

Yet learning and play are often conceptualized as mutually exclusive by young children (see e.g., Shirilla et al., 2019). Children think that play reflects their autonomy, because they can choose to do what they want. Learning, in contrast, involves mandatory activities. Similarly, children think play is fun and enjoyable while learning is serious and dull. Play is also inherently social, even if one sometimes plays alone. Learning, however, is a mostly solitary endeavor (see Beisser et al., 2013; Howard et al., 2006; Karrby, 1990; Keating et al., 2000; King & Howard, 2014; Robson, 1993; Rothlein & Brett, 1987). Indeed, in the study where we asked to children to report on ways that they learned (Sobel & Letourneau, 2015), no child ever mentioned that they learned through play. The closest we ever saw children refer to this idea was in a corpus analysis of children's natural speech, which we used as the title for one of our papers: "They danced around in my head and I learned them" (Sobel, Li & Corriveau, 2007).

But these studies might not capture the extent to which children conceptualize learning through play, or play in general. In particular, many of the studies on the relation between play and learning present children with stimuli (photos or vignettes) and ask children to categorize whether the individuals under discussion are playing or learning (or whether they are playing or working). That forced-choice question does not allow for a third response, which is that individuals may be both learning and playing.

To look at whether children conceptualize play and learning in similar ways, we first wanted to have children define and reflect on "playing," much in the same way that they defined and reflected on "science" and "learning" in our previous interviews. We[2] asked 70 children between the ages of 4 and 10 (33 girls, 37 boys; mean age=83.04 months; age range 48–131 months) to define "playing."

We coded how children responded to this question into five categories (see figure 9.2). Similar to when children were asked to define "learning," they also sometimes generated Identity responses to the question of what playing is (e.g., "Playing is when you play"). Children also sometimes generated Content responses that focused on the kinds of things that happen in

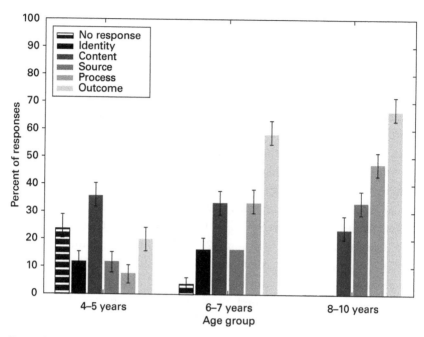

Figure 9.2
Percentage of children who generated definitions of "playing" in each category.

play (e.g., "Using your toys and playing games"). They also talked about who they played with (e.g., "Hanging out with your friends," coded as Source), how they played (e.g., "chasing each other," "building things," "pretending," coded as Process), and the result of playing ("having fun" or "being happy," coded as Outcome). Much like in the learning interviews, we combined these last three responses together, because they all indicated a more process-based understanding of what play is and how it works. As children got older, they were more likely to generate these kinds of definitions, even when controlling for the length of their responses.

Children were then asked to provide examples of things that they had played before and of how they engaged in that play. Children provided between 1 and 8 examples of play (M = 3.47, SD = 1.59). Each example was categorized using one of six mutually exclusive codes: (1) Physical Activity, including sports, playground activities, outdoor games (like tag), and unstructured activities (like climbing or running); (2) Structured Indoor Activities, including board games, card games, video games, educational games, puzzles, and mazes; (3) Creative Activities, including construction activities, drawing, arts and crafts, and building with blocks or LEGO; (4) Pretense, including role play and other pretense activities that specifically involved object representation; (5) Functional Object Play, including playing with toys, dolls, or pretense that involved giving toy objects specific functions, like pretending to cook with toy pots and pans; and (6) No Content, such as simply mentioning play with another person (e.g. "I play with grandma"). Finally, we also coded all of the examples as being either *solitary*, in which children do not mention other people in their example or specifically state they were alone, or *social*, in which the example involves other people.

Overall, we found that children generated examples of physical activities and of structured indoor activities most frequently. Children generated examples of solitary play and social play with the same frequency: 73% of the children generated at least one example of solitary play and 83% of the children generated at least one example of social play, not a significant difference.[3] There were relatively few reliable effects of age or gender. In terms of age, as children got older, they talked less about functional object play (such as playing with dolls or other toys).[4] Additionally, older children were more likely to generate examples of social play.[5] But generating an example of solitary play did not correlate with age.[6] In terms of gender, girls talked more about pretend play than boys.[7] But there were no other differences between boys and girls. Finally, we considered the relation between the ways in which children defined "play" and the examples of play they generated. Unlike the relation between children's definitions of "learning" and their reflections of how they learned, there were no relations between children's definitions of "playing" and what they did during play or how they played.[8]

There is one other interesting null result here. Before we started interviewing children about play, we thought about how we would code the data. One of the codes that we thought about, a priori, was academic or proto-academic activities (like counting). We thought children might generate some examples of playing that happened at school or that involved learning. But they didn't, so we did not even consider this variable.

This study—much like the studies where children are asked to categorize events as either learning or play—showed that young children can reflect on their own play. But children in these studies did not seem to relate play with learning (or vice versa), despite the great importance placed on this relation from educational and psychological researchers. However, it is possible that we discouraged children from talking about learning through play because we did not mention anything about learning in these interviews. Much like the forced-choice method, then, we might have underestimated the extent to which children recognize that playing can lead to learning or that learning can occur while playing.

To address this possibility, in more recent work, we asked 5- to 8-year-olds to define both "learning" and "play," using these same interview methods (Letourneau & Sobel, 2020). We chose this narrower age range because children's definitions of both "learning" and "play" tended to change during this time, moving from describing particular topics or activities to more process-based definitions, which we took to reflect the emergence of a more metacognitive ability to reflect on the nature of these activities. Asking children about both learning and playing allowed us to directly compare how children talked about each and whether the developmental trajectory of children's definitions of these concepts was similar or different.

In this study, children's process-based definitions of "play" emerged earlier than their process-based definitions of "learning" (see figure 9.3). This provides evidence that children's ability to reflect on different activities (like play and learning) does not emerge in a domain-general way. That is, children describe the process of how play works for them differently from how they do so for learning—and also differently from how they do so for teaching, as we will see in the next section.

We did two other things in this interview. First, we asked children whether they could give examples of cases where they were playing and learning at the same time, and why those activities were both playing and learning. Second, we asked half the children to provide instances of playing that had features that were congruent with play, for example, times when playing was fun, chosen by the child who was playing, or not done with adults. These children were also asked to provide instances of learning that had features that were congruent with learning, for example, times when learning was done in a serious manner, or when learning was guided by an adult or not freely chosen. The other half of the children were asked to provide instances of playing that had features congruent with learning and instances of learning that had features congruent with playing.

We found that children were able to generate at least some instances of playing and learning at the same time in response to the first prompt. Some of the examples were based in physical activities, such as, "When I was doing ice

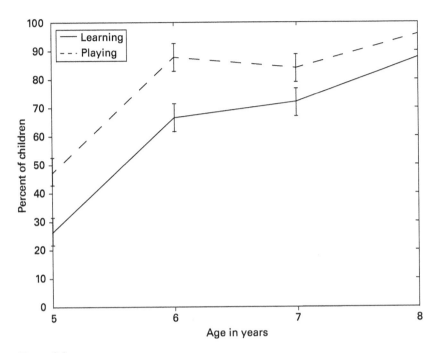

Figure 9.3
Percentage of children who generated process-based definitions of "learning" and "play" across the four age groups in Letourneau and Sobel (2020).

skating I was playing on the ice and I was learning how to skate" (age 7 years, 4 months). Other examples involved children realizing only later that they were engaged in both activities: "When I was in my classroom. I was playing school with my friend and we were actually doing real math" (age 6 years, 10 months). Finally, some of them were based in learning about what one was playing (or what one was playing with): "Well once I was playing with my new American Girl doll and I was reading the tag and it told me to not brush her hair and it taught me how to make her hair curly and keep it curly" (age 8 years, 11 months). Similarly, "Maybe when I first learned how to solve a Rubik's cube, I went on a computer, and I was playing around with it, and I wasn't able to solve a color, and then I went on the internet and I got interested in it and then I solved . . . I tried solving it and I looked it up on the internet and I learned how to solve it" (age 8 years, 2 months).

The number of examples children generated of learning and playing at the same time increased with age. This frequency also significantly correlated with whether children generated a process-oriented definition of "learning" or "playing." Critically, in an analysis designed to isolate the independent contribution of these factors, we found that whether children generated a process-oriented definition of "learning" was the only factor that uniquely explained how frequently they generated examples of playing and learning at the same time. This seems to indicate that children's understanding of how learning happens allows them to connect play with learning, not their understanding of play itself (see Letourneau & Sobel, 2020, for statistical details).

Similarly, and perhaps unsurprisingly, children were able to generate more examples of learning and playing with congruent features than with incongruent features in response to the second prompt. The number of examples that children generated correlated with whether they had generated a process-oriented definition of "learning" and "play," as well as with their age. Again, when we ran a model to isolate the unique contribution of these factors, only whether children generated a process-based definition of "learning" mattered to their ability to answer these questions (beyond the differences explained by the questions we had asked them). In both cases, age did not predict children's ability to understand that learning and playing can happen at the same time. Again, these results indicate that children must have a process-oriented understanding of learning to appreciate that learning can happen through play. In turn, this suggests that children's development of an understanding of learning as a process involving knowledge change might be a bottleneck in their ability to see that learning and play are related.

"What Is Teaching?"

Teaching is often conceptualized as "causing to learn" (Kastovsky, 1973). This suggests that teaching has a cognitive relation to learning. Children might be aware of this as soon as they use words like "learn" and "teach" in their everyday conversation, which is fairly early in development (as reviewed by Bartsch et al., 2003). Teaching has also been described as a "natural cognitive ability" (Strauss, 2005, p. 368; see also Strauss & Ziv, 2012), in that teaching is a universal and basic form of communication that children learn through everyday social interactions rather than through explicit instruction. Children engage in actions designed to teach at early ages (Ashley & Tomasello,

1998; Frye & Ziv, 2005; Kruger & Tomasello, 1998), and school-age children often teach others without prompting (e.g., Brown & Palincsar, 1989; Flynn, 2010; Rogoff, 1990).

However, while children can engage in teaching activities, they might not understand that those activities are teaching. Such a metacognitive understanding of how teaching works and of how to make choices about whom to teach seems reliant on an understanding of other mental states. Specifically, a concept of teaching may need to emerge from a concept of learning, much like the concept of learning seems to emerge from a concept of knowledge (e.g., Astington & Pelletier, 1998; Knutsen et al., 2014; Strauss et al., 2002; Wellman & Lagattuta, 2004; Ziv & Frye, 2004; Ziv et al., 2008; although see Davis-Unger & Carlson, 2008, for an alternative view). This would imply that children's understanding of teaching develops after their understanding of learning, because it relies on children's abilities to think about learning and other related mental states, like knowledge or intention.

With respect to the role of intention, in order to understand teaching as "causing to learn," one must recognize that there is usually an agent doing the causing. That is, teaching might not be "causing to learn" as much as it is "*intentionally* causing to learn." Various studies suggest that children come to an understanding of teaching as requiring some kind of intentionality around the age of 5. For example, Strauss et al. (2002) presented children with stories about a character (A) who learned something from watching another person (B), although person B was unaware that A was watching. Three-and-a-half-year-olds judged that B was teaching, while 5.5-year-olds recognized that B was not teaching. The older children understood the intentional nature of teaching, while the younger children simply conceptualized teaching as demonstrating information.

Similarly, teachers teach individuals whom they believe lack knowledge, regardless of whether that is actually true. But for children to appreciate this relation, they must have at least a rudimentary capacity for reasoning about others' false beliefs. To examine this facet of teaching, Ziv et al. (2008) presented 3.5- and 5.5-year-olds with stories in which teachers either had true or false beliefs about students' knowledge states. For example, the teacher might believe that the student did not know something that the student actually did know, or the teacher might believe that the student knew something that the student actually did not know. Five-and-a-half-year-olds predicted whom the teacher would teach based on the teacher's belief states, while 3.5-year-olds predicted whom the teacher would teach based on the *learner's* belief. More importantly, age was not the crucial factor here—rather, it was performance on a standard false belief task. That is, children's developing understanding of how knowledge works strongly influenced their developing understanding of how teaching works.

Children's understanding of teaching thus relies on their understanding of belief and knowledge, suggesting that this understanding would show a prolonged developmental trajectory. To investigate this, we (Sobel & Letourneau, 2016) asked 4- to 7-year-olds what "teaching" means, and then asked them to reflect on times they were taught and times when others taught them, using a semi-structured interview similar to the investigations of children's understanding of "learning," "play," and "science" that we have already described.

Much like in those other investigations, children's definitions of "teaching" mostly referred to either Content ("To show kids math and science") or to Process ("Showing somebody how to do something so they know how to do it if somebody else asks them to"), although a small number of children simply said that teaching was learning. We found that process-based definitions increased with age: 4- and 5-year-olds generated them only about 17% of the time, while 6- and 7-year-olds did so about 46% of the time. Critically, the frequency with which children generated process-based definitions of "teaching" lagged well behind children's process-based definitions of "learning." Additionally, children who generated process-based definitions of "teaching" were better able to generate examples of when they were taught and when they taught others, controlling for age and for a measure of language production.

Moreover, after children were asked for examples of times they taught others, they were asked to reflect on how they did so. Children who generated process-based definitions were much more detailed in answering these questions. For example, one child who generated a process-based definition (age 7 years, 6 months) responded: "I taught my brother history." When the experimenter prompted this child to explain, "How do you teach them about history?" they said, "Well, I told them about the world and Abraham Lincoln and George Washington and World War II." Critically, these kinds of answers were typical only of children who generated a process-based definition of "teaching," regardless of their age; children who did not generate such definitions simply failed to respond to this question about 75% of the time. These results suggest that, although young children show an intuitive grasp of teaching during their social interactions, they do not really have an explicit understanding of teaching as an intentional action until after the preschool years.

Consistent with this conclusion is the fact that children tended to produce the same kinds of examples of what they had been taught and what they taught others, regardless of whether they had generated a process-based definition of "teaching." Generating a process-based definition of "teaching" thus did not impact whether they understood how to engage in the act of teaching; this only affected children's metacognitive reflections about what teaching is. This result further suggests that there are communicative and situational cues that help children recognize teaching in everyday life, which may also be used when children are thinking of examples of teaching events in response to our interview. The ability to define "teaching" itself as a process that results in knowledge change, however, relies on a different set of cues and cognitive precursors. The act of teaching thus might be a natural form of social interaction, but reflecting on how teaching occurs seems to have a more prolonged developmental trajectory.

More generally, looking at how children's definitions of concepts like "learning," "play," and "teaching" relate to their reflection on these actions provides an interesting parallel to the conclusions we drew about children's understanding of science in chapter 8. Children learn, play, and teach at early ages. Learning, playing, and teaching might all be "natural" and early-emerging aspects of their cognitive development, just as the causal reasoning capacities that they have early in development allow them to engage in rudimentary aspects of scientific thinking. But the ability to explicitly reason about science, learning, playing, and teaching comes later and may involve the kind of metacognitive understanding that we think of as being part of what distinguishes scientific thinking from causal reasoning. While understanding science, learning, playing, and teaching might all have different developmental trajectories, there is a clear distinction between children's ability to perform these actions and to reflect on what these actions mean or how these processes work.

10 Children's Definitions of "Pretending"

This book focuses on the development of scientific thinking out of the foundation of causal reasoning and on how children's concepts of science change over the early elementary-school years. Talking about pretending and imagination, as we do in this chapter, might seem like a radical departure from this focus. Imaginative activities often appear to be in opposition to scientific thinking, involving unfettered flights of fancy that break the rules of reality. Part of growing up, the common wisdom goes, is moving away from imagination in favor of the kind of more serious reasoning needed for activities like science. So what place do imaginative activities like pretending have in this book?

To begin answering that question, we must first define "imagination." The use of this word in everyday language conjures up scenarios that are wildly different from reality; being imaginative is often thought of as equivalent to being creative. While that is certainly something that imagination can do, and that is the part of imagination focused on in our typical discourse, that is not how "imagination" is defined in psychological science.

Imagination is the mental tool that allows us to think about anything outside of current reality. As such, it can be used for any kind of event or process or entity, not just fantastical or creative[1] ones. Understood in this way, most of our imaginative activities are actually quite mundane, like daydreaming about our next vacation, thinking about what could have gone better in a past encounter with our family, or practicing what we are going to say to our boss when asking for a raise. All of those scenarios involve elements that do not reflect the truth of current reality. Imagination is simply the ability to think about those kinds of scenarios. One of the key aspects of scientific thinking is the ability to predict events before observing them. This is what imagination allows us to do.

In this chapter, we continue our investigation of children's category-based intensions by asking how children conceptualize pretending, one kind of imaginative process. We first consider how the development of children's explicit understanding of this abstract concept might parallel the development of the other concepts we have considered so far (science, learning, teaching, and play). Then we broaden our discussion to consider various ways in which pretense and imagination relate to causal reasoning and scientific thinking.

"What Is Pretending?"

Pretending is a kind of imaginative activity. When children pretend, they represent the world not as it is, but as something different: a banana becomes a telephone, a parent becomes a fairy princess, they themselves become grown-ups. While the underlying mental states involved in making these substitutions or in conjuring these invisible entities are the same as in other types of imaginative activities, a key feature that distinguishes pretend play from other kinds of imaginative thought is that pretense is acted out. Unlike daydreaming, counterfactual inference, or making up stories, pretend play does not merely take place in the child's head; it is embodied in some way in the world (see Weisberg, 2015).

Pretend play is also one of the earliest forms of imaginative activity that we can reliably observe.[2] Children begin to engage in simple object-substitution pretense (e.g., pretending to eat a block as if it were a cookie) around 15 to 18 months, and their repertoire of pretend actions and scenarios gradually expands over the course of the preschool years (Piaget, 1962). Social pretense— understanding the pretend worlds of others and being able to share in those worlds—emerges a little later in development, at roughly 2.5 years (Harris & Kavanaugh, 1993). Children can explicitly report on ways that pretense is different from reality by the age of 3 (Morison & Gardner, 1978; Woolley, 1997). Around the age of 4, children start to engage in object representation: Instead of it being necessary to interpret one object as another, children can simply act as if objects that are not there are present (Overton & Jackson, 1973). The emergence of this capacity correlates with children's performance on standard measures of false belief (Taylor & Carlson, 1997), suggesting a relation between pretend play and social-cognitive abilities.

Piaget (1962) posited that the emergence of pretend play was a hallmark of moving from the sensorimotor to the preoperational stages of development; pretense reflects children's abilities to reason about more than what they directly observe. Similarly, Leslie (1988) hypothesized that the emergence of pretense indicated that children possess broader representational mechanisms, which serve as the basis of other developments in their social cognition, such as their understanding of belief and knowledge. In contrast, although Lillard (2001) argued that pretense could be fruitfully understood as a way for children to explore solutions to problems or to enact impossible scenarios in ways that could help them understand reality, she also strongly cautioned against interpreting pretend play as necessary for the development of other cognitive skills (see Lillard et al., 2013).

Regardless of which of these theories of children's pretend play is correct, all of them describe pretending from an external, scientific point of view; they do not provide any information about how children conceptualize their own pretending or the act of pretending in general. But unlike the cases of children's understanding of science and learning covered in the previous chapters, there has been work on this topic. Some of this work has shown that infants recognize that others are pretending (Lillard & Witherington, 2004; Onishi et al., 2007). But most of the studies in this area have focused on whether preschool-age children understand that pretending involves mental states.

In order to make up a scenario or to imbue an object with properties that it does not actually have (i.e., to pretend), one must (1) know what those scenarios or properties are, (2) intentionally think about those scenarios or properties, and (3) be aware that one is engaged in this activity so that one does not come to confuse the pretend representation with reality. Several studies suggest that preschool-age children lack an understanding of all three of these features of pretending. For example, in one set of studies, 4- and 5-year-old children were shown a character, Moe, who was hopping like a kangaroo even though he did not know anything about kangaroos. Children tended to report that Moe was nevertheless pretending to be a kangaroo, seeming to ignore the fact that knowledge of kangaroos is a necessary prerequisite to pretending to be a kangaroo (Lillard, 1993). In another set of studies, preschoolers were shown that Moe was again hopping like a kangaroo, but he was not intending to look like a kangaroo. Again, children tended to report that Moe was pretending to be a kangaroo, even though he lacked the intention to act in this way (Lillard, 1998; Sobel, 2007). In a third set of studies, preschoolers were shown characters who, unbeknownst to them, took on particular appearances (e.g., a boy slipped in the mud and got muddy stripes on his orange shirt, making him look like a tiger, even though he didn't know he looked this way). Again, the majority of preschoolers reported that the character was pretending to be what he looked like (Sobel, 2004a). In each of these cases, young children misunderstand the representational nature of pretense, suggesting that they view pretense merely as a physical activity.

One can examine this conclusion further by asking how children view the relations among the different mental states involved in pretending. Specifically, the causal relation between knowing and intention is an enabling condition: One must first possess a certain kind of knowledge (e.g., what a kangaroo is and how it acts) in order to form the intention to pretend a particular thing (e.g., to jump like a kangaroo). It is impossible to do the latter without the former. In contrast, the causal relation between intention and action is generative: Forming the intention to pretend a particular thing leads to the action, but the surface action is possible without that particular intention. Children might struggle more to understand the former relation than the latter, because they lack a general ability to conceptualize enabling conditions. Indeed, when the enabling condition relation between knowledge and the intention to pretend was made more accessible to preschoolers, their performance on the Moe task improved considerably (Sobel 2009).

Although this line of work using the Moe task generally finds that children think of pretending as a series of actions, we have to contend with the fact that these studies are a bit strange.[3] They all present children with characters and then explicitly tell children about those character's mental states and physical activities. But children rarely, if ever, have access to exactly what another person is thinking or intending. Because of this, we think that it may be more productive—or at least as productive—to ask children to report on what they think pretending is.

Taylor et al. (2003) did this, as part of a larger study on children's understanding of the difference between pretending and lying (if you think about it, these concepts are pretty similar). These researchers categorized children's definitions of "pretending" based on whether they referred to general principles (e.g., "It's where you make something up that can't really happen") or specific examples (e.g., "When you slide and pretend to be a worm"). Younger children (4- and 5-year-olds) generated more of the latter, while older children (6- and 7-year-olds) provided responses that were more evenly distributed between the two categories. Overall, though, the majority of children in their study defined "pretending" in terms of taking the role of someone else or acting like something else.

Unfortunately, that study didn't probe children's understanding of pretending further, nor did it consider how children's definitions of "pretending" related to their knowledge of pretense. To do so, we (Sobel & Letourneau, 2019) first asked 4- to 7-year-olds to define "pretending" (i.e., "What does it mean to pretend?"). A few children said that they did not know or did not want to give a definition. A few other children simply defined "pretending" as pretending ("It means that something is pretend" or "Pretending is when you pretend"). These definitions, like the Identity definitions for "learning" and "teaching" described in chapter 9, were not considered further in our analyses.

The rest of the definitions were coded in three ways, based on some of the definitions of "pretending" described by Lillard (1993) and Austin (1979). We considered whether children talked about pretending being *not real* (e.g., "it means it's not real and you do it," "to make up something"), talked about pretending as involving *agency* (e.g., "it means you're acting," "doing what you want"), or used *mental state* words such as "think," "try," or "believe" (e.g., "you're thinking of something"). These codes were not mutually exclusive, so children's definitions could involve all three or none of them. Table 10.1 gives a few examples taken from this corpus.

Children's age positively correlated with the number of features they used in their definitions, so older children were more likely to define "pretending" using all three types of features. This correlation held when controlling for the length of children's utterances, demonstrating that this relation was not simply a result of the older children talking more. In fact, each of these features alone correlated with children's age. This suggests that, between the ages of 4 and 7, children are more able to articulate pretending in terms of its defining features. What is interesting is that, although all three of these features develop with age, they start at different places and have different developmental trajectories. Most children in the sample generated a definition that included agency, and this feature changed the least over time. Generating a mental state or a reference to something that is not real emerged

Table 10.1

Examples of children's definitions of "pretending" from Sobel & Letourneau (2019) (age in years and gender in brackets)

Response type	Transcribed responses
Not Real Only	Something not real. [5F] Pretending means not real. [7M]
Agency Only	Pretending means that you are playing with toys. [4F] It means when you're sneaky. It means you're a spy. [4M] To play something . . . whatever's part of the game. [6F]
Not Real and Agency	Doing something that's just make-believe. It means it's not real and you do it. [5M] It means you're something else, but you're not really that, but you're playing in a game that you're that. [6M] Pretending means like acting something that is not real. [7M]
Mental State and Agency	To think of your own thing that you like to do. You have to use your imagination and think about strange things and kind things. [7F] Pretending means like you're thinking, and you're thinking and making something happen, but with a costume and stuff. [7M]
Mental State, Not Real, and Agency	Pretending means something's not real and visualizing it and imagining it, I guess, so like if you have a room and then pretending it's a castle. [6F] It means thinking of something that isn't real, thinking of it and learning it. [7M] Using your imagination to make up stuff. [7F]

later and had a steeper developmental trajectory; these features exhibited much more change over time.[4]

In the same study, we asked this group of children to reflect on their own pretending and others' pretending in two ways. First, we gave these children a battery of tasks like the Moe task described above, in which a character is performing an action that appears as though he's pretending to be something, but cannot be pretending because he lacks the appropriate knowledge. We also asked children to reflect on times they pretended by themselves and with others, and how they know others were pretending. These latter questions aim to get at the same information as the Moe task, but in a more naturalistic way.

We found that the number of features that children generated in their definition correlated with their performance on the battery of pretend tasks (including the Moe task), controlling for age and the amount they talked in their definitions. But these definitions did not relate to children's reflections of their own or others' pretending. For the most part, in response to this interview, children emphasized others' actions and appearances when asked how they knew someone else was pretending. For example, one 7-year-old boy was asked how he knew that another person was pretending to be a mommy; he said that he knew because "she [was] pulling up her cell phone." Similarly, a 6-year-old boy said that he knew his cousin was pretending to be Spider-Man because "he was wearing a mask." So young children may understand that someone is pretending given different combinations of appearance, knowledge, and intentions (as probed by the laboratory tasks and the pretending battery we used), but only later begin to spontaneously reference mental states when describing how one pretends. In line with Lillard's arguments, then, young children seem to have an action-based or activity-based understanding of what pretense is, at least for others' pretense (see Lillard 2001; Stich & Tarzia, 2015).

These results bear striking similarity to our findings on children's definitions of "science" and "learning." In all three cases, younger children tend to be more swayed by the outward signs of these activities or by their outcomes: Science is characterized by actions like mixing things together, or outcomes like learning the correct answer; learning does not necessarily involve knowledge change for younger children. As they move through the early elementary-school years, children are more likely to integrate a mentalistic understanding of these activities into their definitions and conceptions. They come to understand science as being a process of learning, and they come to understanding learning and pretending as requiring a certain set of intentions and beliefs.

The Continued Adventures of Moe the Troll

The Moe studies reviewed above shows that children tend to think of pretending as an activity, misunderstanding its reliance on mental states like knowledge and intention. But there was a curious anomaly in the data from the original paper on this task (Lillard, 1993): On one item, children were more likely to correctly report that Moe couldn't be pretending to be something he didn't know about. That item involved Moe running like Simba from The *Lion King*. It was a small effect (about a 15% boost in performance), but it stood out. To investigate this more deeply, we ran a study in which we systematically varied whether the potential pretense was about a fantasy character (e.g., Simba from *The Lion King*) or an ordinary animal (e.g., a cat). Confirming the earlier finding, 4-year-olds in this new study were slightly better at saying that they needed their brains to pretend to be Simba than to pretend to be a cat (Lillard & Sobel, 1999). We also extended this finding to the Moe task. Once again, 4-year-olds were more likely to say that Moe was not pretending to be Simba when he was running like a lion but had no knowledge of lions than to say that Moe was not pretending to be running like a cat even though he had no knowledge of cats (Sobel & Lillard, 2001). As we reviewed in chapter 6, there are some circumstances in which fantastical contexts can bolster children's learning and reasoning; this is another example of that phenomenon.

But why should this be the case? For these data, we believe that we can pinpoint the reason that children were more likely to think maturely about fictional characters than ordinary creatures by drawing a parallel with children's thinking about causality. When children are told that a character is running like Simba, children might think about all of the things that Simba does. For instance, he talks with other animals. Real lions don't do that, and they especially don't spontaneously burst into song with meerkats and warthogs. Mufasa (Simba's father) has a supernatural roar that blows hyenas over;[5] real lions roar, but not like that. Simba also communicates with dead relatives, whereas real lions don't do that—at least as far as we know. Given all of this, even though Simba can run, running is not really representative of what makes Simba special or unique. Pretending to be Simba by running thus does not seem to be a good fit with that task. Put another way, if you are going to pretend to be Spider-Man, you are going to act as if you can swing from a web, shoot webbing at bad guys, climb a wall, or generally try to help people; you are not going to worry about paying your rent (even though Peter Parker does, and this is often represented in the comic—why else would he work at the *Daily Bugle*?). You want your action to be representative of what the character can do.

This is related to an old idea in cognitive psychology: the *representativeness heuristic* (Kahneman & Tversky, 1972). "Heuristic" here refers to a rule of thumb, or a shortcut in reasoning. These rules generally work, but they might not be the most rational rules or the best possible mechanisms to use all the time. "Representativeness" here refers to the idea that we jump to a particular conclusion in thinking that a sample is a pure reflection of a population. To illustrate with a brief example, during the Super Bowl between San Francisco and Kansas City in 2020, San Francisco was winning 20–10 with about seven minutes to go. A reputable website put out a tweet that suggested the 49ers had a 95% probability of winning at that point in the game. But Kansas City came back and won the game 31–20. Many people were angry at this tweet in retrospect, and mocked the website that forecasted victory for San Francisco. But 95% isn't 100%; the website wasn't wrong. This example reveals these fans' use of the representativeness heuristic:[6] People represent numbers that are close to 100% as being 100%, because 100% is just easier to think about.

To test whether something about representativeness was involved in children's better performance with the fantasy characters, we introduced 4-year-olds to a new character in a storybook: Zoltron from the planet Zolnar, who looked like an anthropomorphic purple carrot (Sobel, 2006). Zoltron was described as "being from another planet" to establish the fantastical nature of the scenario (following Dias & Harris, 1988).

In one condition, the story revealed that Zoltron engaged in mostly fantastical actions (i.e., actions that violated real-world causal structure), like walking through walls or never having to sleep. In the other condition, the story revealed that Zoltron engaged in a set of ordinary actions, like walking through a door or sleeping through the night. One of Zoltron's actions (playing in the sandbox) was identical across the two storybooks, and the picture of this action was also identical across the two storybooks.

After children were read the story, they were introduced to Moe the troll, who was acting in the same manner to the one action that was common across both Zoltron stories. That is, Moe was playing in the sandbox, which we represented by having the troll doll play in a cardboard sandbox we created for the experiment. Children were then told that Moe had not read the story, so he had no knowledge of Zoltron. They were asked whether Moe was pretending to be Zoltron. Children routinely were more successful on this test question (i.e., denied that Moe was pretending to be Zoltron) when Zoltron had engaged in mostly fantastical actions in the story as opposed to ordinary ones.

What's going on here? To determine whether Moe is pretending to be Zoltron, children must reason about a particular kind of counterfactual: They must compare whether the action/appearance of Moe (the potential pretender) is representative of the action/appearance of Zoltron (the target of the pretense). The more that children believe that Moe's action and appearance are like Zoltron's, the more likely they should be to say that Moe is pretending. That is, fantasy itself might not be the driver of the benefit in performance for these cases. Rather, the fantastical actions performed by Zoltron in the story indicate that, when a potential pretender (Moe) engages in an ordinary action, that action is insufficient to indicate the character is pretending. But when Zoltron only engages in ordinary actions in the story, Moe's ordinary action is more representative, and thus more likely to be intended as pretense. In general, then, when making judgments about pretending, children must learn to use the representativeness of the pretend action in order to decide whether the character is pretending.

Building Fictional Worlds

When children engage in a pretend game or when they hear or watch a narrative story, they build a representation of the world that this game or story describes. This representation allows them to make judgments about the game or story and the characters within it, like how realistic or fantastical it is. How do children do this?

To answer that question, it is helpful to start with adults, because they also need to create mental representations of the stories they consume. How do adults do this? As an example, consider the world of Star Trek. In the show *Star Trek: The Next Generation*, the starship *Enterprise* has a tractor beam, which "employ[s] superimposed subspace/gravitation force beams" to allow for the "direct manipulation of relatively large objects in proximity to a starship. Such operations can take the form of towing another ship, modifying the speed or trajectory of a small asteroid, or holding a piece of instrumentation at a fixed position relative to the ship" (Sternbach & Okuda, 1991, p. 89). The tractor beam is pretty useful, and there are a number of cases where it could have been used (or at least tried) as a solution to a problem that threatened the *Enterprise*. A journey into the bowels of the internet reveals whole threads dedicated to this idea; we are not alone in suspecting that the writers of the show tended to forget that the *Enterprise* had it.[7] But, when it does make an appearance, consumers of this kind of fiction tend to shrug their shoulders—in a world that already contains transporters, replicators, holodecks, and faster-than-light travel, a tractor beam sounds reasonable. But for all the technological marvels of *Star Trek*, we assume that the *Enterprise* has bathrooms and that the human beings on board the *Enterprise* need to eliminate waste.[8]

In contrast, consider *The Love Boat*, the late-1970s TV show about people finding love aboard a cruise ship. It ran for nine seasons and had 249 episodes, so it must have had some staying power. If you were a consumer of this kind of fiction, and you saw Captain Stubing deploy a tractor beam on the *Pacific Princess*, you likely would not have simply shrugged. In fact, you may have felt the need to completely rethink the nature of the show you were watching.[9]

What these examples show is that adults are quite adept at constructing different types of fictional worlds from sparse information (Weisberg, 2016). Adults understand (and tend to gravitate toward) different genres of fiction, recognizing that the rules of a story in the genre of magical realism will be different from the rules of a detective novel. In support of this idea, we (Weisberg & Goodstein, 2009) showed that adults fill in the gaps of fictional worlds based both on their real-world knowledge and on their understanding of the kind of story that they are consuming. In this study, adult participants were presented with three possibilities for a story: entirely realistic, containing a few impossible events, or containing many impossible events. These participants used the degree of similarity of the fictional world to the real world as the basis for their judgments about whether to extend novel facts into that world. For example, in a story in which characters engaged in no violations of real-world causal structure, adults were likely to say that the world of that story resembled the real world quite a bit. In contrast, when a story did violate several aspects of real-world causal structure, adults were less likely to agree that facts that were true of the real world were also true in the world of that story.

These results demonstrate that understanding a fictional story is quite similar to navigating the real world. Both reality and fiction have gaps: Adults do not know everything that there is to know about the structure of reality, and no story is ever long enough to fully describe the world in which it takes place. Luckily, neither engaging in everyday activities nor engaging with fiction requires knowing everything there is to know about the world. In both cases, we can use our existing knowledge and reasoning abilities to fill in gaps when they arise (see Schacter, 2012). This argument further suggests that, as one's knowledge changes and develops, the way in which one understands fictional stories changes and develops with it. Just as children's causal knowledge constrains the kinds of inferences they make about new real-world causal systems, the same knowledge constrains how they construct fictional worlds.

So how do children think about fictional worlds and make judgments about new elements within them? To begin to answer this question, we followed up our work with adults by investigating preschoolers' inferences about novel stories (Weisberg, Sobel, Goodstein & Bloom, 2013). We presented 4-year-olds with one of two kinds of stories. Both had the same basic structure, but one presented several violations of real-world causal laws (e.g., a character who teleported to the ice cream store) while the other presented no real-world causal violations (e.g., a character who walked to the ice cream store). In each case, children were told that some pages of the story were missing, and they were asked to fill in the blanks of the story with new pages. At the points where they had to fill in the story, children were always given the choice between an entirely realistic event and an event that broke some real-world law. Regardless of which story they saw, children were significantly more likely to choose the realistic events. That is, rather than being indiscriminately attracted to fantasy, as popular ideas of children tend to assume, and rather than matching events to their story contexts, as adults tend to do, children are *reality-prone*.

Critically, children in these studies were not reality-prone under all circumstances. When shown the three pairs of choice pictures from this study without any story context and asked to choose which one they liked, children tended to pick a mix of fantastical and realistic events. Further, in a follow-up study, an experimenter told a new group of 4-year-olds that she liked a set of events that were either all realistic or all fantastical. She then asked children which new event she would also like: a novel realistic one or a novel fantastical one. In this case, children correctly chose the event whose ontological structure matched the experimenter's preference. These control studies demonstrate that children do not have a general preference for real-world events; they only showed this preference when asked about which events belong in a story.

Indeed, in another set of follow-up studies, we asked 4-year-olds to construct stories completely from scratch, choosing each time between a fantastical and realistic element to move the plot forward (Sobel & Weisberg, 2014). We found that children were highly consistent in their choices. If they started with a fantastical element, they tended to pick fantastical elements throughout their story construction; if they started with a realistic element, they tended to pick realistic elements throughout their construction. This indicates an understanding that stories need to be internally consistent in the worlds they describe, as in the *Star Trek/Love Boat* example above. But, critically, 80% of children generated all or mostly realistic stories, while only 20% generated all or mostly fantastical ones, replicating the tendency to choose realistic story events that we had found in prior work. And when asked whether unusual events in stories (like the existence of paint that never dried) should be used as the basis for further inferences about that story, preschoolers tended deny that later events would have the same unusual structure. Instead, they tended to say that the story world would match reality more closely (Weisberg & Hopkins, 2020).

These results do not mean that children lack all sensitivity to the structure of fictional worlds. In the study in which children were conservative about extending unusual premises within the world of a story, they were willing to license some familiar impossible premises, like talking animals (Weisberg & Hopkins, 2020; see also Van de Vondervoort & Friedman, 2014). In addition, when asked to choose between two different types of impossible events to continue a story (fantasy and science fiction), 4- to 6-year-olds were able to correctly match these events by genre, choosing fantasy events to continue fantasy stories and science-fiction events to continue science-fiction stories (Kibbe et al., 2018). Nevertheless, when compared to adults, children are generally much less sensitive to the structure of story worlds and much more reality-prone.

But why should this be the case? Before answering that question, it is important to note that children's tendency to continue fantastical stories with realistic events is not an error. Events that are consistent with real-world causal structure occur in all kinds of stories. While fantastical stories can introduce novel violations of causal structure (like the tractor beam in *Star Trek*), even highly unrealistic fictional stories contain many realistic elements (like characters needing to use the bathroom). One cannot necessarily infer that stories containing many violations of real-world rules should contain additional violations. But children's tendency to use real-world causal structure to construct fictional worlds does reflect a certain kind of conservatism in their thinking about possibility.

So how can we explain this conservatism? It is possible that preschoolers (unlike adults) reason according to the *principle of minimal departure* (Ryan, 1980; see also Lewis, 1978; Walton, 1990). This principle states that the construction of a fictional world involves (by default) everything that is possible in the real world, unless the story explicitly forbids it. Although this is not the only possible way to construct fictional worlds (see Weisberg, 2016), if young children follow this principle, it could explain why they fill in stories the way that they do. Adults, by contrast, have a more nuanced view of how much of reality belongs in a story. They are able to flexibly accept and reject premises based on a more holistic assessment of the kind of fictional world they are encountering, rather than assuming that the entirety of reality belongs within the fictional world as a default. What develops over the course of the lifespan, then, is not just the causal and domain knowledge on which fictional worlds are based, but also a higher-order set of assumptions about how that knowledge should be applied within a given fictional world.

Imagination and Causal Reasoning

How children interpret fictional worlds nicely illustrates an important point of connection between imagination and scientific thinking. One of the main goals of scientific inquiry is to determine causal structure, specifically separating genuine causal relations among events from factors that are merely correlates, as the control of variables strategy allows one to do. The causal graphical model framework that we introduced in part I of this book is a way of describing a representation of such knowledge, and we showed there that children can reason according to its principles in a variety of settings.

One of the appealing facets of this framework is its explanation of counterfactual reasoning (Glymour, 2001; Pearl, 2000; Woodward, 2003). Counterfactuals involve constructing a hypothetical representation of a causal structure that is different in some way from the original representation. One can then reason about that modified structure using one's general inference-making abilities. To do so, one must quarantine the original representation (or decouple it from reality; see Leslie, 1987, 1994). This allows the original representation to be retained independently of a new representation on which a counterfactual inference is performed (via reasoning about interventions on certain nodes in the model). The counterfactual is calculated by reasoning about the results of these (hypothetical) interventions. The actual representation of the world, because it has been quarantined, is restored when the counterfactual inference is complete.

But thinking counterfactually crucially involves imagination. For example, we may be able to observe only cases where two events (A and B) occur together. Nevertheless, we can mentally construct scenarios in which A is removed, or in which A and B are related in different ways than they are in reality. Being able to think about *possible* causal structures and the results of various interventions on them is, at its heart, an imaginative process, and one that is vital to causal reasoning (e.g., Gopnik & Walker, 2013; Weisberg & Gopnik, 2013).

These arguments draw a deep connection between causal reasoning and imagination in general, and between causal reasoning and pretend play in particular (see Leslie, 1987, 1994). When children engage in a pretend scenario, they are demonstrating their ability to copy a representation of the world and modify it, while keeping this representation separate from their original representation of the real world. This allows children to draw inferences within the pretend scenario that do not affect their understanding of reality. For example, when a child engages in simple object substitution to pretend a banana is a telephone, their representation of real-world bananas is maintained. Actions within the pretend game that imply that the banana can be used for communication do not affect children's understanding that, in reality, bananas are not communication devices. This way, children do not confuse telephones with bananas (or pretense with reality, more generally). This description makes clear that pretending and counterfactual reasoning involve the same mental capacities. Both require an individual to take an existing representation, copy it, quarantine the original, modify the copy, and then draw conclusions about the modified representation.

One of our studies directly tested this relation, finding evidence that counterfactual thinking and pretending are indeed linked in young children (Buchsbaum, Bridgers, Weisberg & Gopnik, 2012). Specifically, this study taught 3- and 4-year-olds a new causal relation: Putting a special object called a zando on a machine made the machine play the song "Happy Birthday." Once children learned this causal relation, we found that they were able to think counterfactually about it. To test this, we asked them what would happen if the zando were not a zando and we put it on the machine. Children tended to say that putting it on the machine would not make the machine play music. This is a rather complex piece of reasoning. To answer this question correctly, children must suppress what they know to be true in reality (the object really was a zando) and think about an imagined (non-real) scenario in which the zando did not have its usual causal powers.

Even more impressively, the preschoolers in this study were able to transfer this newly learned causal relation into a pretend game. When they were given the opportunity to pretend that a box was the machine and a block was the zando, they chose to put the pretend zando (rather than a second object) on the pretend machine to make it play music in the pretend game. And these preschoolers' tendencies to respond in this way in the pretend scenario were significantly related to their tendencies to correctly answer the counterfactual questions about the real zando and the real machine, even when controlling for other factors like age and executive function abilities. This evidence suggests that imaginative abilities underlie both children's playful actions in pretend games and their counterfactual reasoning abilities in general.

Counterfactual Thinking in Development

The arguments in the previous section lead to the strong conclusion that pretending and imagining are deeply related to counterfactual reasoning, and hence to both causal reasoning and scientific thinking. Although we believe that this theory is sound, aside from the one study reviewed above, there is little direct evidence for it. Here, we review prior work on children's understanding of counterfactual reasoning in general, aiming to illustrate how it relates to both imaginative activities and causal reasoning.

First, there is a long literature on infants' causal perception, suggesting that infants can perceive causal relations among events based on their spatiotemporal relations sometime during the second half of the first year of life (e.g., Leslie & Keeble, 1987; Oakes & Cohen, 1990; Saxe et al., 2005; Saxe et al., 2007; Sobel & Kirkham, 2007). Perceptual information can be a powerful clue to causality, so much so that even adults can experience illusions of causality, which act much like visual illusions. Take, for example, the blicket detector, which we have discussed throughout this book. Most versions of the detector actually present a causal illusion: The block is not making the machine activate. Rather, the experimenter is pressing a button, hidden to the child, at the same time as the block is placed on top of the machine. The button is depressed as long as the block is on the machine and the experimenter stops pressing the button when the block is lifted off. This gives the appearance that the block is causing the machine to turn on.[10] This spatiotemporal relation gives the illusion of causality, much like visual illusions (such as the Ponzo illusion) afford an illusion of depth. Even if you know how the machine works, the illusion stands: Placing a block on the machine appears to cause it to activate.

However, whether infants are genuinely thinking causally on the basis of these perceptual cues is a matter of interpretation (see, e.g., Sommerville & Woodward, 2005). Because we believe that thinking causally goes hand in hand with the ability to think counterfactually, one way to resolve this issue would be to test whether infants can do so. This is an understandably tricky challenge; how could we tell if preverbal babies are thinking about states of affairs that reflect possibilities rather than reality? For this reason, studies on counterfactual thinking tend to focus on older children.

This body of work suggests that children starting around the age of 3 or 4 can engage in certain kinds of counterfactual inferences (e.g., Beck et al., 2006; Guajardo & Turley-Ames, 2004; Harris et al., 1996; Nyhout & Ganea, 2019; Rafetseder et al., 2010; Rafetseder & Perner, 2014; Riggs et al., 1998; Robinson & Beck, 2000). For example, Beck et al. (2006, Study 2) showed 4-year-olds an apparatus with a forked slide, so that a toy could slide down either the spotted side of the slide or the striped side of the slide. After sending the toy down the top of the slide and seeing on which side of the fork the toy ended up, children were asked, "What if it had gone the other way, where would it be?" Children were able to correctly indicate the bottom of the other slide, suggesting that they could consider this counterfactual situation.

There is disagreement, however, about the role of counterfactual reasoning in causal inference. Some researchers have claimed that children understand causal relations among events by reasoning about the absence of candidate causes—that is, by genuinely reasoning counterfactually. In support of this view, Harris et al. (1996) claimed to show that children were capable of engaging in a form of counterfactual inference at the age of 3. In one of their tasks, they first showed children a representation of a clean white floor. Then, a character comes home and doesn't take her muddy shoes off, so she makes the floor all dirty (represented by placing a set of tracks on the "floor"). Children were asked about what the floor would look like if the character had taken off her shoes before she walked on it. They were successfully able to answer that the floor would be clean in this case.

On this basis, and on the basis of the fact that pretend behavior emerges early in development, Harris and his colleagues suggested that counterfactual reasoning was a natural form of causal inference (following suggestions from Mackie, 1974).

In contrast, most of the other studies cited above suggest that counterfactual inference might be a more difficult form of inference than other kinds of causal inferences, such as making predictions. For example, Riggs et al. (1998, Experiment 4) contrasted preschoolers' ability to make an inference about a future hypothetical and about a counterfactual. In this study, children were shown a box with items that had pictures on them and another box with items that did not have pictures on them and were told about this categorization scheme. In one condition (future hypothetical), children were shown a blank piece of paper and were asked which box it would go into if the experimenter drew on it. In the other condition (counterfactual), they were shown a piece of paper with a picture on it and were asked which box if would go into if the experimenter had not drawn on it. Children were more accurate at answering the future hypothetical question than the counterfactual one.

Interestingly, Riggs et al. (1998) also suggested that the false belief task was a measure of counterfactual reasoning, and that the developmental progression that many researchers have documented between ages 3 and 4 (e.g., Wellman et al., 2001) was due to children's developing counterfactual reasoning capacities. Consider, for example, a standard unexpected transfer task (Wimmer & Perner, 1983). In this type of task, children are introduced to a character (Sally) who has a chocolate. Sally does not want her chocolate right now and puts it in a location (such as a drawer) and then leaves the room to do another activity. While she is away, another character (Anne) finds the chocolate and moves it to another location (such as a cabinet). This sets up a situation where Sally has a false belief: The chocolate is now in the cabinet, but Sally thinks that the chocolate is in the drawer, because that is where she last saw it. After children hear this story, they are asked where Sally will look for her chocolate when she returns (alternately, some researchers ask where Sally thinks the chocolate is). Typically, 3-year-olds do not succeed on this task; they tend to say that Sally will look for the chocolate where it currently is (the cabinet). In contrast, 4-year-olds (and hopefully you as well, dear reader) correctly report that Sally will look for the chocolate where she left it (the drawer).

The way that researchers have typically interpreted this pattern of responses over the past thirty-five or so years is by saying that, in order to give the latter (correct) response, children have to make an inference about the content of Sally's beliefs. In contrast, as noted above, Riggs et al. (1998) observed that this measure is also a test of counterfactual inference: If Anne had not moved the chocolate, then Sally's beliefs would be correct. These researchers argued that the mental state aspect of the inference might not be as relevant as the counterfactual; remove the mental state but preserve the counterfactual, and children should still show the same developmental trajectory.

To test this interpretation, these researchers presented children with a version of an unexpected transfer task in which children could be asked to reason about a character's false belief or about a counterfactual state of affairs that did not involve any mental states. In their procedure, two characters were together in their house. One wasn't feeling well and went to bed. The other went to get medicine for the first. While the second character was away, the first character was called away from the house to help put out a fire. Children were equally accurate at answering a counterfactual question about the character's whereabouts that involved no mental state inference ("If there had been no fire, where would the first character be?") as they were about answering a question about the belief states of the second character ("While the first character is out of the house, where does the second character think the first character is?"). These data imply that counterfactual reasoning is a domain-general cognitive capacity, one that underlies or contributes to many different kinds of inferences, including children's abilities to think about false beliefs.

We agree with the idea that causal inferences are related to and sometimes even performed on the basis of counterfactual inferences. But we would like to make a slightly more nuanced view of the relation between these abilities—a view that can explain why counterfactual inferences are sometimes difficult for children. One of the main features of the causal graphical model framework is its ability to support counterfactual inferences (Pearl, 2000). If children's representations of causal knowledge are well-described by this framework (as we argued in part I of the book), then if they can make causal inferences, they can also make counterfactual inferences. However, the studies reviewed above suggest that this is not always the case, and that some kinds of inferences (e.g., counterfactuals) are more difficult than others (e.g., hypotheticals). One reason that this is sometimes the case (such as in Riggs et al.'s 1998 procedure) might have to do with the development of domain-general cognitive capacities (such as inhibitory control, which appears to correlate with counterfactual reasoning measures; see also Beck et al., 2009; Drayton et al., 2011). As we argued in chapter 1, these other cognitive capacities can constrain children's performance on these tasks.

Similarly, counterfactual and future hypothetical questions are usually asked in different ways. Counterfactual questions typically take the form of "What if X had (or had not) happened, would Y [still] have happened?" Future hypothetical questions typically take the form of "What if X happens, then will Y happen?" The different modal verbs used in these questions might contribute to the different levels of linguistic difficulty that children might have with these questions, because these verbs have different developmental trajectories in their comprehension and production (Coates, 1983; Kuczaj & Daly, 1979).

To investigate this issue, we (Sobel, 2001) constructed a measure of future hypothetical and counterfactual inference that removed some of these demands (particularly the linguistic ones). Specifically, we asked 3- and 4-year-olds to talk about a picture book instead of to answer direct future hypothetical and counterfactual questions. In this study, children were shown the first three pictures of a story that depicted a future hypothetical situation, and then were asked to fill in the fourth picture in the storybook. Or, children were shown all four pictures, and then a change was made to the third picture, and then children were asked to fill in a new fourth picture to represent a counterfactual.[11] We based the initial stories on the measures used by Harris et al. (1996) and Riggs et al. (1998), and we found that children were equally good at reasoning about counterfactuals as they were about future hypotheticals.

We also varied what the stories were about. When we used similar stories to Riggs et al. (1998), we found that there was some development between the ages of 3 and 4, which is similar to the developmental trajectory one sees when one studies performance on the false belief task (Wellman et al., 2001). But when the stories were changed to be about whether a character would be happy or sad based on their desire being fulfilled or unfulfilled, both 3- and 4-year-olds were able to reason accurately about both the future hypotheticals and counterfactuals. In contrast, when the story was about whether someone would be surprised, both 3- and 4-year-olds struggled to reason accurately about both future hypotheticals and counterfactuals; they performed no different from chance on both types of stories.

These findings are reminiscent of the effects of contextualization that we discussed in chapter 6. As a reminder, we found that the way that a problem is framed or described can influence children's and adults' reasoning abilities. We suspect that the same kind of mechanism is at play here, and that the type of causal knowledge that a child possesses could affect their ability to reason. So both 3- and 4-year-olds are good at counterfactuals about desire-based situations because they possess a pretty good understanding of the causal relations involved in reasoning about their own and others' desires (see Repacholi & Gopnik, 1997; Wellman & Woolley, 1990; see also Wellman & Liu, 2004). The mental state of surprise, on the other hand, is a more complex concept, and the causal relations involved inferring when someone is surprised might not be available to children until after the age of 4 (see e.g., Hadwin & Perner, 1991).

Studies like this one support our argument that children's counterfactual reasoning abilities are connected to their causal reasoning capacities. They also begin to suggest an answer to the question of directionality: The existence of context effects suggests that children's causal reasoning abilities and their general knowledge support their counterfactual reasoning, rather than children having a fully domain-general counterfactual ability that operates on any type of causal inference.

Another of our studies (Sobel, 2004b) tried to resolve this question more directly by presenting 3- and 4-year-olds with stories about impossible events. In one set of stories, we asked children whether those events were possible. For example, a character wanted to throw a ball up in the air and have it float—is that possible? In another set of stories, we showed the same children a character who tried to do one of these events and failed. Then, we asked about a counterfactual: Was there anything that the character could have done differently to bring about this event? This study also varied the domain of the story, so sometimes the impossible event broke a physical law (like the floating example above), sometimes a biological law (like wanting to never go to sleep again), and sometimes a psychological law (like causing someone to know the location of an object they have never seen). Children's judgments about whether these events were possible in the first set of stories related to their ability to state that the character could not do anything different because the event was impossible in the second set of stories; if children judged that the events were impossible in the first set of stories, they were more likely to say that nothing could be done differently to accomplish an impossible goal. Importantly, this relation was domain-specific: Judgments that physical events were impossible related to counterfactual inferences about physical events, but not to counterfactual inferences about biological events, for example. Again, this suggests that counterfactual thinking might depend on one's causal knowledge, not the other way around.

This work, however, may still present an incomplete view of counterfactual reasoning. As Rafetseder and Perner (2010, 2014) note, the tasks presented in all these studies could potentially be solved if one were only thinking conditionally (i.e., considering "if . . . then" statements or statements of general rules). That is, the work discussed so far does not provide direct evidence that preschoolers are genuinely thinking counterfactually as opposed to making a series of conditional inferences. To demonstrate this, they presented children a version of the task from Harris et al. (1996), but with two characters. Both of these characters have muddy shoes and they both walk across the clean floor, making it dirty. Children are then asked whether the floor would be dirty or clean if only one of the characters had taken off her shoes before she walked across the floor. Preschoolers tend to incorrectly say that the floor would be clean, and it is not until later in development (age 7 or potentially even adolescence, depending on the nature of the study) that children reliably answer correctly (Rafetseder et al., 2013). Unlike younger children, what these older children seem to be doing is reasoning about the "nearest possible world," as Lewis (1973) suggested, taking more of the causal structure of the antecedents into account (see also Leahy et al., 2014). That is, in this view, the older children in these studies are genuinely reasoning counterfactually, while the younger children seem to be attempting to solve the problem by translating it into a conditional statement.

Interestingly, in the same paper, these researchers also speculate on the relation between the developmental trajectory of causal reasoning (particularly as it relates to counterfactual reasoning) and scientific thinking. They write,

> One could also expect a parallel development between [the kind of counterfactual reasoning that takes causal structure into account] and scientific reasoning abilities. A fundamental element in scientific reasoning is differentiating variables

that are causally relevant for the observed outcome from variables that are not. For example, to determine whether the piece of metal was the only causally relevant factor for the Concorde to crash, one would need to manipulate this factor while holding all other independent variables constant. This amounts to the nearest possible world constraint. (Rafetseder & Perner, 2014, p. 57)

While we believe that counterfactual reasoning does rely on one's existing causal knowledge, as Sobel (2011) suggested, we also accept some of the conclusions from Rafetseder and Perner (2014) about a more prolonged developmental trajectory for counterfactual reasoning, particularly given the complexity of some of their tasks. But it also seems reasonable to conclude that causal knowledge is necessary for the kind of counterfactual reasoning that Rafetseder and Perner tested, as well as the capacity to reason about possible worlds in order to represent a hypothetical intervention on the antecedent.

There's a broader point here. In chapter 2, we introduced the idea that Bayesian inference is a good description of how children might make causal inferences and learn causal structure. Bayesian inference starts with a representation of a hypothesis space, but there are no physical tokens or manifestations of that hypothesis space. That representation is in the mind. This means that making inferences using a Bayesian process requires the ability to imagine. Imagination is thus part of the domain-general mechanism of causal inference, which governs not only counterfactual reasoning, but aspects of causal reasoning as well.

These connections can be seen in some recent empirical work. For example, McCormack et al. (2013) showed that encouraging 5-year-olds to think counterfactually improved their reasoning about a form of blocking, similar to the Gopnik et al. (2001) procedure used to test children's use of the Markov assumption described in chapter 2. Similarly, Engle and Walker (2021) showed that asking counterfactual questions to 5-year-olds helped them determine what data disambiguated competing interpretations. In a synthesis of this literature, Walker and Nyhout (2020) argue that counterfactual questions allow children to "perform mental simulations and interventions on causal models . . . to consider and test alternatives (particularly those with lower prior probability) to an initial hypothesis" (p. 269). So while having a representation of the causal structure of a set of events might allow children to reason counterfactually, it is also the case that the domain-general capacity for imagination helps to guide some forms of causal reasoning. Because children have the capacity to imagine, they can learn not only from statistical regularity, but also from considering what might have been.

There is a corollary to this hypothesis. In chapter 3, we argued that children are not born with the capacity for causal inference, but that it emerges from their statistical learning capacities during the second half of the first year of life, as they begin to integrate their ability to understand statistical regularity with their ability to understand the importance of intervention on the world. The emergence of the capacity to imagine—which seems to begin in earnest in the first half of the second year of life—is another domain-general capacity that plays an important role in the cascading development of causal reasoning and causal knowledge. Describing the developmental process in this way suggests that there is a period of time in development during which children represent causal knowledge, make causal inferences, and learn new pieces of causal structure, but cannot yet engage in counterfactual reasoning. This probably occurs between the second half of the first year of life (when we think children start genuinely reasoning about causality beyond mere statistical regularity) and the middle of the second year of life (when children start pretending). This is an open question for investigation.[12]

Possibility and Probability

Based on the work reviewed above, we conclude that counterfactual thinking and reasoning about imaginary worlds depend both on a domain-general capacity for imagination and domain-specific causal knowledge. This conclusion is supported by another line of work on children's understanding of possibility. For a long time, researchers have documented that even preschoolers have a good understanding of the difference between fantasy and reality (e.g., Corriveau et al., 2009; Morison & Gardner, 1978; Woolley, 1997) or between what is possible and what is impossible (e.g., Schult & Wellman, 1997; see Weisberg, 2013, for review). Although many have taken this as the end of the story, Shtulman and colleagues noted that most of this work contrasts fantastic or impossible events with relatively ordinary ones. Simba the Lion King is fictional; lions are real. But what about events that are neither completely impossible nor totally ordinary, like having a lion or a tiger for a pet? This kind of event is possible, just unlikely (*Tiger King* and Siegfried & Roy notwithstanding). Do children understand that there is a distinction between impossibility and improbability, just as they understand the difference between impossibility and possibility?

To answer this question, Shtulman and Carey (2007) presented 4- to 8-year-olds with ordinary events (eating an apple), impossible events (eating lightning), and improbable events (drinking onion juice). They asked children whether these events could happen in real life. All of the children said that the ordinary events could happen in real life and the impossible events could not, as in previous work using such events. For the improbable events, there was a developmental trajectory. Four-year-olds treated the improbable events like the impossible ones, saying that they could not happen. By the time that children were 8, they correctly judged that the improbable events could happen in real life, like the ordinary ones.

Shtulman and Carey argued that the reason that young children were confused by improbable events was not that they lacked imagination, but rather that they lacked the experience to recognize that such events could be possible (although see Lane et al., 2016, for an alternative interpretation). Given that preschoolers had never experienced drinking onion juice or having a lion for a pet, or any of their other improbable events, the children judged them as impossible. Older children, in contrast, were better able to distinguish between what they had personally experienced and whether an event was possible. One related interpretation of these data is that 4-year-olds genuinely struggle with this kind of question, interpreting "could it happen" as "has it happened," while older children interpret the question in a more mature way, as about whether something is possible in the modal sense.

We (Weisberg & Sobel, 2012) supplemented this interpretation by testing 4-year-olds' understanding of improbable stimuli in two ways. First, we replicated Shtulman and Carey's method, directly asking these children whether improbable events could happen in real life. We found, as they had, that these young children tended to mischaracterize improbable events as impossible. Second, we presented these events in the context of a story, in which a character experienced only improbable events. We then asked children to choose how the story should continue: with a novel improbable event, a novel impossible event, or a novel ordinary event. These three events were presented in pairs, so children's choices were between an improbable and an impossible event or between an improbable and an ordinary event. Children usually chose the improbable event over the impossible one to continue the story, but they were at chance at choosing between the improbable and ordinary events. This suggests that in the context of fiction—where 4-year-olds do not have to rely on their experience—they can accurately distinguish between the improbable and impossible actions. Further, these 4-year-olds correctly treated the improbable actions more like ordinary, possible ones than like impossible ones.

What these data suggest is that, even at these young ages, children are not just responding on the basis of their experience. Rather, children are engaging their causal knowledge to reason about imagined situations. This again suggests that children's causal knowledge is the basis of their counterfactual reasoning capacities. In order to reason about a counterfactual, children must first be able to represent the causal structure they are to reason about. They might use counterfactuality to confirm causal knowledge, but the causal knowledge comes first.

The Role of Inhibition

Before we conclude, we want to consider whether everything we have just said about counterfactual reasoning could be wrong. Another interpretation of the relation between causal and counterfactual reasoning is that counterfactual reasoning actually has little to do with the causal knowledge that one possesses. Rather, making a counterfactual inference is a more domain-general process that is heavily reliant on inhibitory control. In this argument, causal knowledge appears to be relevant to counterfactual reasoning because young children have difficulty suppressing their existing knowledge in order to reason on the basis of counterfactual antecedents rather than on the basis of what they know to be true. Indeed, many studies show that preschoolers' counterfactual reasoning abilities are correlated with their performance on various inhibitory control measures (e.g., Beck et al., 2009; see also Beck, Riggs & Burns, 2011). Moreover, in the adult literature, working memory plays a clear role in counterfactual inferences (e.g., Byrne, 2005).

There is little doubt that inhibition plays a role in counterfactual reasoning, especially because thinking about counterfactuals often requires one to inhibit one's current real-world knowledge. But talking about the role of inhibitory control in development requires unpacking what is specifically meant by the "increasing evidence that domain-general executive functions underpin children's counterfactual thinking" (Beck, Riggs & Burns,

2011, p. 111). Is causal knowledge truly irrelevant to the picture, and can all of children's failures or mistakes be explained by inhibitory processes?

We do not think so. To clarify our position, we borrow an argument from Benson et al. (2013), who discuss the possible ways in which inhibitory control might relate to theory of mind development. One possibility is that children's developing inhibitory control underlies their social cognitive development. The false belief task, for example, requires reasoning not on the basis of one's knowledge of the actual state of the world (there are candles inside a crayon box), but rather on the basis of a different representation (the crayon box should contain crayons, so an ignorant reasoner will think there are crayons inside, not candles). Inhibitory control is necessary for successful performance on this task. Different kinds of inhibitory control, which develop on different timescales based on task complexity and context, allow for the expression of various cognitive and social cognitive capacities throughout development (e.g., the A-not-B task, the false belief task, various measures of understanding pretense; see e.g., Davidson et al., 2006).

In contrast to this standard story that inhibitory control allows for the *expression* of false belief capacities, Benson et al. (2013) suggest that there is another way of interpreting the role of inhibitory control in cognitive development. Specifically, the development of inhibitory control allows for the *emergence* of conceptual change. In this view, children's developing inhibitory control is necessary for reasoning, but not sufficient. To support this argument, these researchers examined the relation between children's developing theory of mind and inhibitory control. They found that 3-year-olds who scored high on inhibitory control measures were not necessarily better at the false belief task than children who scored low on such measures at the outset of their training, but they were more amenable to training on the false belief task. In their view, having better inhibitory control capacities allow children to better process information relevant to learning.

We suggest that something similar is happening with respect to the relation between causal knowledge and counterfactual abilities. Having better inhibitory control capacities does not, on its own, allow children to reason counterfactually. Instead, inhibitory control capacities provide children with improved access to their causal knowledge and greater capacity to reason about the alternate, non-real representations that they construct in order to reason counterfactually.

Imagination and Science

When using a broad definition of "imagination," as any representation that is not meant to reflect the current truth of reality, it is easy to see how imagination has a crucial role to play in scientific thinking (see Weisberg, 2020). A large part of scientific practice involves thinking about possible explanations for patterns of data, alternative ways that experiments could have turned out, and hypotheses for why things happened in a certain way. But these possible explanations and hypotheses are really just imagined scenarios. For example, we may not know exactly why a particular population of frogs suddenly developed a series of deformities on their legs. But we can use the information that we have to imagine an explanation that could fit the facts: A nearby company might have started dumping chemicals in the water, or there might have been unusual fluctuations in temperature recently (similar to what Walker & Nyhout, 2020, call "why else" reasoning). We can think of the process of constructing a hypothesis in parallel to the process of creating a pretend scenario or writing a story. In the science cases, these are stories that are inspired by and connected to real events. Nevertheless, these stories are not necessarily real, and so they require imagination.

We believe that participants in the diagnostic reasoning task we introduced in chapter 5 go through the same sort of imaginative thought process. When they first encounter the blicket detector, they need to come up with some idea of how it works in order to solve the puzzle. Maybe the detector senses number, so it needs four blocks on it to turn green. This hypothesis might or might not be true, so it has to be imagined. As participants gain further information about the machine (e.g., it can turn green with only two blocks on it), they need to update this imagined representation so that it more closely fits the facts.

This example and the work reviewed throughout this chapter make it clear that children's (and adults') causal knowledge constrains their imagination. The framework of causal graphical models that we introduced in chapter 2 provides a way to think about this process. As a quick reminder, causal graphical models are ways of representing different causal structures and different types of relations between causes and effects. In combination with Bayes' rule, or some other formalism for drawing inferences from observed data, they can be used to model how people think about causal systems in the world. More importantly for the current purposes, though, is the fact that these models can be represented and manipulated independently of each other and independently of the structure of the world. This means that they can be used to represent imagined or counterfactual structures as well.

On the flip side, imagination can also be used to manipulate representations of causal structure. In situations where participants do not know the correct underlying structure, they can construct multiple possible models that match the data that they have observed. As new information comes in, they can test out different versions of these models and intervene on them to see which ones can be made to match the new observations (sometimes called "graph surgery"; see Pearl, 2000). Imagination and causal counterfactual reasoning are thus intimately linked in how we think about the world. However, as we have seen across the chapters in this book, explicitly reasoning in this way might have a prolonged developmental trajectory.

To conclude, at the outset of this chapter, we considered the standard view whereby pretending and imagining are unrelated to (perhaps even antithetical to) causal reasoning and scientific thinking. What we hope to have shown is that engaging in and understanding imagination might provide children with insight into the power of possibility—insight that they can use when thinking scientifically. Further, children's causal knowledge provides a foundation on which their cognitive capacities build. Reasoning about pretense, possibility, and counterfactuals thus all emanate from children's developing causal knowledge, and all affect the ways in which children reason about science.

IV Conclusion

11   What Does It Mean to Engage in Scientific Thinking?

We began this book with a question: How do we bridge the gap between children's causal reasoning skills, which develop early, and their scientific thinking skills, which have a longer developmental trajectory? To answer this question, we first reviewed the causal graphical model framework, which provides a more precise way of describing how children represent causal knowledge and engage in causal reasoning. We then focused on two facets of scientific thinking: (1) how causal reasoning, particularly in context and combined with metacognitive development, forms part of the foundation of scientific thinking, and (2) how an explicit understanding of science, which potentially allows children to appreciate that they are doing science, relates to children's scientific thinking. We believe each of these processes is necessary for the development of scientific thinking, that neither alone is sufficient, and that these processes have similar but mostly independent developmental trajectories.

We briefly review those conclusions below, but first we want to note that our original concept for this book was to talk to different audiences—not only to researchers in our home field of cognitive development, but also to researchers interested in science education, to educators, and to the field of psychology more generally. In closing, we want to acknowledge that this book is merely a starting point. For example, many researchers are interested in the causal graphical model framework used to represent causal reasoning, which we sketched out in part I of the book. There are more detailed descriptions of that framework, many of which are specifically geared for researchers who want to build computational models. Our goal in including a brief overview of that work here was to articulate the idea that theories of development could be formalized. This provides great advantages because those formalizations make predictions about how reasoning occurs. But this approach also has limitations: Not everything fits into a nice, neat computational framework. Some of the unanswered questions about rational constructivism are difficult to incorporate into such a computational framework, such as the role of culture or parent-child interaction. These concerns operate in the other direction as well, in that there is a strong possibility that the field of artificial intelligence can be positively influenced by gaining a greater understanding of child development. Knowing more about how children learn might provide insight into how computers or robots could learn.

Similarly, one might object to our approach by claiming that experiments that use a blicket detector or other minimally contextualized methods are not studying scientific thinking. The most extreme version of this view is that scientific thinking can be studied only in a classroom or a lab, and it can include information drawn from only a particular set of topics. While we disagree with this extreme view, we do take to heart this general criticism, which was a primary motivation for the design of the studies presented in part II of the book. Our goal in detailing that work was to articulate the foundations of scientific thinking and to illustrate how one might bridge the gap between research in cognitive development and scientific thinking. The conclusions from those studies can apply to aspects of teaching science, as we discuss below, but they also embrace the idea that scientific thinking is all around us in our daily lives. Additionally, and importantly, we intended that work to serve as just one set of examples of what the two fields can learn from each other; much more work needs to be done to continue building that bridge.

Finally, one might object to our general approach of asking children questions about how science or learning works. These tasks are artificial and strange. We do not disagree; our goal with that line of work was to get researchers, educators, practitioners, and parents (basically, adults) thinking about these concepts. We often read academic research papers on science and learning that define these terms in the ways that adults (particularly academic researchers) think they should be defined; children's behaviors and thoughts are then judged through that lens. Asking children to define a concept like "science" possibly gets them to think about science, but also gets us as consumers of this kind of research to think about what science is as well, which hopefully makes us better, more open-minded consumers of science. But we also acknowledge that science is culturally constructed, and interviewing children across different cultures about science and other concepts related to learning (something we did not investigate) would be an interesting and important future direction.

Most importantly, as reflected in the quote from Carl Sagan at the start of chapter 1, our investigations and interpretations of children's causal reasoning and scientific thinking are fallible. If they weren't, we would be poor scientists. This is all to say that this book is a first step. We assume that few people will be completely satisfied with what we have written or how we have framed these investigations. We certainly are not. Not only because we have so much left to learn, but also because there are so many more ways in which we can try to strengthen the connections between causal reasoning and scientific thinking, and so many more ways in which we can apply what we have learned to help create a more scientifically literate society.

Causal Reasoning as a Foundation for Scientific Thinking

In part II of the book, we reported on a series of empirical studies that provide examples of how to explain the gap between work on young children's understanding of simple, less contextualized causal systems and work on older children's thinking about complex and scientifically rich causal systems. Preschool-age children, and possibly also infants, have some of the foundation for scientific thinking from their causal reasoning capacities, but these capacities undergo further development during the early elementary-school years, and likely beyond. Specifically, young elementary-school children are developing capacities to diagnose the structure of complex causal systems, to account for uncertainty in these systems, to engage in explicit belief revision when confronted with counterevidence, to incorporate aspects of their developing metacognitive abilities, and to recognize and design unconfounded experiments on these systems. Rapid maturation in these capacities seems to occur between the ages of 6 and 8, as children become increasingly able to think about systems with multiple variables, even when the specific efficacies of some of those variables are unknown. These developments occur at the same time as children's developing understanding of interpretive (or advanced) theory of mind, one of the component abilities of which is navigating disagreements between conflicting beliefs.

So while the processes involved in mature scientific thinking have their roots in early causal reasoning abilities, the development of these processes is not complete during the preschool years. Indeed, this development can be influenced by children's experiences in school, as our investigation of the effect of different curricula in the Springfield School District demonstrated (chapter 7). That is, preschoolers may be little scientists, but they also have much science left to learn. This is true not just with respect to acquiring knowledge about scientific content, but also with respect to how to go about doing science.

And this is perfectly fine. Much research in our home field of cognitive development has focused on young children and infants, aiming to chart the origins of various thinking processes. But this work can assume that it is sufficient to show that some abilities are present early in life, hence researchers may underestimate the importance of later developments in these abilities. Infants and young children have the foundational abilities to do science, abilities that allow them to systematically investigate and make sense of their world. As they develop, both their growing knowledge and their growing repertoire of skills shape the trajectory of how they are able to learn and express new scientific thinking abilities.

Explicit Understanding of Science

We have also traced the developmental trajectory of children's explicit understanding of science and related concepts like learning, play, and pretending. This line of work found that younger children tend to hold a content-based view of these concepts, swayed by these activities' visible markers and by the outcomes of the actions involved. As children get older, they incorporate a more mentalistic understanding, recognizing the roles of intentions and beliefs in all of these concepts. For example, younger children tend to view science only as a particular set of topics. Starting between the ages of 6 and 8, they begin to recognize that science is also a way of learning. We see similar development in children's understanding of learning and teaching, although the explicit understanding of each of these concepts matures with different developmental trajectories.

Across the book, but particularly in chapter 1 and chapter 8, we also outlined why we think that children's understanding of science should relate to their scientific thinking. Briefly, inspired by arguments from Kuhn (e.g., Kuhn, 1989, 2007a, 2011; Kuhn & Pearsall, 2000), we suggested that children who conceptualize science as involving reasoning and learning might be better able to marshal their developing scientific thinking skills in service of solving a task than are children who believe that science is merely a set of topics. These latter children might fail to understand the connection between their scientific abilities and the task at hand, and hence might not perform as well on the task. Briefly, if you do not think that a problem requires scientific thinking, you might not think about it scientifically.

The first time we tested this hypothesis, we indeed found a robust relation between children's definitions of "science" as a process of learning and their success on our diagnostic reasoning measure from chapter 5 (as we reported in chapter 8). This inspired us to continue our investigations, aiming to reproduce this initial effect. But as we conducted more research, we found that this relation was weaker and more complex than we had expected.[1] We found only marginal relations between children's definitions of "science" and performance on various diagnostic reasoning measures in a follow-up study, and there were some occasional connections between these definitions and children's performance with other measures of scientific thinking in our museum-based work.

Frankly, we can talk about this case as a process of our own belief revision. The fact that we did not directly reproduce our original result suggests that our hypothesis might be incorrect or that it needs to be modified to say that the effect is weaker than we initially expected. We believe that an explicit understanding of science is an important facet of scientific thinking. But, like belief revision, understanding uncertainty, or learning how to reason within complex contexts, it may have its own unique developmental trajectory, just like the other members of the family of abilities that make up scientific thinking.

For example, aspects of children's definitions of "science" related to their exploratory behaviors with a museum exhibit on gears (Callanan, Legare, Sobel et al., 2020): Children who defined science not as a specific activity and those who defined science as related to learning were more likely to engage in the systematic exploration that related to their causal knowledge. Similarly, these definitions related to how children played with an exhibit on electric circuits: Children who defined science as something that they themselves do or enjoy made more connections or disconnections among components in the process of building a circuit during free play with their parents (Sobel, Letourneau et al., 2021). While the frequency of their own actions during this free play session did not relate to their ability to solve challenges on their own, understanding science in terms of a personal connection did seem to give them a sense of agency in completing play-related goals during play with their parents.

Why this might happen is not clear—after all, neither of these latter studies were designed to investigate the question of how children's explicit understanding of science might relate to their scientific thinking abilities. But it is possible that certain aspects of our hypothesis hold true. One possible way to reconcile all these results is to say that the relation between children's ability to define "science" and their causal reasoning is weak when they are asked to reason about information generated by others. That does not mean that the relation does not exist. It means that there is still an effect worth investigating and describing, but the relation is complicated. When children are in control of their own behavior, their understanding that science is a learning process might indeed provide them with more opportunities to learn from play. Autonomy is important to children's reasoning, and this is something that we as a field need to consider in more detail.

Perhaps we should not be surprised that these relations are complex. As we saw in chapters 9 and 10, an explicit understanding of learning is not necessary for children to learn and an explicit understanding of pretending is not necessary for children to pretend. The explicit understanding of a concept might help children only to engage in metacognitive inferences related to that activity, but not to engage in the activity itself. For instance, as we reported in chapter 10, children's definitions of "pretending" related not to how they recalled engaging in pretend play, but rather to their judgments about pretense based on the metacognitive inferences they had to make about others' mental states. That is, there is likely still an important role for children's developing awareness of science as an active process of learning in their development into fully-fledged scientific thinkers, but this relation may occur for aspects of their scientific thinking other than their first-order abilities to solve diagnostic reasoning problems.

Constraints and Enabling Conditions: Relations among Knowledge and Skills

Another theme that we have considered throughout this book is how different aspects of children's developing knowledge and skills relate to each other and work together to reveal or suppress aspects of children's scientific thinking. We first introduced this idea in chapter 1 with reference to how children's theory of mind constrains their performance on other tasks, such as how they use what they know about others' beliefs to appreciate whether a character has learned something.

A similar example of the role of underlying capacities relating to children's cognition can be found in the potential relations between infants' motor development and their causal reasoning, as reviewed in chapter 3. Infants might initially represent causality as statistical regularities among events, and that statistical learning is motivated by their social interactions and their attentional capacities (e.g., Wu & Kirkham, 2010). As their motor capacities develop, this expands what information becomes available to them and also changes what kind of data they can produce. They can begin to do more than just actively interpret their observations; they can intervene to generate new information (remember how infants knock stacks of objects over to learn about how objects behave, but also possibly about how grownups react?). These developing capacities to intervene on their environments in increasingly complex ways might in turn change the way that infants represent information. This example also illustrates the vital importance of children's active interactions with their environments; learning to act on the world might affect how we represent it. An important facet of the causal graphical model framework as a description of how children represent causal knowledge is precisely that it values the role of action—intervention—in the development of children's reasoning abilities.

To take a more cognitive example, throughout the preschool years, the development of a variety of domain-general capacities can constrain and facilitate aspects of children's causal reasoning. As we discussed in chapter 10, children's pretend play, which emerges during the second year of life, can indicate that children have the representational capacities necessary to put aside one representation and reason about another (e.g., Leslie, 1988). Such representational capacities might be the basis of children's counterfactual reasoning. Counterfactual reasoning requires a representation of the specific causal structure one is reasoning about, as well as more domain-general imaginative capacities to represent a copy of that model, on which a hypothetical intervention is performed to represent the "what if" statement. Possessing domain-specific knowledge about the causal structure and possessing domain-general representational capacities, including the imaginative capacity and the cognitive control to act on the represented structure, are both necessary for counterfactual reasoning, but neither alone is wholly sufficient.

Developments in other aspects of children's cognition thus first constrain and then facilitate children's causal reasoning. What is interesting about this hypothesis is that it offers an analogy for our take on the relation between causal reasoning and scientific thinking. Specifically, having the capacity to engage in causal inference is necessary but not sufficient for fully engaging in scientific thinking. Beyond the ability to represent causal relations, one also needs to be able to change one's beliefs and understand the conditions under which such belief change is warranted. The latter involves not only learning from sets of data, but also being able to recognize when one is wrong; articulating different beliefs in light of new information is at the forefront of scientific thinking. Belief revision also involves more domain-general processes, such as cognitive control, that are also necessary to inhibit the complex or unfamiliar ways in which a problem is phrased in order to recognize its underlying causal structure. But even though the capacity to reason about causes and the ability to represent events in terms of their causal structure do not reflect the entirety of scientific thinking, they serve as its foundation. They allow children to integrate the data they observe and draw conclusions about it, which forms an important part of scientific thinking, even if not its entirety.

As we have documented, the many cognitive and metacognitive capacities involved in scientific thinking have distinct developmental trajectories between the preschool and elementary-school years. But each of these capacities also imposes constraints on children's abilities to progress in their development of scientific thinking abilities. Lacking the knowledge of how a particular domain of science works (e.g., biology) can affect children's abilities to reason about that domain (e.g., whether a character ever needs to sleep), even if they may be perfectly capable of drawing parallel inferences for a different domain (e.g., whether an object can float in the air). On the flip side, possessing such knowledge does not automatically guarantee that children will be able to draw appropriate inferences; it merely sets up the enabling conditions for their potential reasoning.

Further, and somewhat speculatively, the idea that domain-specific knowledge constrains some of our domain-general capacities, like causal reasoning, can also potentially explain why gaining metacognitive insight into these capacities is so difficult. If our causal reasoning abilities and scientific thinking abilities are constrained in different ways at different times by domain knowledge and by the development of other abilities, then it stands to reason that it would be more difficult to understand the inner workings of these capacities than to understand the nature of our content knowledge or of other aspects of our thinking.

It is this intricate interplay between children's knowledge and skills and the ways in which these aspects of their cognitive capacities constrain each other that we wish to ensure is part of the conversation about how children's scientific thinking develops. Children's thinking is not fully domain-general; children are not necessarily able to express their skills fully in all circumstances as these skills are developing. There are effects of context, both cognitive and sociocultural, that influence children's performance on lab-based tasks and in school. Put another way, it may not make sense to ask *when* children become able to succeed at expressing a particular skill. Rather, our claims and conclusions must be sensitive to the larger developmental system within which children exercise their abilities.

Constructing Scientists

The work described in this book has been primarily concerned with charting how children's early abilities to solve simple and relatively decontextualized causal reasoning problems begin to develop into the skills that they will eventually need to engage in more complex scientific thinking. As we conclude, we want to consider how these findings could translate into recommendations for formal and informal science education. One straightforward conclusion from the work presented here is to focus more on children in kindergarten, first grade, and second grade. We have found shifts in children's performance on diagnostic reasoning tasks and in their definitions of "science" between the ages of about 6 and about 8. This period thus seems to be an important time for children (at least in our culture) to develop both their scientific thinking skills and their knowledge of what science is. In turn, this implies that interventions aimed at boosting scientific thinking in this age group may be particularly successful.

Formal Learning Environments

It is crucial to note that children enter school already in possession of powerful causal reasoning mechanisms. This argument has been made before with respect to children's knowledge of the content of scientific theories, because even preschoolers have rich conceptions of the biological, physical, and psychological world (see Carey, 2009; Inagaki & Hatano, 2006; Shtulman, 2017; Vosniadou, 1994). Teachers who view their students as blank slates are thus fighting an uphill battle, as they attempt to get their students to learn things that contradict students' preexisting knowledge. For example, most kindergarteners believe that the Earth is flat, because of the way that it looks to them, and lessons trying to convince them that the Earth is round tend to be generally ineffective; they can even lead to further misconceptions as children try to integrate this new knowledge with their existing beliefs. In general, addressing the existing conceptions and misconceptions that children have, rather than ignoring them, paves the road for a better and more effective educational experience (see work on refutation texts in science education; Sinatra & Broughton, 2011; Tippett, 2010). Children learn with repeated exposure, but this takes effort.

Attempting to teach children about science without understanding their existing beliefs about science might be similarly ineffective. Gaining insight into how children conceptualize this topic can help us reach them where they are, rather than assuming that they will straightforwardly absorb our adult concepts of science. Similarly, children come to school already in possession of some abilities to reason about causal systems, which seem to be present in preschool (e.g., Buchanan & Sobel, 2011; Gopnik & Wellman, 2012; Schulz & Sommerville, 2006). Our work suggests that children in kindergarten and first grade are ready to advance these skills to more fully-fledged scientific thinking abilities.

Our work on children's definitions can also help us to understand how best to structure early science education. It is worth noting that our approach of asking children for explicit definitions of various concepts is a bit outside the norm, both in educational research and in our home field of cognitive development. There is some merit in that argument, as we noted in chapter 8, because children might not have good insight into the content of their own concepts. Nevertheless, we see value in our approach, both for basic developmental science research and for education. For one thing, we believe that this technique could be used fruitfully as a jumping-off point for understanding conceptual development. Using this approach, we have found many connections between children's definitions and their performance on various tasks, particularly within the domain of learning:

Children who understand learning at a more mature level are also better able to reflect on their own learning. But even if such relations do not exist, that null finding provides interesting information about the link between children's abilities and their explicit knowledge in development. We can then ask when children become explicitly able to provide reasonable definitions for a concept and whether that ability seems to impact their performance related to that concept. We can also ask how the abilities that infants and young children possess implicitly come into conscious awareness, such that older children and adults can access and manipulate them explicitly. Finally, we can ask whether providing children with aspects of these definitions facilitate their performance related to the concept.

The answers to these questions have implications far beyond the development of scientific thinking. At the heart of this question is the issue of consciousness: What is the difference between possessing an implicit ability to perform some task and additionally possessing the explicit awareness that one is doing that task, or how one is doing so? Asking about children's definitions can thus be an important tool in the kit that we use to map the trajectory of the development of children's awareness and metacognitive abilities.

With respect to education, our work on children's definitions also suggests a potentially helpful shift in how science is framed in educational settings. Formal and informal learning environments tend to present children with "science classes" or "science activities"—but the mere act of calling some kinds of learning content or activities "science," while not applying this label to other things, might shift how children think about science. Specifically, it might lead children to think that science is only a particular set of content areas or a certain subset of activities. Broader actions that are clearly scientific (like asking questions and searching systematically for answers) might not be categorized as "science" by younger children. Indeed, our interviews found that the younger children we investigated tended to conceptualize science as a narrow set of topics. In turn, this may hamper their abilities to engage in scientific thinking on a wider variety of tasks. Much like telling children to "be scientists" (Rhodes et al., 2019), focusing preschoolers and young elementary-school students on the science corner of their classroom or separating science topics into their own class might discourage them from recognizing that they can engage in scientific thinking in other places and times.

Our suggestion is to integrate science into early educational curricula in a more holistic way. While the practice of science does tend to concern itself with a particular set of topics and questions, at its heart, science is a method of investigating the world. This method involves gathering evidence for one's claims, thinking critically and carefully about this evidence, weighing one's existing beliefs against the strength of incoming information, and objectively navigating situations of uncertainty and disagreement, among other things. These skills should surely not be confined to a single classroom or to a brief experience, only to be left behind when children are engaged in literacy and math and social studies; these skills are ones that children will need to succeed in all aspects of their lives.

This is not necessarily how early science education occurs, at least in the United States. Science is often presented to young children as a topic or activity, and it is confined to a (shockingly brief) period of time in the school day (particularly compared to math or language arts). No wonder many of our younger participants claimed that science was only about dinosaurs or potions. Some schools are beginning to address this issue by including more scientific content in nonscience times in the curriculum, as when literacy lessons focus on reading science texts. We would argue that this is a good beginning, but also that science needs to be more fully and deeply integrated into all aspects of education. More specifically, the skills of scientific thinking should be treated the way that current curricula treat the skills of literacy. Reading is a skill that children must master, not merely for its own sake, but in order to engage with any kind of content across the curriculum (e.g., social studies, math problems) and to succeed in adult life. Scientific thinking is also a set of skills that children must master in order to think clearly about topics across the curriculum and to succeed in adult life. By respecting children's existing abilities in scientific thinking, and then by engaging with both these developing skills and with their explicit knowledge, we can make real progress on teaching science to the youngest learners.

One additional message of our work is to focus science teaching not just on the content of science or on science facts but on the nature and practice of science. This recommendation is at the core of new guidelines for science teaching (e.g., the Three-Dimensional approach in the Next Generation Science Standards; see NGSS Lead States, 2013; National Research Council, 2012), and our work strengthens the basis for this recommendation.

Specifically, we have argued that a broader conception of science (as a process of learning, rather than as merely a set of topics) has links to children's abilities to reason scientifically. While our evidence for this connection in childhood is tentative and requires further investigation, this is more strongly the case for adults: Adults who have a more accurate view of how science is practiced are more likely to accept scientific claims (Weisberg, Landrum et al., 2018, 2021). It is possible that teaching students how science can create knowledge and that science involves active processes of learning and discovery could help underpin their abilities to think scientifically.

In general, then, science is not a topic of study that should be done only in a particular place or at a particular time. Science is a set of skills that apply in myriad circumstances in our lives. Integrating scientific thinking broadly across the curriculum may well be the key to nurturing young children's causal reasoning abilities to fully bloom into mature capacities for scientific thinking.

One other facet of children's understanding is important here. Children's identities—particularly as learners—are constructed over time and based on their experience. Identity-based language, such as "let's be scientists," can suggest to children that "scientist" is a social group, and that only some people are members of that group. This can lead them to interpret any failure at a science task as a sign that they are not members of that group, in turn leading them to persist less on challenging topics and problems. In contrast, the action-based language of "doing science" suggests to children that they are engaged in a set of activities. If they are challenged under these circumstances, instead of interpreting the struggle as a sign that they lack the qualities of a scientist, they may be more likely to interpret this struggle as situational and not specifically about them as learners (Archer et al., 2010; Rhodes et al., 2020). So, when integrating scientific thinking into the classroom environment, introducing this kind of thinking as an action as opposed to an identity might promote greater engagement with the material.

Informal Learning Environments

Many facets of our thinking about children's development have been inspired by our collaborations with informal learning institutions. Although interactions about science can happen anywhere, we focus here on museum settings because they offer the ideal environments to consider how children engage in various aspects of scientific thinking. Museums are designed to promote learning through active experiences, and they integrate a diversity of authentic activities and interactions, providing information that children can use to engage in belief revision and learning about science.

Dave's lab has worked with Providence Children's Museum for about seventeen years, while Deena's lab has worked with the Academy of Natural Sciences for about eight years. In general, academic researchers and informal learning institutions can partner in a number of different ways, often to mutual benefit (Callanan, 2012; Sobel & Jipson, 2016). Critically, working in informal learning environments allows researchers to document how children can learn from everyday experiences, including from their observations within the museum environment and from their interactions with the exhibits and with other people (e.g., caregivers, siblings, peers, museum staff).

The two primary museums with which we collaborate have different pedagogical goals. Providence Children's Museum believes in the importance of play for the developing child—not just their cognitive development, but their social-emotional and identity development as well. Mind Lab, the program that Dave helped establish with the museum staff, focuses on learning about how children (particularly young children) learn, and not about any particular scientific content. Indeed, the majority of the work on children's understanding of learning, teaching, and play described in chapter 9 emerged from a collaboration with members of the museum staff. In contrast, the Academy of Natural Sciences is a natural history museum with the mission of helping visitors to understand the natural world and to feel inspired to care for it. While the Academy has a dedicated children's exhibit, called *Outside In*, the majority of the museum caters to a broad audience across the age spectrum.

To investigate the role of these different missions more closely, we are currently conducting studies that compare children's exploration, reflection, and learning across these two museums. As an example of some of our preliminary results, we (Stricker & Sobel, 2021) tested 120 children between the ages of 6 and 9 (62 girls and 58 boys; mean age=94.92 months; age range: 72–119 months) on their conceptions of learning. Half of the children were tested at Providence Children's Museum after they had engaged in free play at a set of exhibits, where they had the opportunity to choose where to go and what to do. After their exploration and play, children were tested on a short experiment. The other half of the children were tested in the lab, where they did not have an exploratory play session, although they were given a similar experiment. We then introduced both groups of children to a picture of a novel toy (based on Bonawitz et al., 2011). The toy had a number of different manipulanda, and it was not clear what the toy did or how it worked. We told the children that this toy could do a lot of things, and we asked children how they wanted to learn how the toy worked: by playing with it themselves, by asking their grown-up, or by watching someone else play with it.

Overall there was a difference in the distribution of responses between the two environments.[2] In both conditions, the most frequent answer was to play with the toy, but this answer varied between the conditions. Forty-seven percent of the children in the lab chose to learn about the toy by playing with it themselves, while 63% of the children tested at the museum did so. In contrast, 35% of the children in the lab chose to ask their grown-up, while only 20% of the children at the museum did so. Choosing to watch another child didn't differ between the two environments (18% in the lab vs. 17% at the museum).[3] Overall, these data suggest the possibility that the environment in which children are tested affects how they might respond to experiments and interview questions.[4]

Related to this finding, however, is the possibility that how children engage in scientific thinking differs depending on the context they are in. In particular, consider our own findings on the relations between children's definitions of "science" and aspects of their scientific thinking. The first experiment where we asked children to define "science" and then related these definitions to their performance on our diagnostic reasoning measure was done at Providence Children's Museum. Recall that in this sample, we found this hypothesized relation, which held even when controlling for age. We similarly found effects of children's conceptualization of science and facets of their exploration at Providence Children's Museum and in other children's museums around the country. But we did not find this relation when we tested children in schools, in the laboratory, or at the Academy.

It is possible—and this is a hypothesis we would like to investigate further— that children's access to their scientific thinking differs across these contexts. Places where children are explicitly aware that they are doing science, like schools or museums with a scientific focus, implicitly encourage children to bring their developing scientific thinking to bear on tasks. When children are tested in these settings, they might be already primed to do science, thus lessening the impact that their own definitions of "science" might have on their task performance. But when children are tested in environments where science is not necessarily part of their perception of this environment, like a children's museum that does not focus on science content, there might be more scope for children's own conceptions of science to play a role in their task performance. Specifically, they might engage in scientific thinking in these environments only if they conceptualize science as being generally related to learning. Indeed, as noted above, we did not see a relation between children's definitions of "science" and their diagnostic thinking in the data sets that we collected at the Academy or in the Springfield School District. But we did see this relation in environments that were focused more on play and learning, which did not necessarily have the pedagogical goal to teach scientific content.[5]

Taken together, these findings invite us to deepen our understanding of how formal learning environments, informal learning environments, and everyday experiences contribute to the development of children's scientific thinking skills. Cognitive development does not happen just in the laboratory; its findings must generalize to classrooms, museums, and everyday interactions with the world. Similarly, science learning does not happen just in the classroom; its practices and processes can be found throughout children's lives. We hope that our work can help to illuminate how this learning happens, allowing all children to reach their potential as scientific thinkers.

What Constructing Science Means

While we like the play on words in the title of the book, we need to conclude by clarifying a possible misconception that this title might raise. "Constructing science" implies an end state—that there is a point in development where children become capable of scientific thinking. But scientific thinking develops throughout the lifespan, beginning with children's causal reasoning abilities and continuing throughout adulthood. Young children are already on that journey; they are learning the skills necessary for scientific thinking and are learning to apply such skills across many situations. But if even children have access to the rudiments of scientific thinking, and if these skills continue to develop over the course of their lives, then why are we not a more scientifically literate society?

While there are many influences on adults' scientific beliefs and their abilities to apply their own capacities for scientific thinking (e.g., Angier, 2007;

Weisberg et al., 2021), based on our own work, we suspect that one additional answer to this question has to do with the metacognitive awareness that is necessary for scientific thinking. Not only is scientific thinking complex in its own right, requiring coordination with multiple other cognitive skills, scientific thinking also may require insight into one's own thought processes. This alone may make aspects of scientific thinking difficult to achieve. As an analogy, consider the process of learning to drive a car. This is a skill that is somewhat difficult to learn (and that, like scientific thinking, requires the coordination of multiple cognitive skills as well as some social interaction). Eventually, though, it becomes automatic. But this automaticity may get in the way of metacognitive awareness; trying to explain to someone else how to drive a car or trying to become consciously aware of our own actions as we are doing so becomes increasingly difficult. Scientific thinking might be similar. Recall that most adults responded to our diagnostic reasoning measure in the way we had intended (following the logic of interactive causality), but a subset of them could not articulate the reasoning behind their choice.

The need for metacognitive awareness thus might be part of what makes scientific thinking difficult for adults, or at least more difficult than the causal reasoning capacities that even young children seem to possess. A better analogy, then, may be that causal reasoning is like learning to drive a car, while scientific thinking is more like learning to fly a plane. Many adults learn to drive a car, just as young children develop a causal reasoning system, potentially from lower-level mechanisms of statistical regularity. Driving is something that so many people learn to do, it is almost considered a rite of passage in our society. Learning to fly, however, is something that fewer people learn to do and, as such, seems more intimidating. But both skills are built up out of more foundational skills, and both skills, though difficult, can become automatic through practice and experience. The more we become aware that we are engaging in the kinds of reasoning and metacognition necessary for scientific thinking, the easier it will become for us to do so. So, to make one last *Star Trek* reference, let's fly.

Notes

Chapter 1

1. Dave went to graduate school at the University of California at Berkeley. The first time he went to Berkeley, he walked through Sather Gate and onto the main campus. Turning left at the library, he saw the word "Psychology" etched in marble on the side of the Valley Life Sciences building. Turning to his host—a graduate student who was familiar with the campus—he asked whether that was the psychology department, having never seen so nice a building and inspired by the majesty that he would be working in such environs. His host replied "no" with a laugh, and explained that the psychology department used to be housed there, but was moved to Tolman Hall to make room for the other biological sciences etched in the marble around the building. Tolman, a 1960s concrete building, built in a Brutalist architectural style, was something of a letdown. The folk continuum is real even (and maybe especially) inside university structure.

2. Whether those neurological developments are based on biological maturation independent of environmental input is unclear, but seems incredibly unlikely.

3. Because the examples in this section are about psychological knowledge, one might object that these examples are not well-matched to the examples above about physical relations among objects. This argument goes like this: Physics is a science, so understanding the causality of physical relations among objects is clearly "science knowledge." Psychology is "less" of a science, thus understanding relations among mental states is not science knowledge, but merely knowledge about human behavior. Our response, of course, is that psychology is a science; there is every bit as much causal structure in understanding psychology as there is in understanding physics. The main difference is that most of physics involves deterministic relations, which can be mathematically described. Human behavior, in contrast, is stochastic. But its relations are just as causal, as many philosophers have argued (e.g., Campbell, 2007). Just because psychological relations are stochastic does not stop people from writing formulas that attempt to explain human behavior, which basically summarizes a major goal of cognitive science.

4. Frankly, we doubt that most adults know all the whys behind their behavior, but at least adults can try to articulate them.

Chapter 2

1. Gopnik et al. (1999) describe a similar version of this kind of systematic exploration as the "drop the spoon" game: Whereas younger infants (around 8 months) drop items off their high chair to learn that unsupported objects fall, older infants (around 14 months) might do it to learn how many times they can drop an item before a parent takes it away.

2. We are both still wondering how "carrying clipboards" became a defining property of scientists. That said, on the show *Odd Squad*, the child-scientist character who works in the gadget lab (Oscar) does wear a lab coat and carry a clipboard. So maybe we are missing something.

3. There are certain exceptions to this, such as Bullock et al. (1982) and Shultz (1982), who set the groundwork for the contemporary study of causal reasoning in young children.

4. There is an open question as to whether nonhuman animals have the same kinds of causal reasoning capacities as young children. Although it initially seemed as though the answer to this question was "no," over the past twenty years, work in comparative psychology has uncovered that nonhuman animals can make highly sophisticated causal inferences (e.g., Blaisdell et al., 2006; Schloegl & Fischer, 2017) and can perform similarly to human children in some tasks (e.g., Seed et al., 2012). However, the reasoning mechanisms used by human and nonhuman animals may still be different, even if they sometimes result in similar inferences (see Penn & Povinelli, 2007, for a critical review). We do not wish to take a position on this debate, but we will point out that an important difference between human children and nonhuman animals is the number of trials that members of these two groups often require to make an inference. Children usually require only one, or a very small number, whereas nonhuman animals usually require orders of magnitude more. So while there might be some shared evolutionary foundations, the kind of scientific thinking we describe in this book seems more likely to be unique to human reasoning.

5. Although these simple causal systems may be more common in laboratory experiments than in the real world, there are real-world examples of them as well. After Brown University renovated Dave's building, he discovered that his office lights worked on a touch panel that lacked sensitivity, so simply touching it turned the lights on only about half the time. To make sure that he does not work in the dark, he has to stand there and press the switch really hard, usually for about six seconds. It took a while (and at least one call to the facilities department) to determine that the strength of the button push was the key to activating the lights. In addition, the light switch is probabilistic; it works about 85% of the time. So sometimes Dave just

stands there pushing the button over and over. At this point, the most reasonable conclusion may be that it's operated by a Cartesian demon.

6. At the outset, we would like to say that there are more computationally sophisticated descriptions of this framework (e.g., Glymour, 2002; Gopnik et al., 2004; Pearl, 2000). Our goal here is to describe the framework so that readers can understand that rational constructivism can have a more precise computational formalism. Because of this, we are not presenting many computational details here, and refer the interested reader to these or other references.

7. We'd like to say that what faithfulness implies is the assumption that we are not living in the Matrix, but (a) individuals with philosophical inclinations might argue with this statement and (b) there's a fourth Matrix movie in the works, which might change what that phrase means. So we are not going to make this statement.

8. Some might argue that models like the one shown in figure 2.2 are so simplistic and general (X and Z can stand for just about anything) that the faithfulness assumption does not matter. That is, building these models as a representation does not buy the child anything because these models are not realistic models of the world. We disagree for two reasons. First, children's representations of causal structure are presumably simpler than adults' (i.e., they have less content knowledge), and we have to start somewhere to get a handle on what they might be representing. There are numerous cases in which learners are better off with a "less is more" approach (e.g., Newport, 1990). Second, faithfulness is not really a psychological principle. It's an assumption about translating the representation inherent in a model to an understanding of causality. That said, the initial critique does have merit; the models built using this framework do tend to be (perhaps unrealistically) simple. We address this point in chapter 3 when we discuss nonindependence.

9. Studying children's causal reasoning involves reading a lot of comic books and watching a lot of Marvel movies (at least for us).

10. Alison Gopnik led the team of Jennifer Esterly, Greg Robison, and Dave in creating the blicket detector. Dave and Jen went on to use the detector to study various aspects of children's causal and scientific reasoning. Greg pivoted to studying children's humor; we always suspected that Greg had the better time.

11. If you want to build your own blicket detector, follow the instructions here: https://osf.io/5qt2h/.

12. Schwartz and Reisberg (1991) provide an excellent summary of this theoretical approach and take the reader through many of the same calculations as we present here.

13. This is sometimes separated into two variables, alpha and beta, representing the stimulus and the outcome separately. Because alpha and beta are constants, these numbers can be multiplied together and represented as K.

14. McCormack et al. (2009) used different controls than our studies did—controls that were more faithful to the associative reasoning literature. These researchers found a developmental difference, whereby 4-year-olds did not engage in backward blocking, but 5-year-olds did. We revisit this finding in note 17 below.

15. Xu (2019) also describes a view of cognitive development called *rational construc-tivism*, which is slightly different from the description we present here. It is beyond the scope of this chapter to detail all the differences between this view and the one we describe, but we want to acknowledge that this term—and this theory—is still evolving.

16. Here, we describe those priors and how they differ in general terms. A formal description of this modeling for the interested reader can be found in Griffiths and Tenenbaum (2007, 2009).

17. There are two other points we want to make here. First, this mechanism also potentially explains the development observed in McCormack et al. (2009), mentioned in note 14. While the 4-year-olds in our studies were able to relate causal properties to insides, it is possible that the children tested in McCormack's study were not. Given this, these children would have interpreted the backward blocking demonstration dif-ferently, leading them to respond that the B object was efficacious. Second, it is not clear whether children are explicitly representing each and every one of these hypoth-eses. In fact, it is unlikely that they are, because the hypothesis space can be infinite or at least very large. Instead, it seems more likely to us that this is an implicit process, so we are not aware that we are engaging in this kind of representation.

18. Perspective really determines if this is a blessing or a curse.

Chapter 3

1. This, in turn, provides a foundation for thinking about other kinds of causal models, since more complex models can be broken down into three-node models of common causes, common effects, and chains.

2. Ahl and colleagues (see e.g., Ahl et al., 2020) posit a different developmental tra-jectory than the work from our lab. However, they also use more complex stimuli, which might have made it more difficult for children to reason in the same way as they did with our stimuli (see our discussion of contextualization in chapter 6). Nevertheless, we think that these two sets of findings are generally consistent.

3. Carey (2009) clearly disagrees with the rational constructivist framework in other ways, particularly with respect to what other kinds of knowledge might be innate (see also Spelke & Kinzler, 2007) and with respect to the argument that the causal graphical model framework serves as a good representation of knowledge. But regarding the idea that a concept of "cause" is innate, there does seem to be some agreement between these theoretical positions.

4. There is an open question here, which is beyond the scope of this chapter: Do nonhuman animals reason only by associative inference or do they have the same kinds of inferential capacities as human beings? Penn and Povinelli (2007) provide an outstanding review of this issue. They conclude (and we concur) that the answer lies somewhere in the middle. We take this conclusion as evidence that there are some discontinuities between nonhuman animals and human beings, which is one of the bases for our suggesting that causal inference is not innate and starts with associative reasoning mechanisms.

5. Not just for causal concepts, but for conceptual structure in general—it's what Keil (1989) referred to as "original sim."

6. It is not clear whether that acceleration lasts; to our knowledge, no one has given 3-month-olds the sticky mitten experience and then tested them a month later on understanding goal directedness. But it is certainly plausible that this acceleration does last (see Libertus et al., 2016; Wiesen et al., 2016).

7. There is an alternative version of this argument, namely that causality is innate, and what the motor experience does is allow babies to access their existing understanding that event regularity should be processed in terms of causal relations. Although we do not favor this interpretation, as it does not explain the transition between 5- and 8-month-olds that we demonstrated in our previous work (Sobel & Kirkham, 2006, 2007), we suggest that this is an open empirical question.

8. It is tempting to suggest that such accounts of development are more akin to dynamic systems models (e.g., Thelen & Smith, 1996). Indeed, Smith and Thelen (2003) make some suggestions for how sociocultural findings, particularly interaction, might be interpreted in this way. The algorithms that they suggest, however, seem to focus on the kinds of internal, cognitive developmental processes (such as the A-not-B error) that are similar to information-processing algorithms. It is an open question whether such models may be better accounts of the kinds of social learning processes we describe here.

9. After many sociologists and anthropologists disproved of and disavowed Levy-Bruhl's hypotheses, one prominent scholar continued to cite his work: Piaget (see Jahoda, 2000). Piaget compared early childhood to other cultures that showed (according to Levy-Bruhl) "non-logical" thought patterns.

10. These authors would not consider themselves as describing a rational constructivist framework, but there are similarities between these models and the Bayesian algorithms we described in the previous chapter. What is important to note is that we are imagining an extension of these authors' work; Werchan et al. (2016) did not apply their model to cultural differences.

11. One of the ways we tried to consider some of these ideas in Callanan, Legare, Sobel, et al. (2020) was to look at how parents viewed the relation between play and learning and their goals for visiting a children's museum (inspired by work by

Gaskins, 2008). Unfortunately, this part of the study resulted in inconclusive findings. Investigating this issue more systematically is a high priority for our future endeavors.

12. Bronfenbrenner (1979) described developmental psychology as "the science of the strange behavior of children in strange situations with strange adults for the briefest possible periods of time" (p. 513).

13. We assume that some researchers in theory of mind would say that these false belief contrastives are children's initial attempt to verbalize their "implicit" theory of mind capacities, as suggested by looking time studies (e.g., Onishi & Baillargeon, 2005). However, we cannot find a reference in this literature that directly makes this claim. Further, this assumes that such looking time studies are robust; they may not be (see e.g., Sabbagh & Paulus, 2018).

14. There is an open question here as to whether this tendency is culturally universal. Gauvain and colleagues, for example, use cross-cultural investigations to suggest that it is not (e.g., Gauvain et al., 2013, but see Callanan, Solis, et al., 2020, for a different interpretation).

Chapter 4

1. As we write this book, Dave's daughter (age 12) is in seventh grade and is actively writing a report on the control of variables strategy, using a version of the slopes task (Chen & Klahr, 1999) in her science class. This is the first time she has been introduced to the strategy in her public school education.

2. The superpencil detector is actually the heating element of a broken coffee maker, connected via a fake USB port to a computer running a MATLAB script that allows the researcher to control whether "superlead" is detected on a given trial (via a surreptitious mouse click on the part of the experimenter).

3. This study in particular looks at using simpler contexts to ask questions about children's concepts of planets, and finds that simplifying the context facilitates reasoning in elementary-school-age children. We address this issue directly in chapter 6.

4. Indeed, the week they discussed causal and scientific reasoning, Dave polled his advanced seminar on cognitive development. No student thought snake activity was a risk factor for earthquakes a priori. Although this is merely anecdotal evidence, it suggests that even adults have prior knowledge that might influence their reasoning.

5. Critically, these children were all tested before they entered formal schooling environments, so the age-related difference observed here is not the result of only the older children being in school.

6. Unlike the more contemporary studies, which contextualized their findings in terms of the causal structure of a physical system, Reiber (1969) investigated children's hypothesis testing about another's behavior (where they have hidden a candy), which is inherently more probabilistic.

7. At this point, we just want to acknowledge that "little engineers" is a bad description. Most engineers we know (including many we have worked with throughout the years) are thoughtful, scientific in their reasoning, and not purely driven by generating rewarding behavior if it means not understanding causal structures.

Chapter 5

1. In the early days of the blicket detector, this enabling switch was wired to the box. Dave tested children in a room with a one-way mirror; a wire led from the detector, down the floor, up the wall next to the one-way mirror, and into a hole in the wall, which led to the observation gallery. It seemed pretty obvious to us that a person was in that gallery, moving the switch; the wire even sometimes moved slightly during the experiment. Because we wanted to be sure that participants were convinced that the detector really worked as we described, we piloted the first blicket detector experiment (described in chapter 2) on a set of adults. None of them figured out that the machine was controlled by a person behind the mirror. Only one child—a five-year-old—figured out how the machine really worked during the year that we tested in this manner. That was in 1996, so that child is 31 upon this book's publication. We really wonder what they are doing now.

2. For experimental control, we counterbalance the order in which participants see these four events. Half of the participants see the order described here (all four blocks, two combinations of three blocks, the single block), and the other half see the events in the opposite order (the single block, two combinations of three blocks, all four blocks). This variable did not make a difference to any of our results and will not be discussed further.

Chapter 6

1. The real quote uses "magic" instead of "fantasy."

2. This difference is statistically significant, exact proportion test, $p = .049$.

3. Like the other two versions of this study, half of the children saw the dinosaurs introduced in this order, and the other half saw the dinosaurs introduced in the opposite order. Order never mattered to children's performance, so we do not discuss this variable further.

Chapter 7

1. Exact proportions tests; all p-values $< .001$

2. Interpretation: $\chi^2(1) = 12.74$, $p < .001$; Preference: $\chi^2(1) = 5.09$, $p = .02$

3. $\chi^2(1) = 0.25$, $p = .62$

4. $\chi^2(1) = 0.18$, $p = .67$

5. $\chi^2(1) = 7.08$, $p = .008$

6. $\chi^2(1) = 67.53$, $p < .001$

7. Paired t-tests, Fact: $t(76) = -4.47$, $p < .001$; Interpretation: $t(77) = -3.66$, $p < .001$; Preference: $t(77) = -1.76$, $p = .08$

8. $\chi^2(1) = 4.96$, $p = .03$

9. $\chi^2(1) = 2.04$, $p = .15$

10. $\chi^2(1) = 10.45$, $p = .001$

11. Because some children received the same task twice, we changed the colors between the testing sessions. We here refer to the colors used in chapter 5, which were presented to these children in 2015; in 2017 we used yellow, brown, green, and gray blocks that made the machine light up either blue or purple.

12. Exact proportions test, $p = .07$

13. $\chi^2(1) = 0.001$, $p = .97$

14. Exact proportions test, $p = .15$

15. Exact proportions test, $p < .001$

16. Exact proportions test, $p < .001$

17. One student did not respond to the main test question on the blicket task, so results for this task involve 111 students rather than the full 112.

18. Binomial test, $p = .04$

19. Exact proportions test, $p = .89$

20. $\chi^2(1) = 9.03$, $p = .003$

21. Binomial test, $p < .001$

22. $\chi^2(1) = 3.95$, $p = .05$

23. Binomial test, $p < .001$

24. $\chi^2(1) = 2.20$, $p = .14$

25. Chi-squared tests, all χ^2 values < 1.63, all p-values $> .20$

26. $t(169) = -0.03$, $p = .98$

27. $t(115) = -2.79$, $p = .006$

28. MAP scores can range from 100 to 350, and this test is designed to measure performance from first grade to high school on the same scale.

29. As with the MAP, scores on the PSSA are designed to measure performance across multiple grades on the same scale (third grade through eighth grade; minimum score = 600).

30. First graders: reading correlation, $r(214) = 0.26$, $p < .001$; math correlation, $r(214) = 0.31$, $p < .001$

31. Third graders: reading correlation, $r(107) = 0.32$, $p < .001$; math correlation, $r(107) = 0.31$, $p = .001$

32. Mean reading score of first graders who responded correctly on the fact trial = 194.82 (SD = 14.85); mean reading score of first graders who responded incorrectly = 186.09 (SD = 13.81); $t(214) = -4.27$, $p < .001$

33. Mean math score of first graders who responded correctly on the fact trial = 195.49 (SD = 10.59); mean math score of first graders who responded incorrectly = 187.13 (SD = 13.79); $t(214) = -4.99$, $p < .001$

34. Mean reading score of first graders who responded correctly on the interpretation trial = 193.20 (SD = 15.33); mean reading score of first graders who responded incorrectly = 189.59 (SD = 14.51); $t(214) = -1.76$, $p = .08$. Mean reading score of third graders who responded correctly on the interpretation trial = 216.17 (SD = 11.00); mean reading score of third graders who responded incorrectly = 208.14 (SD = 15.17); $t(107) = -3.01$, $p = .003$

35. Mean math score of first graders who responded correctly on the interpretation trial = 193.93 (SD = 12.76); mean math score of first graders who responded incorrectly = 190.49 (SD = 11.95); $t(214) = -2.02$, $p = .04$. Mean math score of third graders who responded correctly on the interpretation trial = 223.64 (SD = 10.33); mean math score of third graders who responded incorrectly = 217.07 (SD = 14.83); $t(107) = -2.58$, $p = .01$

36. Mean reading score of third graders who responded correctly on the preference trial = 214.94 (SD = 12.03); mean reading score of third graders who responded incorrectly = 204.89 (SD = 12.03); $t(107) = -2.33$, $p = .02$

37. Language correlation, $r(105) = .24$, $p = .01$; math correlation, $r(105) = .25$, $p = .009$; science correlation, $r(104) = .30$, $p = .002$

38. Mean reading score of first graders who responded correctly on the blicket task = 193.63 (SD = 16.27); mean reading score of first graders who responded incorrectly = 192.05 (SD = 15.34); $t(153) = -0.62$, $p = .53$

39. Mean math score of first graders who responded correctly on the blicket task = 191.41 (SD = 13.27); mean math score of first graders who responded incorrectly = 195.32 (SD = 10.94); $t(153) = 2.01$, $p = .05$.

40. Mean reading score of third graders who responded correctly on the blicket task = 216.89 (SD = 11.7); mean reading score of third graders who responded incorrectly = 210.58 (SD = 13.0); $t(107) = -2.66$, $p = .009$

41. Mean math score of third graders who responded correctly on the blicket task = 223.41 (SD = 10.48); mean math score of third graders who responded incorrectly = 220.10 (SD = 13.45); $t(107) = -1.44$, $p = .15$

42. Mean PSSA language score of children who responded correctly on the blicket task = 1137.22 (SD = 107.17); mean PSSA language score of children who responded incorrectly = 1088.40 (SD = 90.77); $t(105) = -2.50$, $p = .01$

43. Mean PSSA math score of children who responded correctly on the blicket task = 1106.77 (SD = 121.77); mean PSSA math score of children who responded incorrectly = 1074.81 (SD = 105.88); $t(105) = -1.42$, $p = .16$

44. Mean PSSA science score of children who responded correctly on the blicket task = 1575.37 (SD = 163.97); mean PSSA science score of children who responded incorrectly = 1506.47 (SD = 144.08); $t(104) = -2.27$, $p = .03$

Chapter 8

1. During the writing of this book, Dave's daughter was in seventh grade (age 12). Her science teacher asked her class to draw a scientist. She reported that the majority of the students in her class drew pictures consistent with the features described here.

2. While the reference to potions might have to do with the prevalence of Harry Potter, responses featuring potions were common in the Draw a Scientist task and other measures that pre-date that book series. So we can't say that Harry Potter alone caused an uptick in children's beliefs that potions are part of science.

3. Chi-squared tests, all χ^2 values > 7.70, all p-values < .006

4. Chi-squared tests, all χ^2 values < 0.70, all p-values > .40

5. $\chi^2(4) = 316.69$, $p < .001$

6. $\chi^2(4) = 122.37$, $p < .001$

7. $r_s(929) = .05$, $p = .13$

8. $r_s(929) = -.04$, $p = .19$

9. $r_s(929) = -.01$, $p = .71$

10. Exact proportions test, $p = .22$

11. Exact proportions test, $p = .27$

12. Exact proportions test, $p = .03$

13. $r_s(929) = .26$, $p < .001$

14. Exact proportions test, $p < .001$

15. $r_s(929) = .22$, $p < .001$

16. Exact proportions test, $p < .001$

17. $\chi^2(1) = 3.85$, $p = .05$

18. $\chi^2(1) = 4.60$, $p = .03$

19. $\chi^2(1) = 3.65$, $p = .06$

20. $\chi^2(1) = 5.32$, $p = .02$

21. $\chi^2(1) = 0.93$, $p = .33$

22. $B = -0.02$, $SE = 0.007$, $p = .004$

23. $B = 0.30$, $SE = 0.07$, 95% CI = [0.16, 0.43], Wald $\chi^2(1) = 17.90$, $p < .001$

24. $B = 0.49$, $SE = 0.10$, 95% CI = [0.28, 0.69], Wald $\chi^2(1) = 22.22$, $p < .001$

25. $B = 0.26$, $SE = 0.16$, 95% CI = [−0.05, 0.57], Wald $\chi^2(1) = 2.69$, $p = .11$

26. $B = 0.07$, $SE = 0.13$, 95% CI = [−0.20, 0.33], Wald $\chi^2(1) = 0.25$, $p = .62$

27. $t(1051) = -6.02$, $p < .001$

28. $r(938) = .13$, $p < .001$

29. Children: $r_s(938) = .50$, $p < .001$; Adults: $r_s(111) = .35$, $p < .001$

30. Age effect: $t(929) = -7.14$, $p < .001$; Length effect: $t(929) = -3.67$, $p < .001$

31. Binomial regression predicting receiving a Learning code from the interaction of age and response length: $B = -0.0001$, $SE = 0.0002$, $p = .60$

32. $t(92) = -1.76$, $p = .08$

33. $t(107) = -2.76$, $p = .007$

34. $t(107) = 2.74$, $p = .007$

35. Learning: $t(105) = -1.78$, $p = .08$; Other Process: $t(105) = 1.72$, $p = .09$

36. $t(105) = -2.00$, $p = .05$

37. $t(105) = -1.78$, $p = .08$

38. For these analyses we constructed 2 x 2 tables of children whose responses fit both coding categories under consideration, only one or the other category, or neither. We conducted a series of chi-squared tests on these tables to test for relations; all p-values > .06.

39. $\chi^2(1) = 46.47$, $p < .001$

40. $\chi^2(1) = 7.49$, $p = .006$

41. Learning: $\chi^2(1) = 3.97$, $p = .05$, Other Process: $\chi^2(1) = 11.02$, $p < .001$

42. $r_s(70) = .39, p = .001$

43. $r_s(69) = .37, p = .001$

44. $r_s(69) = .07, p = .08$

45. Exact proportion test, $p = .03$

46. These data were collected by Elena Schiavone as part of her undergraduate honors thesis project.

47. "Puzzle" framing: 48% correct, $SD = 51\%$; "Science" framing: 49% correct, $SD = 50\%$; $t(87) = -0.04, p = .97)$

48. Callanan et al. (2020) used a slightly different coding system than what we used here. However, the change in coding system does not change the results reported in that paper.

49. All $r_s(301)$-values $> |.15|$, all p-values $< .008$.

50. B$=-13.19$, SE$=4.10$, 95% CI $[-21.23, -5.16]$, Wald $\chi^2(1) = 10.36, p = .001$. This is still a negative relation, so generating a definition of science as a specific activity still related to *less* systematic exploration when age is factored into the analysis.

51. B$=14.22$, SE$=4.90$, 95% CI $[4.62, 23.82]$, Wald $\chi^2(1) = 8.43, p = .004$.

52. Engagement: $r_s(107) = -.20, p = .04$; Performance $r_s(107) = -.22, p = .02$. Note that these are negative correlations, which means that if children generated a definition of science that included a specific activity, they were less engaged by the challenges and less likely to complete challenges on their own.

53. Engagement: $r_s(106) = -.12, p = .21$; Performance $r_s(106) = .14, p = .14$

54. Correlations with age: Engagement: $r_s(109) = .49$, p$< .001$; Performance: $r_s(109) = .47$, $p < .001$; Generating a definition of science that was specific: $r_s(107) = -.20, p = .04$.

55. We analyzed this via a set of Generalized Estimating Equations, controlling for children's age and the type of circuit that they constructed during free play (because not every parent-child dyad built every type of circuit we considered during free play), looking at child and parent action as a count variable (Poisson distribution). Child action: B$=0.28$, SE$=0.12$, 95% CI $[0.05, 0.50]$, Wald $\chi^2(1) = 5.65, p = .02$; Parent action: B$=-1.03$, SE$=0.45$, 95% CI $[-1.91, -0.16]$, Wald $\chi^2(1) = 5.35, p = .02$

56. Chi-squared tests, all χ^2 values < 6.60, all p-values $> .16$

57. $\chi^2(4) = 6.49, p = .17$

58. Chi-squared tests, all χ^2 values > 10.58, all p-values $< .04$

59. Chi-squared tests, all χ^2 values > 7.80, all p-values $< .10$

60. $r(86) = .52, p < .001$

61. This was confirmed by a univariate ANOVA on these scores with age group as a fixed factor, which revealed a significant effect of age group: $F(3, 84) = 8.83$, $p < .001$, $\eta^2 = 0.24$

62. Post-hoc test with Tukey correction, $p = .99$

63. Post-hoc test with Tukey correction, $p = .97$

64. Post-hoc test with Tukey correction, $p = .005$

65. $t(84) = -2.20$, $p = .04$; confirmed with ordinal regression, odds ratio $= 3.20$, $p = .02$

66. $t(84) = -2.88$, $p = .005$; confirmed with ordinal regression, odds ratio $= 3.34$, $p = .007$

67. Exact proportions tests, all p-values $< .001$

68. $\chi^2(1) = 14.43$, $p < .001$

69. Exact proportion tests, all p-values $< .04$

70. Exact proportion tests, all p-values $> .07$

71. $\chi^2(1) = 2.67$, $p = .10$

72. Psychology: $\chi^2(1) = 19.67$, $p < .001$; chemistry: $\chi^2(1) = 13.65$, $p < .001$

73. Exact proportion tests, all p-values $< .015$

74. Exact proportion test, $p = .14$

75. $\chi^2(1) = 5.68$, $p = .02$

76. Biology: $\chi^2(1) = 2.85$, $p = .09$; chemistry: $\chi^2(1) = 2.85$, $p = .09$

77. Exact proportion tests, all p-values $< .04$

78. Exact proportion tests, chemistry: $p = .05$, psychology: $p = .005$, biology: $p = .19$

79. Exact proportions tests, all p-values $< .08$

80. Exact proportions tests, all p-values $< .04$

81. Exact proportions tests, all p-values $< .04$

82. Exact proportions test, $p = .28$

83. Exact proportions test, $p < .001$

84. Exact proportions test, $p < .001$

85. Exact proportions test, $p = .23$

86. Exact proportions test, $p = .23$

87. Exact proportions test, $p = .43$

88. Exact proportions test, $p = .005$

89. Chemistry appropriate method, $t(97) = -2.50$, $p = .01$; chemistry inappropriate method, $t(97) = -2.34$, $p = .02$

90. Biology appropriate method, $t(99) = -3.63$, $p < .001$; biology inappropriate method, $t(99) = -0.56$, $p = .58$

91. Psychology appropriate method, $t(99) = 0.77$, $p = .44$; psychology inappropriate method, $t(99) = -2.41$, $p = .02$

92. The semantics of the word "just" are complicated (Lee, 1987). On Lee's analysis, we seem to be following a "restrictive" account, but sometimes children used words like "only," and we suspect that the intention of many of their utterances was to be exclusive (i.e., the character is not doing science). When words like "just" or "only" were applied to cases where the child said the character was doing science, then the word might be more elaborative (Warstadt, 2019).

93. $\chi^2(2) = 3.74$, $p = .15$

94. $\chi^2(2) = 0.51$, $p = .77$

95. $\chi^2(2) = 14.77$, $p < .001$

96. Exact proportion tests, chemistry $p < .001$; biology $p < .001$; psychology $p = .008$

97. Exact proportion tests, chemistry $p = .04$; biology $p = .64$; psychology $p = .18$

Chapter 9

1. Wald $\chi^2(4) = 9.38$, $p = 05$. Specifically, we build a Generalized Estimating Equation assuming a binomial logistic response, with the response to the test question (yes or no) as the dependent measure and with Question Type (Fact vs. Skill), Intention, Outcome, and age (in months) as independent factors, testing for these main effects and this interaction.

2. This work was done collaboratively with Susan Letourneau.

3. McNemar $\chi^2(1) = 1.16$, $p = .28$

4. $r_s(68) = -.24$, $p = .04$

5. $r_s(68) = .28$, $p = .02$

6. $r_s(68) = -.14$, $p = .22$

7. Fisher Exact Test, $p = .02$

8. Factoring out children's age and the mean length of utterance of the definition (a gross measure of language capacity, which we have used throughout these studies), all $r(66)$-values $< |.16|$, all p-values $> .20$.

Chapter 10

1. Much like science, learning, teaching, play, and (as we discuss below) pretending, we have also asked children to define "creativity." This work focuses mostly on 5- to 10-year-olds' understanding that creativity involves novelty and innovation. Understanding the former emerges earlier in development (around age 8) than the latter (sometime in adolescence; see Stricker & Sobel, 2020).

2. It is conceivable that longer-looking time in studies that use the violation of expectation method also involve a certain kind of imagination, given that infants are reacting to what they actually observe based on their expectations of how the world should be, and expectations are imagined representations. We are agnostic as to whether differences in looking-time behavior in this paradigm indicate the same kind of imaginative capacities we discuss here (see e.g., Wellman & Liu, 2007).

3. Since Dave started working on children's understanding of pretense in 1996, he has revised his belief about just about everything related the topic, except for the fact that these studies are just plain strange. He loves them. But they are strange.

4. Overall, the difference among the three codes was significant, Cochran's $Q(2, N=66)=19.60$, $p<.001$. Regression models suggest that the best model that captures generating a definition with agency and age in months is defined by the equation $y=-6.13+0.08x$. For the Not Real code, that equation is $y=-10.29+0.13x$, and for the Mental State code, that equation is $y=-13.47+0.15x$.

5. In *The Lion Guard*, it's revealed (retconned, really) that the "Roar of the Guard" skips a generation. Simba does not have this ability, but it is passed on to his son, Kion.

6. And that trolls are not just limited to studies of pretend play.

7. It's worth noting that it was the use of the tractor beam that kept the *Enterprise* inside a time loop in the episode "Cause and Effect" (S5, E18), so it's possible that the crew was a bit shy of deploying it after that.

8. Bathrooms are never shown on the *Enterprise* in the original *Star Trek* series, but there is a bathroom on a shuttlecraft in *Star Trek: The Motion Picture*. There is also a reference to characters being in a bathroom on the *Enterprise-D* in the *Star Trek: The Next Generation* episode "Home Soil" (S1, E17).

9. That said, *The Love Boat*, *Charlie's Angels*, and *Fantasy Island* all seemed to exist in the same shared universe, so the ontological structure of that show might have been more complex than meets the eye.

10. The original blicket detector did not work this way. It was based on a pressure plate, which was designed by the engineers in the UC Berkeley Psychology Department. They did good work, because no one could replicate their design. That original

detector was built in 1995 and functioned until 2017 (twenty-two years!). We moved to this newer design only when we conceptualized some experiments in which we wanted the detector to activate without any objects on it. No box that we have built subsequently has yet to last more than ten years.

11. Alternatively, a change was made to the fourth picture, and children were asked to fill in a new third picture, which represents a different kind of counterfactual inference (see also Guarjado & Turley-Ames, 2004).

12. It is worth noting, however, that if looking-time studies using violation of expectation provide evidence for counterfactual reasoning abilities (see note 2), then children might have this domain-general capacity much earlier in development.

Chapter 11

1. This is, by the way, the nature of science, and it highlights the importance of reproducibility.

2. This was analyzed via a multinomial logistic regression, treating response as a nominal response, and age and location as independent variables. The overall model was significant, Likelihood Ratio $\chi^2(4) = 9.28$, $p = .05$. The difference between the "ask a grown-up" response compared with the "play with it myself" response was significant when location was contrasted, $B = 0.91$, $SE = 0.44$, Wald $\chi^2(1) = 4.16$, $p = .04$, Odds Ratio $= 2.47$, 95% CI [1.04, 5.90].

3. In a parallel sample of 51 children interviewed at the Academy (22 girls and 29 boys; mean age = 95.97 months, age range = 72–122 months), 53% of children said that they would play with the toy, 22% said that they would ask a grown-up, and 25% said that they would watch another child. This distribution is similar to what we obtained at Providence Children's Museum, but data collection for this sample was interrupted by COVID. Further investigations should consider whether children's exploration differs between the two museums and whether children's reflections on what they learn from their exploration differs. We are currently working on these projects.

4. We also gave these children the same "What is learning?" interview that was described in chapter 9. Their responses on that interview did not differ between the two settings, only their beliefs about how they would want to go about learning something new. Our next step is to reproduce this finding by working with children outside of the museum, giving them the two interviews either immediately before or immediately after their visit.

5. Incidentally, in support of this interpretation, the three museums where we tested the relation between children's definitions of "science" and their exploratory behaviors were Providence Children's Museum in Providence, RI, Children's Discovery Museum in San Jose, CA, and Thinkery in Austin, TX. Although Thinkery does not have the term

"science" in its name, it was rebranded to Thinkery from Austin Children's Museum to reflect more of a focus on STEAM content (Science, Technology, Engineering, Arts, and Math) and on a more inquiry-rich, play-based learning experience. It is, frankly, much more of a children's science museum than the other two environments. When the data from the three museums were considered separately, the relation between children's exploration and their definitions of "science" held in the Providence sample and in the sample from Children's Discovery Museum individually, but not in the sample from Thinkery.

References

Ahl, R. E., Amir, D., & Keil, F. C. (2020). The world within: Children are sensitive to internal complexity cues. *Journal of Experimental Child Psychology, 200,* 104932. https://doi.org/10.1016/j.jecp.2020.104932

Ahl, R. E., & Keil, F. C. (2017). Diverse effects, complex causes: Children use information about machines' functional diversity to infer internal complexity. *Child Development, 88*(3), 828–845. https://doi.org/10.1111/cdev.12613

Aikenhead, G. S., & Ryan, A. G. (1992). The development of a new instrument: "Views on Science-Technology-Society" (VOSTS). *Science Education, 76*(5), 477–491. https://doi.org/10.1002/sce.3730760503

Alfieri, L., Brooks, P. J., Aldrich, N. J., & Tenenbaum, H. R. (2011). Does discovery-based instruction enhance learning? *Journal of Educational Psychology, 103*(1), 1–18. https://doi.org/10.1037/a0021017

Allan, L. G. (1980). A note on measurement of contingency between two binary variables in judgment tasks. *Bulletin of the Psychonomic Society, 15*(3), 147–149. https://doi.org/10.3758/BF03334492

Alter, A. L., Oppenheimer, D. M., & Zemla, J. C. (2010). Missing the trees for the forest: A construal level account of the illusion of explanatory depth. *Journal of Personality and Social Psychology, 99*(3), 436–451. https://doi.org/10.1037/a0020218

Amsel, E., & Brock, S. (1996). The development of evidence evaluation skills. *Cognitive Development, 11*(4), 523–550. https://doi.org/10.1016/S0885-2014(96)90016-7

Angier, N. (2007). *The canon: A whirligig tour of the beautiful basics of science.* Boston: Houghton Mifflin.

Anglin, J. M. (1977). *Word, object and conceptual development.* New York: Norton.

Archer, L., Dewitt, J., Osborne, J., Dillon, J., Willis, B., & Wong, B. (2010). "Doing" science versus "being" a scientist: Examining 10/11-year-old schoolchildren's constructions of science through the lens of identity. *Science Education, 94*(4), 617–639. https://doi.org/10.1002/sce.20399

Asch, S. E. (1956). Studies of independence and conformity: I. A minority of one against a unanimous majority. *Psychological Monographs: General and Applied, 70*(9), 1–70. https://doi.org/10.1037/h0093718

Ashley, J., & Tomasello, M. (1998). Cooperative problem-solving and teaching in preschoolers. *Social Development, 7*(2), 143–163. https://doi.org/10.1111/1467-9507.00059

Astington, J. W., Harris, P. L., & Olson, D. R. (Eds.). (1988). *Developing theories of mind*. Cambridge, UK: Cambridge University Press.

Astington, J. W., & Pelletier, J. (1998). The language of mind: Its role in teaching and learning. In D. R. Olson & N. Torrance (Eds.), *The handbook of education and human development: New models of learning, teaching and schooling* (pp. 569–593). Oxford: Blackwell. https://doi.org/10.1111/b.9780631211860.1998.00027.x

Austin, J. L. (1979). Pretending. In J. O. Urmson & G. J. Warnock (Eds.), *Philosophical papers* (3rd ed., pp. 253–271). Oxford: Oxford University Press.

Baillargeon, R. (1994). How do infants learn about the physical world? *Current Directions in Psychological Science, 3*(5), 133–140. https://doi.org/10.1111/1467-8721.ep10770614

Baillargeon, R., Needham, A., & Devos, J. (1992). The development of young infants' intuitions about support. *Early Development and Parenting, 1*(2), 69–78. https://doi.org/10.1002/edp.2430010203

Bang, M., Warren, B., Rosebery, A. S., & Medin, D. (2012). Desettling expectations in science education. *Human Development, 55*(5–6), 302–318. https://doi.org/10.1159/000345322

Baron-Cohen, S., Wheelwright, S. J., Hill, J., Raste, Y., & Plumb, I. (2001). The "Reading the Mind in the Eyes" Test revised version: A study with normal adults, and adults with Asperger syndrome or high-functioning autism. *Journal of Child Psychology and Psychiatry, 42*(2), 241–251. https://doi.org/10.1111/1469-7610.00715

Bartsch, K., Horvath, K., & Estes, D. (2003). Young children's talk about learning events. *Cognitive Development, 18*(2), 177–193. https://doi.org/10.1016/S0885-2014(03)00019-4

Bartsch, K., & Wellman, H. M. (1995). *Children talk about the mind*. New York: Oxford University Press.

Barzilai, S., & Eshet-Alkalai, Y. (2015). The role of epistemic perspectives in comprehension of multiple author viewpoints. *Learning and Instruction, 36*, 86–103. https://doi.org/10.1016/j.learninstruc.2014.12.003

Barzilai, S., & Weinstock, M. (2015). Measuring epistemic thinking within and across topics: A scenario-based approach. *Contemporary Educational Psychology, 42*, 141–158. https://doi.org/10.1016/j.cedpsych.2015.06.006

Beck, S. R., Riggs, K. J., & Burns, P. (2011). Multiple developments in counterfactual thinking. In C. Hoerl, T. McCormack & S. R. Beck (Eds.), *Understanding counterfactuals, understanding causation: Issues in philosophy and psychology* (pp. 110–122). Oxford: Oxford University Press. https://doi.org/10.1093/acprof:oso/9780199590698.003.0006

Beck, S. R., Riggs, K. J., & Gorniak, S. L. (2009). Relating developments in children's counterfactual thinking and executive functions. *Thinking & Reasoning, 15*(4), 337–354. https://doi.org/10.1080/13546780903135904

Beck, S. R., Robinson, A. N., Ahmed, S., & Abid, R. (2011). Children's understanding that ambiguous figures have multiple interpretations. *European Journal of Developmental Psychology, 8*(4), 403–422. https://doi.org/10.1080/17405629.2010.515885

Beck, S. R., Robinson, E. J., Carroll, D. J., & Apperly, I. A. (2006). Children's thinking about counterfactuals and hypotheticals as possibilities. *Child Development, 77*(2), 413–423.

Beisser, S. R., Gillespie, C. W., & Thacker, V. M. (2013). An investigation of play: From the voices of fifth- and sixth-grade talented and gifted students. *Gifted Child Quarterly, 57*(1), 25–38. https://doi.org/10.1177/0016986212450070

Bemis, R. H., Leichtman, M. D., & Pillemer, D. B. (2011). "I remember when I learned that!" Developmental and gender differences in children's memories of learning episodes. *Infant and Child Development, 20*(4), 387–399. https://doi.org/10.1002/icd.700

Bemis, R. H., Leichtman, M. D., & Pillemer, D. B. (2013). I remember when you taught me that! Preschool children's memories of realistic learning episodes. *Infant and Child Development, 22*(6), 603–621. https://doi.org/10.1002/icd.1807

Benson, J. E., Sabbagh, M. A., Carlson, S. M., & Zelazo, P. D. (2013). Individual differences in executive functioning predict preschoolers' improvement from theory-of-mind training. *Developmental Psychology, 49*(9), 1615–1627. https://doi.org/10.1037/a0031056

Benton, D. T., Rakison, D. H., & Sobel, D. M. (2021). When correlation equals causation: A behavioral and computational account of second-order correlation learning in children. *Journal of Experimental Child Psychology, 202*, 105008. https://doi.org/10.1016/j.jecp.2020.105008

Bergstrom, B., Moehlmann, B., & Boyer, P. (2006). Extending the testimony problem: Evaluating the truth, scope, and source of cultural information. *Child Development.* https://doi.org/10.1111/j.1467-8624.2006.00888.x

Bian, L., Leslie, S.-J., & Cimpian, A. (2017). Gender stereotypes about intellectual ability emerge early and influence children's interests. *Science, 355*(6323), 389–391. https://doi.org/10.1126/science.aah6524

Birch, S. A. J., Severson, R. L., & Baimel, A. (2020). Children's understanding of when a person's confidence and hesitancy is a cue to their credibility. *PLOS One, 15*(1), e0227026. https://doi.org/10.1371/journal.pone.0227026

Blaisdell, A. P. (2008). Cognitive dimension of operant learning. In R. Menzel (Ed.), *Learning and memory: A comprehensive reference, Volume 1: Learning theory and behavior* (pp. 173–196). Oxford: Elsevier.

Blaisdell, A. P., Sawa, K., Leising, K. J., & Waldmann, M. R. (2006). Causal reasoning in rats. *Science, 311*(5763), 1020–1022. https://doi.org/10.1126/science.1121872

Blewitt, P., Rump, K. M., Shealy, S. E., & Cook, S. A. (2009). Shared book reading: When and how questions affect young children's word learning. *Journal of Educational Psychology, 101*(2), 294–304.

Bonawitz, E. B., Denison, S., Gopnik, A., & Griffiths, T. L. (2014). Win-stay, lose-sample: A simple sequential algorithm for approximating Bayesian inference. *Cognitive Psychology, 74*, 35–65. https://doi.org/10.1016/j.cogpsych.2014.06.003

Bonawitz, E. B., Fischer, A., & Schulz, L. E. (2012). Teaching 3.5-year-olds to revise their beliefs given ambiguous evidence. *Journal of Cognition and Development, 13*(2), 266–280. https://doi.org/10.1080/15248372.2011.577701

Bonawitz, E. B., & Lombrozo, T. (2012). Occam's rattle: Children's use of simplicity and probability to constrain inference. *Developmental Psychology, 48*(4), 1156–1164. https://doi.org/10.1037/a0026471

Bonawitz, E. B., Shafto, P., Gweon, H., Goodman, N. D., Spelke, E. S., & Schulz, L. E. (2011). The double-edged sword of pedagogy: Instruction limits spontaneous exploration and discovery. *Cognition, 120*(3), 322–330. https://doi.org/10.1016/j.cognition.2010.10.001

Bourdeau, V. D., & Arnold, M. E. (2009). *The science process skills inventory.* Corvallis: 4-H Youth Development Education, Oregon State University.

Bronfenbrenner, U. (1979). *The ecology of human development: Experiments by nature and design.* Cambridge, MA: Harvard University Press.

Brosseau-Liard, P., Penney, D., & Poulin-Dubois, D. (2015). Theory of mind selectively predicts preschoolers' knowledge-based selective word learning. *British Journal of Developmental Psychology, 33*(4), 464–475. https://doi.org/10.1111/bjdp.12107

Brown, A. L., & Palincsar, A. S. (1989). Guided cooperative learning and individual knowledge acquisition. In L. B. Resnick (Ed.), *Knowing, learning, and instruction: Essays in honor of Robert Glaser* (pp. 393–451). Hillsdale, NJ: Erlbaum.

Buchanan, D. W., & Sobel, D. M. (2011). Mechanism-based causal reasoning in young children. *Child Development, 82*(6), 2053–2066. https://doi.org/10.1111/j.1467-8624.2011.01646.x

Buchanan, D. W., & Sobel, D. M. (2014). Edge replacement and minimality as models of causal inference in children. In J. B. Benson (Ed.), *Advances in child development and behavior* (Vol. 46, pp. 183–213). https://doi.org/10.1016/B978-0-12-800285-8.00007-8

Buchanan, D. W., Tenenbaum, J. B., & Sobel, D. M. (2010). Edge replacement and nonindependence in causation. In S. Ohlsson & R. Catrambone (Eds.), *Proceedings of the 32nd Annual Meeting of the Cognitive Science Society* (pp. 919–924). Austin, TX: Cognitive Science Society.

Buchsbaum, D., Bridgers, S., Weisberg, D. S., & Gopnik, A. (2012). The power of possibility: Causal learning, counterfactual reasoning, and pretend play. *Philosophical Transactions of the Royal Society B: Biological Sciences, 367*(1599), 2202–2212. https://doi.org/10.1098/rstb.2012.0122

Bullock, M., Gelman, R., & Baillargeon, R. (1982). The development of causal reasoning. In W. J. Friedman (Ed.), *The developmental psychology of time* (Vol. 3, pp. 209–254). New York: Academic Press. https://doi.org/10.1002/wcs.1160

Bullock, M., Sodian, B., & Koerber, S. (2009). Doing experiments and understanding science: Development of scientific reasoning from childhood to adulthood. In W. Schneider & M. Bullock (Eds.), *Human development from early childhood to early adulthood: Findings from a 20 year longitudinal study* (pp. 173–197). Mahwah, NJ: Erlbaum.

Bullock, M., & Ziegler, A. (1999). Scientific reasoning: Developmental and individual differences. In F. E. Weinert & W. Schneider (Eds.), *Individual development from 3 to 12: Findings from the Munich longitudinal study* (pp. 38–54). New York: Cambridge University Press.

Burden, R. (1998). Assessing children's perceptions of themselves as learners and problem-solvers: The construction of the Myself-as-Learner Scale (MALS). *School Psychology International, 19*(4), 291–305. https://doi.org/10.1177/0143034398194002

Butler, L. P., & Markman, E. M. (2012). Preschoolers use intentional and pedagogical cues to guide inductive inferences and exploration. *Child Development, 83*(4), 1416–1428. https://doi.org/10.1111/j.1467-8624.2012.01775.x

Butler, L. P., & Markman, E. M. (2014). Preschoolers use pedagogical cues to guide radical reorganization of category knowledge. *Cognition, 130*(1), 116–127. https://doi.org/10.1016/j.cognition.2013.10.002

Byrne, R. M. J. (2005). *The rational imagination: How people create alternatives to reality.* Cambridge, MA: MIT Press.

Callanan, M. A. (2012). Conducting cognitive developmental research in museums: Theoretical issues and practical considerations. *Journal of Cognition and Development, 13*(2), 137–151. https://doi.org/10.1080/15248372.2012.666730

Callanan, M. A., Castañeda, C. L., Luce, M. R., & Martin, J. L. (2017). Family science talk in museums: Predicting children's engagement from variations in talk and activity. *Child Development, 88*(5), 1492–1504. https://doi.org/10.1111/cdev.12886

Callanan, M. A., & Jipson, J. L. (2001). Children's developing scientific literacy. In K. Crowley, C. D. Schunn & T. Okada (Eds.), *Designing for science: Implications from everyday, classroom, and professional settings* (pp. 19–43). Mahwah, NJ: Lawrence Erlbaum Associates.

Callanan, M. A., Legare, C. H., Sobel, D. M., Jaeger, G. J., Letourneau, S. M., McHugh, S. R., ... Watson, J. (2020). Exploration, explanation, and parent-child interaction in museums. *Monographs of the Society for Research in Child Development, 85*(1), 7–137. https://doi.org/10.1111/mono.12412

Callanan, M. A., & Oakes, L. M. (1992). Preschoolers' questions and parents' explanations: Causal thinking in everyday activity. *Cognitive Development, 7*(2), 213–233. https://doi.org/10.1016/0885-2014(92)90012-G

Callanan, M. A., Shrager, J., & Moore, J. L. (1995). Parent-child collaborative explanations: Methods of identification and analysis. *Journal of the Learning Sciences, 4*(1), 105–129. https://doi.org/10.1207/s15327809jls0401_3

Callanan, M., Solis, G., Castañeda, C., & Jipson, J. (2020). Children's question-asking across cultural communities. In L. P. Butler, S. Ronfard & K. H. Corriveau (Eds.), *The questioning child: Insights from psychology and education* (pp. 73–88). Cambridge, UK: Cambridge University Press. https://doi.org/10.1017/9781108553803.005

Callanan, M. A., & Valle, A. (2008). Co-constructing conceptual domains through family conversations and activities. In B. Ross (Ed.), *Psychology of learning and motivation* (Vol. 49, pp. 147–165). Elsevier.

Campbell, J. (2007). An interventionist approach to causation in psychology. In A. Gopnik & L. E. Schulz (Eds.), *Causal learning: Psychology, philosophy, and computation* (pp. 58–66). Oxford: Oxford University Press. https://doi.org/10.1093/acprof:oso/9780195176803.003.0005

Caplan, L. J., & Barr, R. A. (1989). On the relationship between category intensions and extensions in children. *Journal of Experimental Child Psychology, 47*(3), 413–429. https://doi.org/10.1016/0022-0965(89)90022-2

Carey, S. (1985). *Conceptual change in childhood*. Cambridge, MA: MIT Press.

Carey, S. (2009). *The origin of concepts*. New York: Oxford University Press.

Carpendale, J. I., & Chandler, M. J. (1996). On the distinction between false belief understanding and subscribing to an interpretive theory of mind. *Child Development, 67*(4), 1686–1706. https://doi.org/10.1111/j.1467-8624.1996.tb01821.x

Chambers, D. W. (1983). Stereotypic images of the scientist: The Draw-a-Scientist test. *Science Education, 67*(2), 255–265. https://doi.org/10.1002/sce.3730670213

Chavajay, P. (2006). How Mayan mothers with different amounts of schooling organize a problem-solving discussion with children. *International Journal of Behavioral Development, 30*(4), 371–382. https://doi.org/10.1177/0165025406066744

Chen, Z., & Klahr, D. (1999). All other things being equal: Acquisition and transfer of the control of variables strategy. *Child Development, 70*(5), 1098–1120. https://doi.org/10.1111/1467-8624.00081

Cheng, P. W. (1997). From covariation to causation: A causal power theory. *Psychological Review, 104*(2), 367–405. https://doi.org/10.1037/0033-295X.104.2.367

Cheng, P. W., & Holyoak, K. J. (1985). Pragmatic reasoning schemas. *Cognitive Psychology, 17*(4), 391–416. https://doi.org/10.1016/0010-0285(85)90014-3

Cheng, P. W., & Novick, L. R. (1992). Covariation in natural causal induction. *Psychological Review, 99*(2), 365–382. https://doi.org/10.1037/0033-295X.99.2.365

Chi, M. T. H., De Leeuw, N., Chiu, M.-H., & Lavancher, C. (1994). Eliciting self-explanations improves understanding. *Cognitive Science, 18*(3), 439–477. https://doi.org/10.1207/s15516709cog1803_3

Chinn, C. A., & Brewer, W. F. (1993). The role of anomalous data in knowledge acquisition: A theoretical framework and implications for science instruction. *Review of Educational Research, 63*(1), 1–49. https://doi.org/10.3102/00346543063001001

Chlebuch, N., Bodas, A., & Weisberg, D. S. (2022). What does the Cat in the Hat know about that? An analysis of the educational and unrealistic content of children's narrative science media. *Psychology of Popular Media*, manuscript in press.

Chouinard, M. M. (2007). Children's questions: A mechanism for cognitive development. *Monographs of the Society for Research in Child Development, 72*(1), 1–112. https://doi.org/10.1111/j.1540-5834.2007.00412.x

Clément, F., Koenig, M. A., & Harris, P. L. (2004). The ontogenesis of trust. *Mind & Language, 19*(4), 360–379. https://doi.org/10.1111/j.0268-1064.2004.00263.x

Coates, J. (1983). *The semantics of the modal auxiliaries*. London: Croom Helm.

Cobb, P. (1994). Where is the mind? Constructivist and sociocultural perspectives on mathematical development. *Educational Researcher, 23*(7), 13–20. https://doi.org/10.3102/0013189X023007013

Coenen, A., Rehder, B., & Gureckis, T. M. (2015). Strategies to intervene on causal systems are adaptively selected. *Cognitive Psychology, 79*, 102–133. https://doi.org/10.1016/j.cogpsych.2015.02.004

Cohen, L. B., Rundell, L. J., Spellman, B. A., & Cashon, C. H. (1999). Infants' perception of causal chains. *Psychological Science, 10*(5), 412–418. https://doi.org/10.1111/1467-9280.00178

Conner, D. B., & Cross, D. R. (2003). Longitudinal analysis of the presence, efficacy and stability of maternal scaffolding during informal problem-solving interactions. *British Journal of Developmental Psychology, 21*(3), 315–334. https://doi.org/10.1348/026151003322277720

Conrad, M., Kim, E., Blacker, K.-A., Walden, Z., & LoBue, V. (2020). Using storybooks to teach children about illness transmission and promote adaptive health behavior—A pilot study. *Frontiers in Psychology, 11.* https://doi.org/10.3389/fpsyg.2020.00942

Cook, C., Goodman, N. D., & Schulz, L. E. (2011). Where science starts: Spontaneous experiments in preschoolers' exploratory play. *Cognition, 120*(3), 341–349. https://doi.org/10.1016/j.cognition.2011.03.003

Coplan, R. J., & Arbeau, K. A. (2009). Peer interactions and play in early childhood. In K. H. Rubin, W. M. Bukowski & B. Laursen (Eds.), *Handbook of peer interactions, relationships, and groups* (pp. 143–161). New York: Guilford Press.

Correa-Chávez, M., & Rogoff, B. (2009). Children's attention to interactions directed to others: Guatemalan Mayan and European American patterns. *Developmental Psychology, 45*(3), 630–641. https://doi.org/10.1037/a0014144

Corriveau, K. H., & Harris, P. L. (2009). Preschoolers continue to trust a more accurate informant 1 week after exposure to accuracy information. *Developmental Science, 12*(1), 188–193. https://doi.org/10.1111/j.1467-7687.2008.00763.x

Corriveau, K. H., Kim, A. L., Schwalen, C. E., & Harris, P. L. (2009). Abraham Lincoln and Harry Potter: Children's differentiation between historical and fantasy characters. *Cognition, 113*(2), 213–225. https://doi.org/10.1016/j.cognition.2009.08.007

Corriveau, K. H., & Kurkul, K. E. (2014). "Why does rain fall?": Children prefer to learn from an informant who uses noncircular explanations. *Child Development*, n/a-n/a. https://doi.org/10.1111/cdev.12240

Cramer, R. E., Weiss, R. F., William, R., Reid, S., Nieri, L., & Manning-Ryan, B. (2002). Human agency and associative learning: Pavlovian principles govern social process in causal relationship detection. *Quarterly Journal of Experimental Psychology Section B, 55*(3b), 241–266. https://doi.org/10.1080/02724990143000289

Croker, S., & Buchanan, H. (2011). Scientific reasoning in a real-world context: The effect of prior belief and outcome on children's hypothesis-testing strategies. *British Journal of Developmental Psychology, 29*(3), 409–424. https://doi.org/10.1348/0261510 10X496906

Csibra, G., & Gergely, G. (2009). Natural pedagogy. *Trends in Cognitive Sciences, 13*(4), 148–153. https://doi.org/10.1016/j.tics.2009.01.005

Cuevas, K., Rovee-Collier, C., & Learmonth, A. E. (2006). Infants form associations between memory representations of stimuli that are absent. *Psychological Science, 17*(6), 543–549. https://doi.org/10.1111/j.1467-9280.2006.01741.x

Cushman, F. (2008). Crime and punishment: Distinguishing the roles of causal and intentional analyses in moral judgment. *Cognition, 108*, 353–380. https://doi.org/10.1016/j.cognition.2008.03.006

Daehler, M. W., & Chen, Z. (1993). Protagonist, theme, and goal object: Effects of surface features on analogical transfer. *Cognitive Development, 8,* 211–229. https://doi .org/10.1016/0885-2014(93)90015-W

Danovitch, J. H., Mills, C. M., Duncan, R. G., Williams, A. J., & Girouard, L. N. (2021). Developmental changes in children's recognition of the relevance of evidence to causal explanations. *Cognitive Development, 58,* 101017. https://doi.org/10.1016/j.cogdev.2021 .101017

Davidson, M. C., Amso, D., Anderson, L. C., & Diamond, A. (2006). Development of cognitive control and executive functions from 4 to 13 years: Evidence from manipulations of memory, inhibition, and task switching. *Neuropsychologia, 44*(11), 2037–2078. https://doi.org/10.1016/j.neuropsychologia.2006.02.006

Davis, H. E., Stieglitz, J., Kaplan, H., & Gurven, M. (2021). *School quality augments differences in children's abstract reasoning, driving educational inequalities: Evidence from a naturally occurring quasi-experiment in Amazonia, Bolivia.* Unpublished manuscript on PsyArXiv. https://doi.org/10.31234/osf.io/d3sgq

Davis-Unger, A. C., & Carlson, S. M. (2008). Development of teaching skills and relations to theory of mind in preschoolers. *Journal of Cognition and Development, 9*(1), 26–45. https://doi.org/10.1080/15248370701836584

Denison, S., Reed, C., & Xu, F. (2013). The emergence of probabilistic reasoning in very young infants: Evidence from 4.5- and 6-month-olds. *Developmental Psychology, 49*(2), 243–249. https://doi.org/10.1037/a0028278

Denison, S., & Xu, F. (2010). Integrating physical constraints in statistical inference by 11-month-old infants. *Cognitive Science, 34*(5), 885–908. https://doi.org/10.1111/j .1551-6709.2010.01111.x

Diamond, A. (1991a). Frontal lobe involvement in cognitive changes during the first year of life. In K. R. Gibson & A. C. Petersen (Eds.), *Brain maturation and cognitive development: Comparative and cross-cultural perspectives* (pp. 127–180). New York: Routledge.

Diamond, A. (1991b). Neuropsychological insights into the meaning of object concept development. In S. Carey & R. Gelman (Eds.), *The Jean Piaget Symposium series. The epigenesis of mind: Essays on biology and cognition* (pp. 67–110). Mahwah, NJ: Lawrence Erlbaum.

Diamond, A., & Gilbert, J. (1989). Development as progressive inhibitory control of action: Retrieval of a contiguous object. *Cognitive Development, 4*(3), 223–249. https:// doi.org/10.1016/0885-2014(89)90007-5

Dias, M. G., & Harris, P. L. (1988). The effect of make believe play on deductive reasoning. *British Journal of Developmental Psychology, 6,* 207–221.

Dias, M. G., & Harris, P. L. (1990). The influence of the imagination on reasoning by young children. *British Journal of Developmental Psychology, 8*(4), 305–318. https://doi .org/10.1111/j.2044-835X.1990.tb00847.x

Dickinson, A. (2001). Causal learning: Association versus computation. *Current Directions in Psychological Science, 10*(4), 127–132. https://doi.org/10.1111/1467-8721.00132

Dickinson, A., & Shanks, D. R. (1995). Instrumental action and causal representation. In *Symposia of the Fyssen Foundation. Causal cognition: A multidisciplinary debate* (pp. 5–25). New York: Oxford University Press.

DiSessa, A. A. (1993). Toward an epistemology of physics. *Cognition and Instruction, 10*(2–3), 105–225. https://doi.org/10.1080/07370008.1985.9649008

DiYanni, C. J., Corriveau, K. H., Kurkul, K., Nasrini, J., & Nini, D. (2015). The role of consensus and culture in children's imitation of inefficient actions. *Journal of Experimental Child Psychology, 137*, 99–110. https://doi.org/10.1016/j.jecp.2015.04 .004

Drayton, S., Turley-Ames, K. J., & Guajardo, N. R. (2011). Counterfactual thinking and false belief: The role of executive function. *Journal of Experimental Child Psychology, 108*(3), 532–548. https://doi.org/10.1016/j.jecp.2010.09.007

Dunbar, K. N. (1995). How scientists really reason: Scientific reasoning in real-world laboratories. In R. J. Sternberg & J. E. Davidson (Eds.), *The nature of insight* (pp. 365–395). Cambridge, MA: MIT Press.

Dunbar, K. N. (2000). How scientists think in the real world: Implications for science education. *Journal of Applied Developmental Psychology, 21*(1), 49–58. https://doi.org /10.1016/S0193-3973(99)00050-7

Dunbar, K. N. (2002). Understanding the role of cognition in science: The Science as Category framework. In P. Carruthers, S. Stich & M. Siegal (Eds.), *The cognitive basis of science* (pp. 154–170). New York: Cambridge University Press.

Dunbar, K. N., & Fugelsang, J. A. (2005). Scientific thinking and reasoning. In K. J. Holyoak & R. G. Morrison (Eds.), *The Cambridge handbook of thinking and reasoning* (pp. 702–725). Cambridge, UK: Cambridge University Press.

Dündar-Coecke, S., & Tolmie, A. (2020). Nonverbal ability and scientific vocabulary predict children's causal reasoning in science better than generic language. *Mind, Brain, and Education, 14*(2), 130–138. https://doi.org/10.1111/mbe.12226

Dweck, C. S. (2006). *Mindset: How we can learn to fulfill our potential.* New York: Ballentine Books.

Dweck, C. S., & Leggett, E. L. (1988). A social-cognitive approach to motivation and personality. *Psychological Review, 95*(2), 256–273. https://doi.org/10.1037/0033-295X .95.2.256

Dyer, H. T. (2018, May 2). I watched an entire Flat Earth Convention for my research—here's what I learnt. *The Conversation*. https://theconversation.com/i-watched-an-entire-flat-earth-convention-for-my-research-heres-what-i-learnt-95887

Eccles, J. S., Wigfield, A., & Schiefele, U. (1998). Motivation to succeed. In W. Damon & N. Eisenberg (Eds.), *Handbook of child psychology: Social, emotional, and personality development* (pp. 1017–1095). Hoboken, NJ: Wiley.

Edson, M. T. (2013). *Starting with science: Strategies for introducing young children to inquiry*. Portland, ME: Stenhouse Publishers.

Einav, S., & Robinson, E. J. (2011). When being right is not enough: Four-year-olds distinguish knowledgeable informants from merely accurate informants. *Psychological Science, 22*(10), 1250–1253. https://doi.org/10.1177/0956797611416998

Eisbach, A. O. (2004). Children's developing awareness of diversity in people's trains of thought. *Child Development, 75*(6), 1694–1707. https://doi.org/10.1111/j.1467-8624.2004.00810.x

Elkind, D. (1976). *Child development and education: A Piagetian perspective*. New York: Oxford University Press.

Emmons, N. A., Lees, K., & Kelemen, D. (2017). Young children's near and far transfer of the basic theory of natural selection: An analogical storybook intervention. *Journal of Research in Science Teaching*. https://doi.org/10.1002/tea.21421

Engle, J., & Walker, C. M. (2021). Thinking counterfactually supports children's evidence evaluation in causal learning. *Child Development, 92(4), 1636–1651*. https://doi.org/10.1111/cdev.13518

Erb, C. D., Buchanan, D. W., & Sobel, D. M. (2013). Children's developing understanding of the relation between variable causal efficacy and mechanistic complexity. *Cognition, 129*(3), 494–500. https://doi.org/10.1016/j.cognition.2013.08.002

Erb, C. D., & Sobel, D. M. (2014). The development of diagnostic reasoning about uncertain events between ages 4–7. *PLOS One, 9*(3), e92285. https://doi.org/10.1371/journal.pone.0092285

Esbensen, B. M., Taylor, M., & Stoess, C. (1997). Children's behavioral understanding of knowledge acquisition. *Cognitive Development, 12*(1), 53–84. https://doi.org/10.1016/S0885-2014(97)90030-7

Evans, E. M. (2000). The emergence of beliefs about the origins of species in school-age children. *Merrill-Palmer Quarterly, 46*(2), 221–254.

Evans, E. M., Weiss, M., Lane, J. D., & Palmquist, S. (2016). The spiral model: Integrating research and exhibit development to foster conceptual change. In D. M. Sobel & J. L. Jipson (Eds.), *Cognitive development in museum settings: Relating research and practice* (pp. 36–64). New York: Routledge.

Evans, J. St. B. T., Barston, J. L., & Pollard, P. (1983). On the conflict between logic and belief in syllogistic reasoning. *Memory & Cognition, 11*(3), 295–306. https://doi.org/10.3758/BF03196976

Faucher, L., Mallon, R., Nazer, D., Nichols, S., Ruby, A., Stich, S., & Weinberg, J. (2002). The baby in the lab-coat: Why child development is not an adequate model for understanding the development of science. In P. Carruthers, S. Stich & M. Siegrist (Eds.), *The cognitive basis of science* (pp. 335–362). New York: Cambridge University Press.

Fender, J. G., & Crowley, K. (2007). How parent explanation changes what children learn from everyday scientific thinking. *Journal of Applied Developmental Psychology, 28*(3), 189–210. https://doi.org/10.1016/j.appdev.2007.02.007

Fernandez-Duque, D., Evans, J., Christian, C., & Hodges, S. D. (2015). Superfluous neuroscience information makes explanations of psychological phenomena more appealing. *Journal of Cognitive Neuroscience, 27*(5), 926–944. https://doi.org/10.1162/jocn_a_00750

Fernbach, P. M., Macris, D. M., & Sobel, D. M. (2012). Which one made it go? The emergence of diagnostic reasoning in preschoolers. *Cognitive Development, 27*(1), 39–53. https://doi.org/10.1016/j.cogdev.2011.10.002

Fischer, F., Kollar, I., Ufer, S., Sodian, B., Hussmann, H., Pekrun, R., . . . Eberle, J. (2014). Scientific reasoning and argumentation: Advancing an interdisciplinary research agenda in education. *Frontline Learning Research, 5*, 28–45. https://doi.org/10.14786/flr.v.2i2.96

Fiser, J., & Aslin, R. N. (2002). Statistical learning of new visual feature combinations by infants. *Proceedings of the National Academy of Sciences, 99*(24), 15822–15826. https://doi.org/10.1073/pnas.232472899

Fitneva, S. A., Lam, N. H. L., & Dunfield, K. A. (2013). The development of children's information gathering: To look or to ask? *Developmental Psychology, 49*(3), 533–542. https://doi.org/10.1037/a0031326

Flavell, J. H., Green, F. L., & Flavell, E. R. (1993). Children's understanding of the stream of consciousness. *Child Development, 64*, 387–398.

Flavell, J. H., Green, F. L., & Flavell, E. R. (1995). Young children's knowledge about thinking. *Monographs of the Society for Research in Child Development, 60*(1).

Flavell, J. H., Mumme, D. L., Green, F. L., & Flavell, E. R. (1992). Young children's understanding of different types of beliefs. *Child Development, 63*(4), 960–977. https://doi.org/10.1111/j.1467-8624.1992.tb01675.x

Flynn, E. (2010). Underpinning collaborative learning. In B. Sokol, U. Muller, J. Carpendale, A. Young & G. Iarocci (Eds.), *Self and social regulation: Social interaction and the development of social understanding and executive function* (pp. 312–336). New York: Oxford University Press. https://doi.org/10.1093/acprof:oso/9780195327694.001.0001

Fodor, J. A. (1983). *The modularity of mind.* Cambridge, MA: MIT Press.

Foster-Hanson, E., Cimpian, A., Leshin, R. A., & Rhodes, M. (2020). Asking children to "be helpers" can backfire after setbacks. *Child Development, 91*(1), 236–248. https://doi .org/10.1111/cdev.13147

Frazier, B. N., Gelman, S. A., & Wellman, H. M. (2009). Preschoolers' search for explanatory information within adult-child conversation. *Child Development, 80*(6), 1592–1611.

Frye, D., & Ziv, M. (2005). Teaching and learning as intentional activities. In B. D. Homer & C. S. Tamis-LeMonda (Eds.), *The development of social cognition and communication* (pp. 231–258). Mahwah, NJ: Lawrence Erlbaum.

Garcia, J., Kimeldorf, D. J., & Koelling, R. A. (1955). Conditioned aversion to saccharin resulting from exposure to gamma radiation. *Science, 122*(3160), 157–158. https://doi.org/10.1126/science.122.3160.157

Gaskins, S. (2008). Designing exhibitions to support families' cultural understandings. *Exhibitionist, 27*, 10–19.

Gauvain, M., Munroe, R. L., & Beebe, H. (2013). Children's questions in cross-cultural perspective. *Journal of Cross-Cultural Psychology, 44*(7), 1148–1165. https://doi.org/10 .1177/0022022113485430

Gauvain, M., & Perez, S. (2015). Cognitive development and culture. In *Handbook of child psychology and developmental science* (pp. 1–43). Hoboken, NJ: John Wiley & Sons. https://doi.org/10.1002/9781118963418.childpsy220

Gauvain, M., & Rogoff, B. (1989). Collaborative problem solving and children's planning skills. *Developmental Psychology, 25*(1), 139–151. https://doi.org/10.1037/0012 -1649.25.1.139

Gelman, S. A. (2003). *The essential child: Origins of essentialism in everyday thought.* New York: Oxford University Press.

Gentner, D. (1983). Structure-mapping: A theoretical framework for analogy. *Cognitive Science, 7*(2), 155–170. https://doi.org/10.1016/S0364-0213(83)80009-3

Gentner, D. (2003). *Language in mind: Advances in the study of language and thought.* Cambridge, MA: MIT Press.

Ginsburg, K. R., Committee on Communications, & Committee on Psychosocial Aspects of Child and Family Health. (2007). The importance of play in promoting healthy child development and maintaining strong parent-child bonds. *Pediatrics, 119*(1), 182–191. https://doi.org/10.1542/peds.2006-2697

Glymour, C. (2001). *The mind's arrows: Bayes nets and graphical causal models in psychology*. Cambridge, MA: MIT Press.

Glymour, C., & Cheng, P. W. (1998). Causal mechanism and probability: A normative approach. In M. Oaksford & N. Chater (Eds.), *Rational models of cognition* (pp. 295–313). Oxford: Oxford University Press. https://doi.org/10.1184/R1/6491081

Godfrey-Smith, P. (2003). *Theory and reality: An introduction to philosophy of science*. Chicago: University of Chicago Press.

Goldstein, M. H., Schwade, J. A., & Bornstein, M. H. (2009). The value of vocalizing: Five-month-old infants associate their own noncry vocalizations with responses from caregivers. *Child Development, 80*(3), 636–644. https://doi.org/10.1111/j.1467-8624.2009.01287.x

Goldstein, T. R., & Alperson, K. (2020). Dancing bears and talking toasters: A content analysis of supernatural elements in children's media. *Psychology of Popular Media, 9*(2), 214–233. https://doi.org/10.1037/ppm0000222

Good, I. J. (2000). Turing's anticipation of empirical Bayes in connection with the cryptanalysis of the naval enigma. *Journal of Statistical Computation and Simulation, 66*(2), 101–111. https://doi.org/10.1080/00949650008812016

Goodman, N. D., Ullman, T. D., & Tenenbaum, J. B. (2011). Learning a theory of causality. *Psychological Review, 118*(1), 110–119. https://doi.org/10.1037/a0021336

Gopnik, A. (1998). Explanation as orgasm. *Minds and Machines, 8*(1), 101–118. https://doi.org/10.1023/A:1008290415597

Gopnik, A., & Astington, J. W. (1988). Children's understanding of representational change and its relation to the understanding of false belief and the appearance-reality distinction. *Child Development, 59*, 26–37.

Gopnik, A., & Bonawitz, E. B. (2015). Bayesian models of child development. *Wiley Interdisciplinary Reviews: Cognitive Science, 6*(2), 75–86. https://doi.org/10.1002/wcs.1330

Gopnik, A., Glymour, C., Sobel, D. M., Schulz, L. E., Kushnir, T., & Danks, D. (2004). A theory of causal learning in children: Causal maps and Bayes nets. *Psychological Review, 111*(1), 3–32. https://doi.org/10.1037/0033-295X.111.1.3

Gopnik, A., & Graf, P. (1988). Knowing how you know: Young children's ability to identify and remember the sources of their beliefs. *Child Development, 59*(5), 1366. https://doi.org/10.2307/1130499

Gopnik, A., & Meltzoff, A. N. (1997). *Words, thoughts, and theories*. Cambridge, MA: MIT Press.

Gopnik, A., Meltzoff, A. N., & Kuhl, P. K. (1999). *The scientist in the crib: What early learning tells us about the mind*. New York: Harper Collins.

Gopnik, A., & Schulz, L. (Eds.). (2007). *Causal learning: Psychology, philosophy, and computation*. Oxford: Oxford University Press.

Gopnik, A., & Slaughter, V. (1991). Young children's understanding of changes in their mental states. *Child Development, 62*, 98–110.

Gopnik, A., & Sobel, D. M. (2000). Detecting blickets: How young children use information about novel causal powers in categorization and induction. *Child Development, 71*(5), 1205–1222. https://doi.org/10.1111/1467-8624.00224

Gopnik, A., Sobel, D. M., Schulz, L. E., & Glymour, C. (2001). Causal learning mechanisms in very young children: Two-, three-, and four-year-olds infer causal relations from patterns of variation and covariation. *Developmental Psychology, 37*(5), 620–629. https://doi.org/10.1037/0012-1649.37.5.620

Gopnik, A., & Tenenbaum, J. B. (2007). Bayesian networks, Bayesian learning and cognitive development. *Developmental Science, 10*(3), 281–287. https://doi.org/10.1111/j.1467-7687.2007.00584.x

Gopnik, A., & Walker, C. M. (2013). Considering counterfactuals: The relationship between causal learning and pretend play. *American Journal of Play, 6*(1), 15–28.

Gopnik, A., & Wellman, H. M. (1994). The theory theory. In L. A. Hirschfeld & S. A. Gelman (Eds.), *Mapping the mind: Domain specificity in cognition and culture* (pp. 257–293). Cambridge: UK: Cambridge University Press.

Gopnik, A., & Wellman, H. M. (2012). Reconstructing constructivism: Causal models, Bayesian learning mechanisms, and the theory theory. *Psychological Bulletin, 138*(6), 1085–1108. https://doi.org/10.1037/a0028044

Griffiths, T. L., Sobel, D. M., Tenenbaum, J. B., & Gopnik, A. (2011). Bayes and blickets: Effects of knowledge on causal induction in children and adults. *Cognitive Science, 35*(8), 1407–1455. https://doi.org/10.1111/j.1551-6709.2011.01203.x

Griffiths, T. L., & Tenenbaum, J. B. (2007). Two proposals for causal grammars. In A. Gopnik & L. E. Schulz (Eds.), *Causal learning: Psychology, philosophy, and computation* (pp. 323–345). Oxford: Oxford University Press.

Griffiths, T. L., & Tenenbaum, J. B. (2009). Theory-based causal induction. *Psychological Review, 116*(4), 661–716. https://doi.org/10.1037/a0017201

Griggs, R. A., & Cox, J. R. (1982). The elusive thematic-materials effect in Wason's selection task. *British Journal of Psychology, 73*(3), 407–420. https://doi.org/10.1111/j.2044-8295.1982.tb01823.x

Guajardo, N. R., & Turley-Ames, K. J. (2004). Preschoolers' generation of different types of counterfactual statements and theory of mind understanding. *Cognitive Development, 19*, 53–80.

Gureckis, T. M., & Markant, D. B. (2012). Self-directed learning: A cognitive and computational perspective. *Perspectives on Psychological Science, 7*(5), 464–481. https://doi .org/10.1177/1745691612454304

Gurven, M., Fuerstenberg, E., Trumble, B., Stieglitz, J., Beheim, B., Davis, H., & Kaplan, H. (2017). Cognitive performance across the life course of Bolivian forager-farmers with limited schooling. *Developmental Psychology, 53*(1), 160–176. https://doi .org/10.1037/dev0000175

Gutiérrez, K. D., & Rogoff, B. (2003). Cultural ways of learning: Individual traits or repertoires of practice. *Educational Researcher, 32*(5), 19–25. https://doi.org/10.3102 /0013189X032005019

Gweon, H., Pelton, H., Konopka, J. A., & Schulz, L. E. (2014). Sins of omission: Children selectively explore when teachers are under-informative. *Cognition, 132*(3), 335–341. https://doi.org/10.1016/j.cognition.2014.04.013

Gweon, H., & Schulz, L. E. (2011). 16-month-olds rationally infer causes of failed actions. *Science, 332*(6037), 1524–1524. https://doi.org/10.1126/science.1204493

Gweon, H., Tenenbaum, J. B., & Schulz, L. E. (2010). Infants consider both the sample and the sampling process in inductive generalization. *Proceedings of the National Academy of Sciences, 107*(20), 9066–9071. https://doi.org/10.1073/pnas.1003095107

Haber, A. S., Sobel, D. M., & Weisberg, D. S. (2019). Fostering children's reasoning about disagreements through an inquiry-based curriculum. *Journal of Cognition and Development, 20*(4), 592–610. https://doi.org/10.1080/15248372.2019.1639713

Hadwin, J., & Perner, J. (1991). Pleased and surprised: Children's cognitive theory of emotion. *British Journal of Developmental Psychology, 9*(2), 215–234. https://doi.org /10.1111/j.2044-835X.1991.tb00872.x

Haith, M. M., Wentworth, N., & Canfield, R. L. (1993). The formation of expectations in early infancy. *Advances in Infancy Research, 8*, 251–297.

Halloun, I. (2001). *Student views about science: A comparative survey.* Beirut: Phoenix Series / Educational Research Center, Lebanese University.

Hanus, D. (2016). Causal reasoning versus associative learning: A useful dichotomy or a strawman battle in comparative psychology? *Journal of Comparative Psychology, 130*(3), 241–248. https://doi.org/10.1037/a0040235

Harris, P. L. (2002). What do children learn from testimony? In P. Carruthers, S. Stich & M. Siegal (Eds.), *The cognitive basis of science* (pp. 316–334). New York: Cambridge University Press.

Harris, P. L. (2012). *Trusting what you're told: How children learn from others.* Cambridge, MA: Harvard University Press.

Harris, P. L., German, T. P., & Mills, P. (1996). Children's use of counterfactual thinking in causal reasoning. *Cognition, 61,* 233–259.

Harris, P. L., & Kavanaugh, R. D. (1993). Young children's understanding of pretense. *Monographs of the Society for Research in Child Development, 58*(1).

Harris, P. L., & Koenig, M. A. (2006). Trust in testimony: How children learn about science and religion. *Child Development, 77*(3), 505–524. https://doi.org/10.1111/j .1467-8624.2006.00886.x

Harris, P. L., Koenig, M. A., Corriveau, K. H., & Jaswal, V. K. (2018). Cognitive foundations of learning from testimony. *Annual Review of Psychology, 69*(1), 251–273. https://doi.org/10.1146/annurev-psych-122216-011710

Harris, P. L., Pasquini, E. S., Duke, S., Asscher, J. J., & Pons, F. (2006). Germs and angels: The role of testimony in young children's ontology. *Developmental Science, 9*(1), 76–96. https://doi.org/10.1111/j.1467-7687.2005.00465.x

Harris, P. L., Yang, B., & Cui, Y. (2017). "I don't know": Children's early talk about knowledge. *Mind & Language, 32*(3), 283–307. https://doi.org/10.1111/mila.12143

Hawkins, J., Pea, R. D., Glick, J., & Scribner, S. (1984). "Merds that laugh don't like mushrooms": Evidence for deductive reasoning by preschoolers. *Developmental Psychology, 20*(4), 584–594. https://doi.org/10.1037/0012-1649.20.4.584

Heiphetz, L., Spelke, E. S., Harris, P. L., & Banaji, M. R. (2013). The development of reasoning about beliefs: Fact, preference, and ideology. *Journal of Experimental Social Psychology, 49*(3), 559–565. https://doi.org/10.1016/j.jesp.2012.09.005

Heiphetz, L., Spelke, E. S., Harris, P. L., & Banaji, M. R. (2014). What do different beliefs tell us? An examination of factual, opinion-based, and religious beliefs. *Cognitive Development, 30*(1), 15–29. https://doi.org/10.1016/j.cogdev.2013.12.002

Henrich, J., Heine, S. J., & Norenzayan, A. (2010). The weirdest people in the world? *Behavioral and Brain Sciences, 33*(2–3), 61–83. https://doi.org/10.1017/S01405 25X0999152X

Hermes, J., Behne, T., & Rakoczy, H. (2018). The development of selective trust: Prospects for a dual-process account. *Child Development Perspectives, 12*(2), 134–138. https://doi.org/10.1111/cdep.12274

Heyes, C. (2012). Simple minds: A qualified defence of associative learning. *Philosophical Transactions of the Royal Society B: Biological Sciences, 367*(1603), 2695–2703. https://doi.org/10.1098/rstb.2012.0217

Heyneman, S. P., & Loxley, W. A. (1983). The effect of primary-school quality on academic achievement across twenty-nine high- and low-income countries. *American Journal of Sociology, 88*(6), 1162–1194. https://doi.org/10.1086/227799

Hickling, A. K., & Wellman, H. M. (2001). The emergence of children's causal explanations and theories: Evidence from everyday conversation. *Developmental Psychology, 37*(5), 668–683. https://doi.org/10.1037/0012-1649.37.5.668

Hirsh-Pasek, K., Golinkoff, R. M., Berk, L. E., & Singer, D. G. (2009). *A mandate for playful learning in preschool: Applying the scientific evidence.* New York: Oxford University Press.

Hogrefe, G.-J., Wimmer, H., & Perner, J. (1986). Ignorance versus false belief: A developmental lag in attribution of epistemic states. *Child Development, 57*(3), 567. https://doi.org/10.2307/1130337

Holyoak, K. J., Junn, E. N., & Billman, D. O. (1984). Development of analogical problem-solving skill. *Child Development, 55*(6), 2042–2055. https://doi.org/10.2307/1129778

Hood, L., & Bloom, L. (1979). What, when, and how about why: A longitudinal study of early expressions of causality. *Monographs of the Society for Research in Child Development, 44*(6). https://doi.org/10.2307/1165989

Hopkins, E. J., & Weisberg, D. S. (2017). The youngest readers' dilemma: A review of children's learning from fictional sources. *Developmental Review, 43*, 48–70. https://doi.org/10.1016/j.dr.2016.11.001

Hopkins, E. J., & Weisberg, D. S. (2021). Investigating the effectiveness of fantasy stories for teaching scientific principles. *Journal of Experimental Child Psychology, 203*, 105047. https://doi.org/10.1016/j.jecp.2020.105047

Hopkins, E. J., Weisberg, D. S., & Taylor, J. C. V. (2016). The seductive allure is a reductive allure: People prefer scientific explanations that contain logically irrelevant reductive information. *Cognition, 155*, 67–76. https://doi.org/10.1016/j.cognition.2016.06.011

Hopkins, E. J., Weisberg, D. S., & Taylor, J. C. V. (2019). Does expertise moderate the seductive allure of reductive explanations? *Acta Psychologica, 198*(July), 102890. https://doi.org/10.1016/j.actpsy.2019.102890

Horner, J. R., & Goodwin, M. B. (2006). Major cranial changes during Triceratops ontogeny. *Proceedings of the Royal Society B: Biological Sciences, 273*(1602), 2757–2761. https://doi.org/10.1098/rspb.2006.3643

Howard, J., Jenvey, V., & Hill, C. (2006). Children's categorisation of play and learning based on social context. *Early Child Development and Care, 176*(3–4), 379–393. https://doi.org/10.1080/03004430500063804

Huang, J., Gates, A. J., Sinatra, R., & Barabási, A.-L. (2020). Historical comparison of gender inequality in scientific careers across countries and disciplines. *Proceedings of the National Academy of Sciences, 117*(9), 4609–4616. https://doi.org/10.1073/pnas.1914221117

Hudicourt-Barnes, J. (2003). The use of argumentation in Haitian Creole science classrooms. *Harvard Educational Review, 73*(1), 73–93. https://doi.org/10.17763/haer.73.1.hnq801u574001877

Inagaki, K., & Hatano, G. (2006). Young children's conception of the biological world. *Current Directions in Psychological Science, 15*(4), 177–181.

Inhelder, B., & Piaget, J. (1958). *The growth of logical thinking from childhood to adolescence.* London: Basic Books.

Jahoda, G. (2000). Piaget and Lévy-Bruhl. *History of Psychology, 3*(3), 218–238. https://doi.org/10.1037/1093-4510.3.3.218

James, K. H., & Swain, S. N. (2011). Only self-generated actions create sensori-motor systems in the developing brain. *Developmental Science, 14*(4), 673–678. https://doi.org/10.1111/j.1467-7687.2010.01011.x

Jant, E. A., Haden, C. A., Uttal, D. H., & Babcock, E. (2014). Conversation and object manipulation influence children's learning in a museum. *Child Development, 85*(5), 2029–2045. https://doi.org/10.1111/cdev.12252

Jaswal, V. K. (2010). Believing what you're told: Young children's trust in unexpected testimony about the physical world. *Cognitive Psychology, 61*(3), 248–272. https://doi.org/10.1016/j.cogpsych.2010.06.002

Jaswal, V. K., Croft, A. C., Setia, A. R., & Cole, C. A. (2010). Young children have a specific, highly robust bias to trust testimony. *Psychological Science, 21*(10), 1541–1547. https://doi.org/10.1177/0956797610383438

Jaswal, V. K., Pérez-Edgar, K., Kondrad, R. L., Palmquist, C. M., Cole, C. A., & Cole, C. E. (2014). Can't stop believing: Inhibitory control and resistance to misleading testimony. *Developmental Science, 17*(6), 965–976. https://doi.org/10.1111/desc.12187

Jenkins, H. M., & Ward, W. C. (1965). Judgment of contingency between responses and outcomes. *Psychological Monographs: General and Applied, 79*(1), 1–17. https://doi.org/10.1037/h0093874

Jirout, J., & Klahr, D. (2012). Children's scientific curiosity: In search of an operational definition of an elusive context. *Developmental Review, 32*(2), 125–160. https://doi.org/10.1016/j.dr.2012.04.002

Johnson, C. N., & Wellman, H. M. (1982). Children's developing conceptions of the mind and brain. *Child Development, 53*(1), 222–234. https://doi.org/10.2307/1129656

Johnson-Laird, P. N., Legrenzi, P., & Legrenzi, M. S. (1972). Reasoning and a sense of reality. *British Journal of Psychology, 63*(3), 395–400. https://doi.org/10.1111/j.2044-8295.1972.tb01287.x

Jones, M., & Love, B. C. (2011). Bayesian fundamentalism or enlightenment? On the explanatory status and theoretical contributions of Bayesian models of cognition. *Behavioral and Brain Sciences*, *34*(4), 169–188. https://doi.org/10.1017/S0140525X 10003134

Kahle, J. B., Meece, J., & Scantlebury, K. (2000). Urban African-American middle school science students: Does standards-based teaching make a difference? *Journal of Research in Science Teaching*, *37*(9), 1019–1041.

Kahneman, D., & Tversky, A. (1972). Subjective probability: A judgment of representativeness. *Cognitive Psychology*, *3*(3), 430–454. https://doi.org/10.1016/0010-0285 (72)90016-3

Kamin, L. J. (1969). Predictability, surprise, attention, and conditioning. In B. A. Campbell & R. M. Church (Eds.), *Punishment and aversive behavior* (pp. 279–296). New York: Appleton-Century-Crofts.

Karmiloff-Smith, A., & Inhelder, B. (1974). If you want to get ahead, get a theory. *Cognition*, *3*(3), 195–212. https://doi.org/10.1016/0010-0277(74)90008-0

Karrby, G. (1990). Children's conceptions of their own play. *Early Child Development and Care*, *58*(1), 81–85. https://doi.org/10.1080/0300443900580110

Kastovsky, D. (1973). Causatives. *Foundations of Language*, *10*(2), 255–315.

Keating, I., Fabian, H., Jordan, P., Mavers, D., & Roberts, J. (2000). "Well, I've not done any work today. I don't know why I came to school". Perceptions of play in the reception class. *Educational Studies*, *26*(4), 437–454. https://doi.org/10.1080/03055690020003638

Keil, F. C. (1989). *Concepts, kinds, and cognitive development.* Cambridge, MA: MIT Press.

Keil, F. C. (2010). The feasibility of folk science. *Cognitive Science*, *34*(5), 826–862. https://doi.org/10.1111/j.1551-6709.2010.01108.x

Keil, F. C., & Batterman, N. (1984). A characteristic-to-defining shift in the development of word meaning. *Journal of Verbal Learning and Verbal Behavior*, *23*(2), 221–236. https://doi.org/10.1016/S0022-5371(84)90148-8

Keil, F. C., Lockhart, K. L., & Schlegel, E. (2010). A bump on a bump? Emerging intuitions concerning the relative difficulty of the sciences. *Journal of Experimental Psychology: General*, *139*(1), 1–15. https://doi.org/10.1037/a0018319

Kelemen, D., Emmons, N. A., Seston Schillaci, R., & Ganea, P. A. (2014). Young children can be taught basic natural selection using a picture-storybook intervention. *Psychological Science*, *25*(4), 893–902. https://doi.org/10.1177/0956797613516009

Kibbe, M. M., Kreisky, M., & Weisberg, D. S. (2018). Young children distinguish between different unrealistic fictional genres. *Psychology of Aesthetics, Creativity, and the Arts*, *12*(2), 228–235. https://doi.org/10.1037/aca0000115

Kim, E., Song, G., Corriveau, K. H., & Harris, P. L. (2013). Young children's deference to a consensus varies by culture and judgment setting. *Journal of Cognition and Culture, 13*(3–4), 367–381. https://doi.org/10.1163/15685373-12342099

King, P., & Howard, J. (2014). Children's perceptions of choice in relation to their play at home, in the school playground and at the out-of-school club. *Children & Society, 28*(2), 116–127. https://doi.org/10.1111/j.1099-0860.2012.00455.x

Kirkham, N. Z., Slemmer, J. A., & Johnson, S. P. (2002). Visual statistical learning in infancy: Evidence for a domain general learning mechanism. *Cognition, 83*(2), B35–B42. https://doi.org/10.1016/S0010-0277(02)00004-5

Klahr, D. (2000). *Exploring science: The cognition and development of discovery processes.* Cambridge, MA: MIT Press.

Klahr, D., & Dunbar, K. N. (1988). Dual space search during scientific reasoning. *Cognitive Science, 12*(1), 1–48. https://doi.org/10.1207/s15516709cog1201_1

Klahr, D., Fay, A. L., & Dunbar, K. N. (1993). Heuristics for scientific experimentation: A developmental study. *Cognitive Psychology, 25*(1), 111–146. https://doi.org/10.1006/cogp.1993.1003

Klahr, D., & Nigam, M. (2004). The equivalence of learning paths in early science education: Effects of direct instruction and discovery learning. *Psychological Science, 15*(10), 661–667. https://doi.org/10.1111/j.0956-7976.2004.00737.x

Klahr, D., Zimmerman, C., & Jirout, J. (2011). Educational interventions to advance children's scientific thinking. *Science, 333*, 971–975.

Kline, M. A. (2015). How to learn about teaching: An evolutionary framework for the study of teaching behavior in humans and other animals. *Behavioral and Brain Sciences, 38*, e31. https://doi.org/10.1017/S0140525X14000090

Klopfer, L. E., & Cooley, W. (1961). *Test on understanding science, Form W.* Princeton, NJ: Educational Testing Services.

Knutsen, J., Frye, D., & Sobel, D. M. (2014). Theory of learning, theory of teaching, and theory of mind. In O. N. Saracho (Ed.), *Contemporary perspectives on research in theory of mind in early childhood education* (pp. 269–290). Charlotte, NC: Information Age Publishing.

Koenig, M. A., Clément, F., & Harris, P. L. (2004). Trust in testimony: Children's use of true and false statements. *Psychological Science, 15*(10), 694–698. https://doi.org/10.1111/j.0956-7976.2004.00742.x

Koenig, M. A., & Harris, P. L. (2005). Preschoolers mistrust ignorant and inaccurate speakers. *Child Development, 76*(6), 1261–1277. https://doi.org/10.1111/j.1467-8624.2005.00849.x

Koenig, M. A., & Woodward, A. L. (2010). 24-month-olds' sensitivity to the prior inaccuracy of the source: Possible mechanisms. *Developmental Psychology, 46*(4), 815–826. https://doi.org/10.1037/a0019664.24-month-olds

Köksal-Tuncer, Ö., & Sodian, B. (2018). The development of scientific reasoning: Hypothesis testing and argumentation from evidence in young children. *Cognitive Development, 48*, 135–145. https://doi.org/10.1016/j.cogdev.2018.06.011

Kominsky, J. F., Strickland, B., Wertz, A. E., Elsner, C., Wynn, K., & Keil, F. C. (2017). Categories and constraints in causal perception. *Psychological Science, 28*(11), 1649–1662. https://doi.org/10.1177/0956797617719930

Korman, J., & Malle, B. F. (2016). Grasping for traits or reasons? How people grapple with puzzling social behaviors. *Personality and Social Psychology Bulletin, 42*(11), 1451–1465. https://doi.org/10.1177/0146167216663704

Koslowski, B. (1996). *Theory and evidence: The development of scientific reasoning*. Cambridge, MA: MIT Press.

Kripke, S. A. (1972). Naming and necessity. In *Semantics of natural language* (pp. 253–355). Dordrecht: Springer Netherlands. https://doi.org/10.1007/978-94-010-2557-7_9

Krogh-Jespersen, S., & Echols, C. H. (2012). The influence of speaker reliability on first versus second label learning. *Child Development, 83*(2), 581–590. https://doi.org/10.1111/j.1467-8624.2011.01713.x

Kruger, A. C., & Tomasello, M. (1998). Cultural learning and learning culture. In D. R. Olson & N. Torrance (Eds.), *The handbook of education and human development: New models of learning, teaching and schooling* (pp. 353–372). Oxford, UK: Blackwell. https://doi.org/10.1111/b.9780631211860.1998.00018.x

Kruschke, J. K., & Blair, N. J. (2000). Blocking and backward blocking involve learned inattention. *Psychonomic Bulletin & Review, 7*(4), 636–645. https://doi.org/10.3758/BF03213001

Kuczaj, S. A., & Daly, M. J. (1979). The development of hypothetical reference in the speech of young children. *Journal of Child Language, 6*(3), 563–579. https://doi.org/10.1017/S0305000900002543

Kuhn, D. (1989). Children and adults as intuitive scientists. *Psychological Review, 96*(4), 674–689. https://doi.org/10.1037/0033-295X.96.4.674

Kuhn, D. (2002). What is scientific reasoning and how does it develop? In U. Goswami (Ed.), *Handbook of childhood cognitive development* (pp. 371–393). Oxford: Blackwell.

Kuhn, D. (2007a). Jumping to conclusions. *Scientific American Mind, 18*(1), 44–51.

Kuhn, D. (2007b). Reasoning about multiple variables: Control of variables is not the only challenge. *Science Education, 91*(5), 710–726. https://doi.org/10.1002/sce.20214

Kuhn, D., Amsel, E., & O'Loughlin, M. (1988). *The development of scientific thinking skills*. Orlando, FL: Academic Press.

Kuhn, D., Cheney, R., & Weinstock, M. (2000). The development of epistemological understanding. *Cognitive Development, 15*(3), 309–328. https://doi.org/10.1016/S0885 -2014(00)00030-7

Kuhn, D., & Dean, D. (2004). Metacognition: A bridge between cognitive psychology and educational practice. *Theory into Practice, 43*(4), 268–274. https://doi.org/10 .1207/s15430421tip4304

Kuhn, D., & Dean, D. (2005). Is developing scientific thinking all about learning to control variables? *Psychological Science, 16*(11), 866–870. https://doi.org/10.1111/j .1467-9280.2005.01628.x

Kuhn, D., Garcia-Mila, M., Zohar, A., & Andersen, C. (1995). Strategies of knowledge acquisition. *Monographs of the Society for Research in Child Development, 60*(4). https:// doi.org/10.2307/1166059

Kuhn, D., & Pearsall, S. (2000). Developmental origins of scientific thinking. *Journal of Cognition and Development, 1*(1), 113–129. https://doi.org/10.1207/S15327647JCD0101N_11

Kuhn, D., Pease, M., & Wirkala, C. (2009). Coordinating the effects of multiple variables: A skill fundamental to scientific thinking. *Journal of Experimental Child Psychology, 103*(3), 268–284. https://doi.org/10.1016/j.jecp.2009.01.009

Kuhn, D., Schauble, L., & Garcia-Mila, M. (1992). Cross-domain development of scientific reasoning. *Cognition and Instruction, 9*(4), 285–327. https://doi.org/10.1207 /s1532690xci0904_1

Kushnir, T., & Gopnik, A. (2005). Young children infer causal strength from probabilities and interventions. *Psychological Science, 16*(9), 678–683. https://doi.org/10 .1111/j.1467-9280.2005.01595.x

Kushnir, T., & Gopnik, A. (2007). Conditional probability versus spatial contiguity in causal learning: Preschoolers use new contingency evidence to overcome prior spatial assumptions. *Developmental Psychology, 43*(1), 186–196. https://doi.org/10.1037/0012 -1649.43.1.186

Kushnir, T., Wellman, H. M., & Gelman, S. A. (2008). The role of preschoolers' social understanding in evaluating the informativeness of causal interventions. *Cognition, 107*(3), 1084–1092. https://doi.org/10.1016/j.cognition.2007.10.004

Lagattuta, K. H., & Wellman, H. M. (2001). Thinking about the past: Early knowledge about links between prior experience, thinking, and emotion. *Child Development, 72*(1), 82–102. https://doi.org/10.1111/1467-8624.00267

Lagattuta, K. H., & Wellman, H. M. (2002). Differences in early parent-child conversations about negative versus positive emotions: Implications for the development of psychological understanding. *Developmental Psychology, 38*(4), 564–580. https://doi.org/10.1037/0012-1649.38.4.564

Lagnado, D. A., & Sloman, S. A. (2004). The advantage of timely intervention. *Journal of Experimental Psychology: Learning, Memory, and Cognition, 30*(4), 856–876. https://doi.org/10.1037/0278-7393.30.4.856

Lagnado, D. A., & Sloman, S. A. (2006). Time as a guide to cause. *Journal of Experimental Psychology: Learning, Memory, and Cognition, 32*(3), 451–460. https://doi.org/10.1037/0278-7393.32.3.451

Lake, B. M., Ullman, T. D., Tenenbaum, J. B., & Gershman, S. J. (2017). Ingredients of intelligence: From classic debates to an engineering roadmap. *Behavioral and Brain Sciences, 40*, e281. https://doi.org/10.1017/S0140525X17001224

Lakoff, G. (1987). *Women, fire, and dangerous things: What categories reveal about the mind*. Chicago: University of Chicago Press.

Lancy, D. F. (2007). Accounting for variability in mother-child play. *American Anthropologist, 109*(2), 273–284. https://doi.org/10.1525/aa.2007.109.2.273

Lancy, D. F. (2016). Playing with knives: The socialization of self-initiated learners. *Child Development, 87*(3), 654–665. https://doi.org/10.1111/cdev.12498

Lane, J. D., Ronfard, S., Francioli, S. P., & Harris, P. L. (2016). Children's imagination and belief: Prone to flights of fancy or grounded in reality? *Cognition, 152*, 127–140. https://doi.org/10.1016/j.cognition.2016.03.022

Lapidow, E., & Walker, C. M. (2020). Informative experimentation in intuitive science: Children select and learn from their own causal interventions. *Cognition, 201*, 104315. https://doi.org/10.1016/j.cognition.2020.104315

Larivière, V., Ni, C., Gingras, Y., Cronin, B., & Sugimoto, C. R. (2013). Bibliometrics: Global gender disparities in science. *Nature, 504*(7479), 211–213. https://doi.org/10.1038/504211a

Latour, B., & Woolgar, S. (1979). *Laboratory life: The construction of scientific facts*. Thousand Oaks, CA: SAGE Publications.

Lave, J. (1988). *Cognition in practice: Mind, mathematics and culture in everyday life*. Cambridge, UK: Cambridge University Press.

Leahy, B., Rafetseder, E., & Perner, J. (2014). Basic conditional reasoning: How children mimic counterfactual reasoning. *Studia Logica, 102*(4), 793–810. https://doi.org/10.1007/s11225-013-9510-7

Leckey, S., Selmeczy, D., Kazemi, A., Johnson, E. G., Hembacher, E., & Ghetti, S. (2020). Response latencies and eye gaze provide insight on how toddlers gather evidence under uncertainty. *Nature Human Behaviour, 4*(9), 928–936. https://doi.org/10.1038/s41562-020-0913-y

Lederman, J. S., & Lederman, N. G. (2004, April). *Early elementary students' and teachers' understandings of nature of science and scientific inquiry: Lessons learned from Project ICAN*. Paper presented at the Annual Meeting of the National Association of Research in Science Teaching, Vancouver, BC.

Lederman, N. G., Abd-El-Khalick, F., Bell, R. L., & Schwartz, R. S. (2002). Views of nature of science questionnaire: Toward valid and meaningful assessment of learners' conceptions of nature of science. *Journal of Research in Science Teaching, 39*(6), 497–521. https://doi.org/10.1002/tea.10034

Lee, D. (1987). The semantics of just. *Journal of Pragmatics, 11*(3), 377–398. https://doi.org/10.1016/0378-2166(87)90138-X

Legare, C. H. (2012). Exploring explanation: Explaining inconsistent evidence informs exploratory, hypothesis-testing behavior in young children. *Child Development, 83*(1), 173–185. https://doi.org/10.1111/j.1467-8624.2011.01691.x

Legare, C. H., Evans, E. M., Rosengren, K. S., & Harris, P. L. (2012). The coexistence of natural and supernatural explanations across cultures and development. *Child Development, 83*(3), 779–793. https://doi.org/10.1111/j.1467-8624.2012.01743.x

Legare, C. H., & Gelman, S. A. (2008). Bewitchment, biology, or both: The coexistence of natural and supernatural explanatory frameworks across development. *Cognitive Science, 32*(4), 607–642. https://doi.org/10.1080/03640210802066766

Legare, C. H., Sobel, D. M., & Callanan, M. A. (2017). Causal learning is collaborative: Examining explanation and exploration in social contexts. *Psychonomic Bulletin & Review, 24*(5), 1548–1554. https://doi.org/10.3758/s13423-017-1351-3

Legare, C. H., Wellman, H. M., & Gelman, S. A. (2009). Evidence for an explanation advantage in naive biological reasoning. *Cognitive Psychology, 58*(2), 177–194. https://doi.org/10.1016/j.cogpsych.2008.06.002

Lei, R. F., Green, E. R., Leslie, S.-J., & Rhodes, M. (2019). Children lose confidence in their potential to "be scientists," but not in their capacity to "do science." *Developmental Science, 22*(6), e12837. https://doi.org/10.1111/desc.12837

Leiserowitz, A., Smith, N., & Marlon, J. R. (2010). *Americans' knowledge of climate change*. New Haven, CT: Yale Project on Climate Change Communication.

Leslie, A. M. (1988). Some implications of pretense for mechanisms underlying the child's theory of mind. In J. W. Astington, P. L. Harris & D. R. Olson (Eds.), *Developing theories of mind* (pp. 19–46). Cambridge, UK: Cambridge University Press.

Leslie, A. M. (1994). Pretending and believing: Issues in the theory of ToMM. *Cognition, 50*(1–3), 211–238.

Leslie, A. M., & Keeble, S. (1987). Do six-month-old infants perceive causality? *Cognition, 25*(3), 265–288. https://doi.org/10.1016/S0010-0277(87)80006-9

Lester, S., & Russell, W. (2010). *Children's right to play: An examination of the importance of play in the lives of children worldwide*. Working paper no. 57. Working Papers in Early Child Development. The Hague. https://doi.org/10.4135/9781473907850.n25

Letourneau, S. M., Meisner, R., & Sobel, D. M. (2021). Effects of facilitation vs. exhibit labels on caregiver-child interactions at a museum exhibit. *Frontiers in Psychology*, 12, 709. https://doi.org/10.3389/fpsyg.2021.637067

Letourneau, S. M., & Sobel, D. M. (2020). Children's descriptions of playing and learning as related processes. *PLOS One*, *15*(4), e0230588. https://doi.org/10.1371/journal.pone.0230588

Levy-Bruhl, L. (1926). *How natives think*. Crow's Nest, NSW: Allen & Unwin.

Lewis, D. (1973). Counterfactuals and comparative possibility. *Journal of Philosophical Logic*, *4*, 418–446.

Lewis, D. (1978). Truth in fiction. *American Philosophical Quarterly*, *15*, 37–46.

Li, J. (2004). Learning as a task or a virtue: U.S. and Chinese preschoolers explain learning. *Developmental Psychology*, *40*(4), 595–605. https://doi.org/10.1037/0012-1649.40.4.595

Libertus, K., Joh, A. S., & Needham, A. W. (2016). Motor training at 3 months affects object exploration 12 months later. *Developmental Science*, *19*(6), 1058–1066. https://doi.org/10.1111/desc.12370

Lillard, A. S. (1993). Young children's conceptualization of pretense: Action or mental representational state? *Child Development*, *64*, 372–386.

Lillard, A. S. (1998). Wanting to be it: Children's understanding of intentions underlying pretense. *Child Development*, *69*(4), 981–993. https://doi.org/10.1111/j.1467-8624.1998.tb06155.x

Lillard, A. S. (2001). Pretend play as twin earth: A social-cognitive analysis. *Developmental Review*, *21*(4), 495–531. https://doi.org/10.1006/drev.2001.0532

Lillard, A. S. (2004). *Montessori: The science behind the genius*. New York: Oxford University Press.

Lillard, A. S., Lerner, M. D., Hopkins, E. J., Dore, R. A., Smith, E. D., & Palmquist, C. M. (2013). The impact of pretend play on children's development: A review of the evidence. *Psychological Bulletin*, *139*(1), 1–34. https://doi.org/10.1037/a0029321

Lillard, A. S., & Sobel, D. M. (1999). Lion Kings or puppies: The influence of fantasy on children's understanding of pretense. *Developmental Science*, *2*(1), 75–80. https://doi.org/10.1111/1467-7687.00057

Lillard, A. S., & Witherington, D. C. (2004). Mothers' behavior modifications during pretense and their possible signal value for toddlers. *Developmental Psychology*, *40*(1), 95–113.

Liquin, E. G., & Gopnik, A. (2022). Children are more exploratory and learn more than adults in an approach-avoid task. *Cognition*, 218, 104940. https://doi.org/10 .1016/j.cognition.2021.104940

Lucas, A. J., Lewis, C., Pala, F. C., Wong, K., & Berridge, D. (2013). Social-cognitive processes in preschoolers' selective trust: Three cultures compared. *Developmental Psychology*, *49*(3), 579–590. https://doi.org/10.1037/a0029864

Lucas, C. G., Bridgers, S., Griffiths, T. L., & Gopnik, A. (2014). When children are better (or at least more open-minded) learners than adults: Developmental differences in learning the forms of causal relationships. *Cognition*, *131*(2), 284–299. https://doi .org/10.1016/j.cognition.2013.12.010

Luchkina, E., Morgan, J. L., Williams, D. J., & Sobel, D. M. (2020). Questions can answer questions about mechanisms of preschoolers' selective word learning. *Child Development*, *91*(5), e1119–e1133. https://doi.org/10.1111/cdev.13395

Luchkina, E., Sobel, D. M., & Morgan, J. L. (2018). Eighteen-month-olds selectively generalize words from accurate speakers to novel contexts. *Developmental Science*, *21*(6), e12663. https://doi.org/10.1111/desc.12663

Luchkina, E., Sommerville, J. A., & Sobel, D. M. (2018). More than just making it go: Toddlers effectively integrate causal efficacy and intentionality in selecting an appropriate causal intervention. *Cognitive Development*, *45*, 48–56. https://doi.org/10 .1016/j.cogdev.2017.12.003

Mackie, J. L. (1974). *The cement of the universe: A study of causation*. Oxford: Oxford University Press.

Mackintosh, N. J. (1974). *The psychology of animal learning*. London: Academic Press.

Macris, D. M., & Sobel, D. M. (2017). The role of evidence diversity and explanation in 4- and 5-year-olds' resolution of counterevidence. *Journal of Cognition and Development*, *18*(3), 358–374. https://doi.org/10.1080/15248372.2017.1323755

Madole, K. L., & Cohen, L. B. (1995). The role of object parts in infants' attention to form-function correlations. *Developmental Psychology*, *31*(4), 637–648. https://doi.org /10.1037/0012-1649.31.4.637

Maguire, M. J., Hirsh-Pasek, K., Golinkoff, R. M., & Brandone, A. C. (2008). Focusing on the relation: Fewer exemplars facilitate children's initial verb learning and extension. *Developmental Science*, *11*(4), 628–634. https://doi.org/10.1111/j.1467-7687.2008 .00707.x

Mangardich, H., & Sabbagh, M. A. (2018). Children remember words from ignorant speakers but do not attach meaning: Evidence from event-related potentials. *Developmental Science*, *21*(2), e12544. https://doi.org/10.1111/desc.12544

Mantzicopoulos, P., Patrick, H., & Samarapungavan, A. (2008). Young children's motivational beliefs about learning science. *Early Childhood Research Quarterly*, *23*, 378–394.

Marcus, G. F., & Davis, E. (2013). How robust are probabilistic models of higher-level cognition? *Psychological Science, 24*(12), 2351–2360. https://doi.org/10.1177/0956797613495418

Markovits, H., & Nantel, G. (1989). The belief-bias effect in the production and evaluation of logical conclusions. *Memory & Cognition, 17*(1), 11–17. https://doi.org/10.3758/BF03199552

Markson, L., & Bloom, P. (1997). Evidence against a dedicated system for word learning in children. *Nature, 385*(6619), 813–815. https://doi.org/10.1038/385813a0

Marr, D. (1982). *Vision: A computational investigation into the human representation and processing of visual information.* Cambridge, MA: MIT Press.

Mascaro, O., & Sperber, D. (2009). The moral, epistemic, and mindreading components of children's vigilance towards deception. *Cognition, 112*(3), 367–380. https://doi.org/10.1016/j.cognition.2009.05.012

Mayer, R. E. (2004). Should there be a three-strikes rule against pure discovery learning? *American Psychologist, 59*(1), 14–19. https://doi.org/10.1037/0003-066X.59.1.14

McClelland, J. L., & Thompson, R. M. (2007). Using domain-general principles to explain children's causal reasoning abilities. *Developmental Science, 10*(3), 333–356. https://doi.org/10.1111/j.1467-7687.2007.00586.x

McCloskey, M. (1983). Intuitive physics. *Scientific American, 248*(4), 122–131.

McCormack, T., Bramley, N., Frosch, C., Patrick, F., & Lagnado, D. (2016). Children's use of interventions to learn causal structure. *Journal of Experimental Child Psychology, 141*, 1–22. https://doi.org/10.1016/j.jecp.2015.06.017

McCormack, T., Butterfill, S. A., Hoerl, C., & Burns, P. (2009). Cue competition effects and young children's causal and counterfactual inferences. *Developmental Psychology, 45*(6), 1563–1575. https://doi.org/10.1037/a0017408

McCormack, T., Simms, V., McGourty, J., & Beckers, T. (2013). Encouraging children to think counterfactually enhances blocking in a causal learning task. *Quarterly Journal of Experimental Psychology, 66*(10), 1910–1926. https://doi.org/10.1080/17470218.2013.767847

McLoughlin, N., Finiasz, Z., Sobel, D. M., & Corriveau, K. H. (2021). Children's developing capacity to calibrate the verbal testimony of others with observed evidence when inferring causal relations. *Journal of Experimental Child Psychology, 210*, 105183. https://doi.org/10.1016/j.jecp.2021.105183

McTighe, J., & Wiggins, G. (2013). *Essential questions: Opening doors to student understanding.* Alexandria, VA: ASCD.

Medin, D., Bennis, W., & Chandler, M. (2010). Culture and the home-field disadvantage. *Perspectives on Psychological Science, 5*(6), 708–713. https://doi.org/10.1177/1745691610388772

Meins, E. (1997). *Security of attachment and the social development of cognition.* Hove, UK: Psychology Press.

Mejía-Arauz, R., Rogoff, B., & Paradise, R. (2005). Cultural variation in children's observation during a demonstration. *International Journal of Behavioral Development, 29*(4), 282–291. https://doi.org/10.1177/01650250544000062

Meng, Y., Bramley, N., & Xu, F. (2018). Children's causal interventions combine discrimination and confirmation. In C. Kalish, M. Rau, J. Zhu & T. T. Rogers (Eds.), *Proceedings of the 40th Annual Conference of the Cognitive Science Society* (pp. 762–767). Madison, WI: Cognitive Science Society.

Mercier, H., Bernard, S., & Clément, F. (2014). Early sensitivity to arguments: How preschoolers weight circular arguments. *Journal of Experimental Child Psychology, 125*(1), 102–109. https://doi.org/10.1016/j.jecp.2013.11.011

Mermelshtine, R. (2017). Parent-child learning interactions: A review of the literature on scaffolding. *British Journal of Educational Psychology, 87*(2), 241–254. https://doi.org/10.1111/bjep.12147

Metz, K. E. (1995). Reassessment of developmental constraints on children's science instruction. *Review of Educational Research, 65*(2), 93–127. https://doi.org/10.3102/00346543065002093

Miller, G. A. (1956). The magical number seven, plus or minus two: Some limits on our capacity for processing information. *Psychological Review, 63*(2), 81–97. https://doi.org/10.1037/h0043158

Miller, J. D., Scott, E. C., & Okamoto, S. (2006). Public acceptance of evolution. *Science, 313*(5788), 765–766. https://doi.org/10.1126/science.1126746

Miller, S. A., Hardin, C. A., & Montgomery, D. E. (2003). Young children's understanding of the conditions for knowledge acquisition. *Journal of Cognition and Development, 4*(3), 325–356. https://doi.org/10.1207/S15327647JCD0403_05

Mills, C. M. (2013). Knowing when to doubt: Developing a critical stance when learning from others. *Developmental Psychology, 49*(3), 404–418. https://doi.org/10.1037/a0029500

Mills, C. M., & Keil, F. C. (2004). Knowing the limits of one's understanding: The development of an awareness of an illusion of explanatory depth. *Journal of Experimental Child Psychology, 87*(1), 1–32. https://doi.org/10.1016/j.jecp.2003.09.003

Mills, C. M., Legare, C. H., Bills, M., & Mejias, C. (2010). Preschoolers use questions as a tool to acquire knowledge from different sources. *Journal of Cognition and Development, 11*(4), 533–560. https://doi.org/10.1080/15248372.2010.516419

Mills, C. M., Legare, C. H., Grant, M. G., & Landrum, A. R. (2011). Determining who to question, what to ask, and how much information to ask for: The development of inquiry in young children. *Journal of Experimental Child Psychology, 110*(4), 539–560. https://doi.org/10.1016/j.jecp.2011.06.003

Mills, C. M., Sands, K. R., Rowles, S. P., & Campbell, I. L. (2019). "I want to know more!": Children are sensitive to explanation quality when exploring new information. *Cognitive Science, 43*(1), e12706. https://doi.org/10.1111/cogs.12706

Mitroff, S. R., Sobel, D. M., & Gopnik, A. (2006). Reversing how to think about ambiguous figure reversals: Spontaneous alternating by uninformed observers. *Perception, 35*(5), 709–715. https://doi.org/10.1068/p5520

Moeller, A. C., Sobel, D. M., & Sodian, B. (2021). *Young children's understanding of controlled tests for diagnostic inference.* Manuscript under review.

Moeller, A. C., & Sodian, B. (2019, March 21). *Preschoolers' understanding of the control of variables strategy.* Paper presented at the Biennial Meeting of the Society for Research in Child Development, Baltimore, MD.

Morison, P., & Gardner, H. (1978). Dragons and dinosaurs: The child's capacity to differentiate fantasy from reality. *Child Development, 49*(3), 642. https://doi.org/10.2307/1128231

Morris, B. J., Croker, S., Masnick, A. M., & Zimmerman, C. (2012). The emergence of scientific reasoning. In H. Kloos, B. Morris & J. Amaral (Eds.), *Current Topics in Children's Learning and Cognition.* Rijeka, Croatia: InTech. https://doi.org/10.5772/53885

Moss-Racusin, C. A., Dovidio, J. F., Brescoll, V. L., Graham, M. J., & Handelsman, J. (2012). Science faculty's subtle gender biases favor male students. *Proceedings of the National Academy of Sciences, 109*(41), 16474–16479. https://doi.org/10.1073/pnas.1211286109

Muentener, P., & Carey, S. (2010). Infants' causal representations of state change events. *Cognitive Psychology, 61*(2), 63–86. https://doi.org/10.1016/j.cogpsych.2010.02.001

Murphy, G. L., & Medin, D. L. (1985). The role of theories in conceptual coherence. *Psychological Review, 92*(3), 289–316. https://doi.org/10.1037/0033-295X.92.3.289

National Academy of Sciences. (2008). *Science, evolution, and creationism.* Washington, DC: National Academies Press.

National Research Council. (2009). *Learning science in informal environments: People, places, and pursuits.* (P. Bell, B. Lewenstein, A. W. Shouse, & M. A. Feder, Eds.). Washington, DC: National Academies Press.

National Research Council. (2012). *A framework for K-12 science education: Practices, crosscutting concepts, and core ideas.* Washington, DC: National Academies Press.

Nazzi, T., & Gopnik, A. (2000). A shift in children's use of perceptual and causal cues to categorization. *Developmental Science, 3*(4), 389–396. https://doi.org/10.1111/1467-7687.00133

Needham, A., & Baillargeon, R. (1997). Object segregation in 8-month-old infants. *Cognition, 62*(2), 121–149. https://doi.org/10.1016/S0010-0277(96)00727-5

Needham, A., Barrett, T., & Peterman, K. (2002). A pick-me-up for infants' exploratory skills: Early simulated experiences reaching for objects using "sticky mittens" enhances young infants' object exploration skills. *Infant Behavior and Development, 25*(3), 279–295. https://doi.org/10.1016/S0163-6383(02)00097-8

Newman, G. E., Keil, F. C., Kuhlmeier, V. A., & Wynn, K. (2010). Early understandings of the link between agents and order. *Proceedings of the National Academy of Sciences, 107*(40), 17140–17145. https://doi.org/10.1073/pnas.0914056107

Newport, E. L. (1990). Maturational constraints on language learning. *Cognitive Science, 14*(1), 11–28. https://doi.org/10.1016/0364-0213(90)90024-Q

Next Generation Science Standards. (2013, April 9). *Final next generation science standards released.* https://www.nextgenscience.org/news/final-next-generation-science-standards-released

NGSS Lead States. (2013). *Next generation science standards: For states, by states.* Washington, DC: National Research Council.

Nickerson, R. S. (1998). Confirmation bias: A ubiquitous phenomenon in many guises. *Review of General Psychology, 2*(2), 175–220. https://doi.org/10.1037/1089-2680.2.2.175

Nosek, B. A., Smyth, F. L., Sriram, N., Lindner, N. M., Devos, T., Ayala, A., . . . Greenwald, A. G. (2009). National differences in gender-science stereotypes predict national sex differences in science and math achievement. *Proceedings of the National Academy of Sciences, 106*(26), 10593–10597. https://doi.org/10.1073/pnas.0809921106

Novaes, C. D. (2013). A dialogical account of deductive reasoning as a case study for how culture shapes cognition. *Journal of Cognition and Culture, 13*(5), 459–482. https://doi.org/10.1163/15685373-12342104

Novick, L. R., & Cheng, P. W. (2004). Assessing interactive causal influence. *Psychological Review, 111*(2), 455–485. https://doi.org/10.1037/0033-295X.111.2.455

Nurmsoo, E., & Robinson, E. J. (2009). Identifying unreliable informants: Do children excuse past inaccuracy? *Developmental Science, 12*(1), 41–47. https://doi.org/10.1111/j.1467-7687.2008.00750.x

Nussenbaum, K., Cohen, A. O., Davis, Z. J., Halpern, D. J., Gureckis, T. M., & Hartley, C. A. (2020). Causal information-seeking strategies change across childhood and adolescence. *Cognitive Science, 44*(9). https://doi.org/10.1111/cogs.12888

Nyhout, A., & Ganea, P. A. (2019). Mature counterfactual reasoning in 4- and 5-year-olds. *Cognition, 183*(April 2018), 57–66. https://doi.org/10.1016/j.cognition.2018.10.027

Oakes, L. M., & Cohen, L. B. (1990). Infant perception of a causal event. *Cognitive Development, 5*(2), 193–207. https://doi.org/10.1016/0885-2014(90)90026-P

Offit, P. A. (2011). *Deadly choices: How the anti-vaccine movement threatens us all.* New York: Basic Books.

ojalehto, b. l., & Medin, D. L. (2015). Perspectives on culture and concepts. *Annual Review of Psychology, 66*(1), 249–275. https://doi.org/10.1146/annurev-psych-010814 -015120

Onishi, K. H., & Baillargeon, R. (2005). Do 15-month-old infants understand false beliefs? *Science, 308*(5719), 255–258. https://doi.org/10.1126/science.1107621

Onishi, K. H., Baillargeon, R., & Leslie, A. M. (2007). 15-month-old infants detect violations in pretend scenarios. *Acta Psychologica, 124*(1), 106–128.

Osterhaus, C., Koerber, S., & Sodian, B. (2015). Experimentation skills in primary school: An inventory of children's understanding of experimental design. *Frontline Learning Research, 3*(4), 56–94. https://doi.org/10.14786/flr.v3i4.220

Osterhaus, C., Koerber, S., & Sodian, B. (2016). Scaling of advanced theory-of-mind tasks. *Child Development, 87*(6), 1971–1991. https://doi.org/10.1111/cdev.12566

Osterhaus, C., Koerber, S., & Sodian, B. (2017). Scientific thinking in elementary school: Children's social cognition and their epistemological understanding promote experimentation skills. *Developmental Psychology, 53*(3), 450–462. https://doi .org/10.1037/dev0000260

Overton, W. F., & Jackson, J. P. (1973). The representation of imagined objects in action sequences: A developmental study. *Child Development, 44*(2), 309–314. https://doi.org/10.2307/1128052

Packer, M. J., & Goicoechea, J. (2000). Sociocultural and constructivist theories of learning: Ontology, not just epistemology. *Educational Psychologist, 35*(4), 227–241. https://doi.org/10.1207/S15326985EP3504_02

Panagiotaki, G., Nobes, G., & Potton, A. (2009). Mental models and other misconceptions in children's understanding of the Earth. *Journal of Experimental Child Psychology, 104*(1), 52–67. https://doi.org/10.1016/j.jecp.2008.10.003

Park, J., & Sloman, S. A. (2013). Mechanistic beliefs determine adherence to the Markov property in causal reasoning. *Cognitive Psychology, 67*(4), 186–216. https:// doi.org/10.1016/j.cogpsych.2013.09.002

Pearce, J. M., & Hall, G. (1980). A model for Pavlovian learning: Variations in the effectiveness of conditioned but not of unconditioned stimuli. *Psychological Review, 87*(6), 532–552. https://doi.org/10.1037/0033-295X.87.6.532

Pearl, J. (1988). *Probabilistic reasoning in intelligent systems: Networks of plausible inference.* San Francisco: Morgan Kauffman.

Pearl, J. (2000). *Causality: Models, reasoning, and inference.* Cambridge, UK: Cambridge University Press.

Pellegrini, A. D., & Boyd, B. (1993). The role of play in early childhood development and education: Issues in definition and function. In O. N. Saracho & B. Spodek (Eds.), *Handbook of research on the education of young children* (pp. 105–121). Routledge. https://doi.org/10.4324/9780203841198

Penn, D. C., & Povinelli, D. J. (2007). Causal cognition in human and nonhuman animals: A comparative, critical review. *Annual Review of Psychology, 58*(1), 97–118. https://doi.org/10.1146/annurev.psych.58.110405.085555

Perner, J. (1991). *Understanding the representational mind: Learning, development, and conceptual change.* Cambridge, MA: MIT Press.

Perner, J., Leekam, S. R., & Wimmer, H. (1987). Three-year-olds' difficulty with false belief: The case for a conceptual deficit. *British Journal of Developmental Psychology, 5*(2), 125–137. https://doi.org/10.1111/j.2044-835X.1987.tb01048.x

Perner, J., & Wimmer, H. (1985). "John thinks that Mary thinks that . . ." Attribution of second-order beliefs by 5- to 10-year-old children. *Journal of Experimental Child Psychology, 39*(3), 437–471. https://doi.org/10.1016/0022-0965(85)90051-7

Piaget, J. (1929). *The child's conception of the world.* London: Routledge.

Piaget, J. (1930). *The child's conception of physical causality.* London: Routledge & Kegan Paul.

Piaget, J. (1962). *Play, dreams, and imitation in childhood.* New York: Norton.

Piaget, J., & Inhelder, B. (1951). *The origin of the idea of chance in children.* New York: Psychology Press.

Piekny, J., Grube, D., & Maehler, C. (2014). The development of experimentation and evidence evaluation skills at preschool age. *International Journal of Science Education, 36*(2), 334–354. https://doi.org/10.1080/09500693.2013.776192

Piekny, J., & Maehler, C. (2013). Scientific reasoning in early and middle childhood: The development of domain-general evidence evaluation, experimentation, and hypothesis generation skills. *British Journal of Developmental Psychology, 31*(2), 153–179. https://doi.org/10.1111/j.2044-835X.2012.02082.x

Pratt, C., & Bryant, P. (1990). Young children understand that looking leads to knowing (so long as they are looking into a single barrel). *Child Development, 61*(4), 973–982. https://doi.org/10.1111/j.1467-8624.1990.tb02835.x

Pratt, M. W., Kerig, P., Cowan, P. A., & Cowan, C. P. (1988). Mothers and fathers teaching 3-year-olds: Authoritative parenting and adult scaffolding of young children's learning. *Developmental Psychology, 24*(6), 832–839. https://doi.org/10.1037/0012-1649.24.6.832

Putnam, H. (1975). The meaning of "meaning." In *Minnesota studies in the philosophy of science* (Vol. 7, pp. 131–193). Minneapolis: University of Minnesota Press.

Quine, W. van O. (1956). Quantifiers and propositional attitudes. *Journal of Philosophy, 53*(5), 177. https://doi.org/10.2307/2022451

Rafetseder, E., Cristi-Vargas, R., & Perner, J. (2010). Counterfactual reasoning: Developing a sense of "nearest possible world." *Child Development, 81*(1), 376–389.

Rafetseder, E., & Perner, J. (2010). Is reasoning from counterfactual antecedents evidence for counterfactual reasoning? *Thinking & Reasoning, 16*(2), 131–155. https://doi.org/10.1080/13546783.2010.488074

Rafetseder, E., & Perner, J. (2014). Counterfactual reasoning: Sharpening conceptual distinctions in developmental studies. *Child Development Perspectives, 8*(1), 54–58. https://doi.org/10.1111/cdep.12061

Rafetseder, E., Schwitalla, M., & Perner, J. (2013). Counterfactual reasoning: From childhood to adulthood. *Journal of Experimental Child Psychology, 114*(3), 389–404. https://doi.org/10.1016/j.jecp.2012.10.010

Rakison, D. H., & Benton, D. T. (2019). Second-order correlation learning of dynamic stimuli: Evidence from infants and computational modeling. *Infancy, 24*(1), 57–78. https://doi.org/10.1111/infa.12274

Randi, J. (2009). I think I can: Developing children's concept of themselves as self-regulated learners. *New England Reading Association Journal, 45*(1), 55–56.

Rehder, B., & Burnett, R. C. (2005). Feature inference and the causal structure of categories. *Cognitive Psychology, 50*(3), 264–314. https://doi.org/10.1016/j.cogpsych.2004.09.002

Repacholi, B. M., & Gopnik, A. (1997). Early reasoning about desires: Evidence from 14- and 18-month-olds. *Developmental Psychology, 33*(1), 12–21. https://doi.org/10.1037/0012-1649.33.1.12

Rescorla, R. A., & Wagner, A. R. (1972). A theory of Pavlovian conditioning: Variations in the effectiveness of reinforcement and nonreinforcement. *Classical Conditioning II: Current Research and Theory, 2*, 64–99.

Rhodes, M., Cardarelli, A., & Leslie, S.-J. (2020). Asking young children to "do science" instead of "be scientists" increases science engagement in a randomized field experiment. *Proceedings of the National Academy of Sciences, 117*(18), 9808–9814. https://doi.org/10.1073/pnas.1919646117

Rhodes, M., Leslie, S.-J., Yee, K. M., & Saunders, K. (2019). Subtle linguistic cues increase girls' engagement in science. *Psychological Science, 30*(3), 455–466. https://doi.org/10.1177/0956797618823670

Richert, R. A., Shawber, A. B., Hoffman, R. E., & Taylor, M. (2009). Learning from fantasy and real characters in preschool and kindergarten. *Journal of Cognition and Development, 10*(1–2), 41–66. https://doi.org/10.1080/15248370902966594

Richert, R. A., & Smith, E. I. (2011). Preschoolers' quarantining of fantasy stories. *Child Development, 82*(4), 1106–1119. https://doi.org/10.1111/j.1467-8624.2011.01 603.x

Rieber, M. (1969). Hypothesis testing in children as a function of age. *Developmental Psychology, 1*(4), 389–395. https://doi.org/10.1037/h0027697

Riggs, K. J., Peterson, D. M., Robinson, E. J., & Mitchell, P. (1998). Are errors in false belief tasks symptomatic of a broader difficulty with counterfactuality? *Cognitive Development, 13*, 73–90. https://doi.org/10.1016/S0885-2014(98)90021-1

Robinson, E. J., & Beck, S. R. (2000). What is difficult about counterfactual reasoning? In P. Mitchell & K. J. Riggs (Eds.), *Children's reasoning and the mind* (pp. 101–119). Hove, UK: Psychology Press.

Robson, S. (1993). "Best of all I like Choosing Time": Talking with children about play and work. *Early Child Development and Care, 92*(1), 37–51. https://doi.org/10 .1080/0030443930920106

Rogers, T., & McClelland, J. L. (2004). *Semantic cognition: A parallel distributed processing approach.* Cambridge, MA: MIT Press.

Rogoff, B. (1990). *Apprenticeship in thinking: Cognitive development in social context.* New York: Oxford University Press.

Rogoff, B. (2003). *The cultural nature of human development.* New York: Oxford University Press.

Rogoff, B., Paradise, R., Mejía-Arauz, R., Correa-Chávez, M., & Angelillo, C. (2003). Firsthand learning through intent participation. *Annual Review of Psychology, 54*(1), 175–203. https://doi.org/10.1146/annurev.psych.54.101601.145118

Ronfard, S., Brown, S., Doncaster, E., & Kelemen, D. (2021). Inhibiting intuition: Scaffolding children's theory construction about species evolution in the face of competing explanations. *Cognition, 211*(February), 104635. https://doi.org/10.1016 /j.cognition.2021.104635

Ronfard, S., Chen, E. E., & Harris, P. L. (2018). The emergence of the empirical stance: Children's testing of counterintuitive claims. *Developmental Psychology, 54*(3), 482–493. https://doi.org/10.1037/dev0000455

Ronfard, S., Chen, E. E., & Harris, P. L. (2021). Testing what you're told: Young children's empirical investigation of a surprising claim. *Journal of Cognition and Development, 22*(3), 426–447. https://doi.org/10.1080/15248372.2021.1891902

Rothlein, L., & Brett, A. (1987). Children's, teachers, and parents' perceptions of play. *Early Childhood Research Quarterly, 2*(1), 45–53. https://doi.org/10.1016/0885 -2006(87)90012-3

Rozenblit, L., & Keil, F. C. (2002). The misunderstood limits of folk science: An illusion of explanatory depth. *Cognitive Science, 26*(5), 521–562. https://doi.org/10.1016 /S0364-0213(02)00078-2

Rubba, P. A., & Andersen, H. O. (1978). Development of an instrument to assess secondary school students understanding of the nature of scientific knowledge. *Science Education, 62*(4), 449–458. https://doi.org/10.1002/sce.3730620404

Rubin, B. C. (2007). Learner identity amid figured worlds: Constructing (in)competence at an urban high school. *Urban Review, 39*(2), 217–249. https://doi.org/10 .1007/s11256-007-0044-z

Rubin, K. H., Fein, G. G., & Vandenberg, B. (1983). Play. In P. Mussen & E. M. Hetherington (Eds.), *Handbook of child psychology, Volume 4: Socialization, personality, and social development* (pp. 693–774). New York: Wiley.

Ruffman, T. K., Perner, J., Olson, D. R., & Doherty, M. (1993). Reflecting on scientific thinking: Children's understanding of the hypothesis-evidence relation. *Child Development, 64*(6), 1617–1636. https://doi.org/10.1111/j.1467-8624.1993.tb04203.x

Ruggeri, A., & Lombrozo, T. (2015). Children adapt their questions to achieve efficient search. *Cognition, 143*, 203–216. https://doi.org/10.1016/j.cognition.2015.07.004

Ruggeri, A., Lombrozo, T., Griffiths, T. L., & Xu, F. (2016). Sources of developmental change in the efficiency of information search. *Developmental Psychology, 52*(12), 2159–2173. https://doi.org/10.1037/dev0000240

Ruggeri, A., Sim, Z. L., & Xu, F. (2017). "Why is Toma late to school again?" Preschoolers identify the most informative questions. *Developmental Psychology, 53*(9), 1620–1632. https://doi.org/10.1037/dev0000340

Ryan, M.-L. (1980). Fiction, non-factuals, and the principle of minimal departure. *Poetics, 9*(4), 403–422.

Sabbagh, M. A., & Baldwin, D. A. (2001). Learning words from knowledgeable versus ignorant speakers: Links between preschoolers' theory of mind and semantic development. *Child Development, 72*(4), 1054–1070. https://doi.org/10.1111/1467-8624.00334

Sabbagh, M. A., & Callanan, M. A. (1998). Metarepresentation in action: 3-, 4-, and 5-year-olds' developing theories of mind in parent–child conversations. *Developmental Psychology, 34*(3), 491–502. https://doi.org/10.1037/0012-1649.34.3.491

Sabbagh, M. A., & Paulus, M. (2018). Replication studies of implicit false belief with infants and toddlers. *Cognitive Development, 46*, 1–3. https://doi.org/10.1016/j .cogdev.2018.07.003

Sabbagh, M. A., & Shafman, D. (2009). How children block learning from ignorant speakers. *Cognition, 112*(3), 415–422. https://doi.org/10.1016/j.cognition.2009.06.005

Saffran, J. R., Aslin, R. N., & Newport, E. L. (1996). Statistical learning by 8-month-olds. *Science, 274*(5294), 1926–1928.

Sanborn, A. N., & Chater, N. (2016). Bayesian brains without probabilities. *Trends in Cognitive Sciences, 20*(12), 883–893. https://doi.org/10.1016/j.tics.2016.10.003

Sandoval, W. A., Sodian, B., Koerber, S., & Wong, J. (2014). Developing children's early competencies to engage with science. *Educational Psychologist, 49*(2), 139–152. https://doi.org/10.1080/00461520.2014.917589

Saracho, O. N., & Spodek, B. (2013). *Handbook of research on the education of young children.* Routledge. https://doi.org/10.4324/9780203841198

Saxe, R., Tenenbaum, J. B., & Carey, S. (2005). Secret agents: Inferences about hidden causes by 10- and 12-month-old infants. *Psychological Science, 16*(12), 995–1001. https://doi.org/10.1111/j.1467-9280.2005.01649.x

Saxe, R., Tzelnic, T., & Carey, S. (2007). Knowing who dunnit: Infants identify the causal agent in an unseen causal interaction. *Developmental Psychology, 43*(1), 149–158. https://doi.org/10.1037/0012-1649.43.1.149

Schacter, D. L. (2012). Adaptive constructive processes and the future of memory. *American Psychologist, 67*(8), 603–613. https://doi.org/10.1037/a0029869

Schauble, L. (1990). Belief revision in children: The role of prior knowledge and strategies for generating evidence. *Journal of Experimental Child Psychology, 49*(1), 31–57. https://doi.org/10.1016/0022-0965(90)90048-D

Schauble, L. (1996). The development of scientific reasoning in knowledge-rich contexts. *Developmental Psychology, 32*(1), 102–119. https://doi.org/10.1037/0012-1649.32.1.102

Schauble, L., Glaser, R., Duschl, R. A., Schulze, S., & John, J. (1995). Students' understanding of the objectives and procedures of experimentation in the science classroom. *Journal of the Learning Sciences, 4*(2), 131–166. https://doi.org/10.1207/s15327809jls0402_1

Schauble, L., Klopfer, L. E., & Raghavan, K. (1991). Students' transition from an engineering model to a science model of experimentation. *Journal of Research in Science Teaching, 28*(9), 859–882. https://doi.org/10.1002/tea.3660280910

Schloegl, C., & Fischer, J. (2017). Causal reasoning in non-human animals. In M. R. Waldmann (Ed.), *The Oxford handbook of causal reasoning* (pp. 699–715). New York: Oxford University Press. https://doi.org/10.1093/oxfordhb/9780199399550.013.36

Schult, C. A., & Wellman, H. M. (1997). Explaining human movements and actions: Children's understanding of the limits of psychological explanation. *Cognition, 62,* 291–324.

Schulz, L. E., & Bonawitz, E. B. (2007). Serious fun: Preschoolers engage in more exploratory play when evidence is confounded. *Developmental Psychology, 43*(4), 1045–1050. https://doi.org/10.1037/0012-1649.43.4.1045

Schulz, L. E., Bonawitz, E. B., & Griffiths, T. L. (2007). Can being scared cause tummy aches? Naive theories, ambiguous evidence, and preschoolers' causal inferences. *Developmental Psychology, 43*(5), 1124–1139. https://doi.org/10.1037/0012-1649.43.5.1124

Schulz, L. E., & Gopnik, A. (2004). Causal learning across domains. *Developmental Psychology, 40*(2), 162–176. https://doi.org/10.1037/0012-1649.40.2.162

Schulz, L. E., Gopnik, A., & Glymour, C. (2007). Preschool children learn about causal structure from conditional interventions. *Developmental Science, 10*(3), 322–332. https://doi.org/10.1111/j.1467-7687.2007.00587.x

Schulz, L. E., & Sommerville, J. A. (2006). God does not play dice: Causal determinism and preschoolers' causal inferences. *Child Development, 77*(2), 427–442. https://doi.org/10.1111/j.1467-8624.2006.00880.x

Schwartz, B., & Reisberg, D. (1991). *Learning and memory.* New York: Norton.

Schwartz, R. S., Lederman, N. G., & Lederman, J. S. (2008, March 30). *An instrument to assess views of scientific inquiry: The VOSI questionnaire.* Paper presented at the International Conference of the National Association for Research in Science Teaching, Baltimore, MD.

Schwichow, M., Croker, S., Zimmerman, C., Höffler, T., & Härtig, H. (2016). Teaching the control-of-variables strategy: A meta-analysis. *Developmental Review, 39*, 37–63. https://doi.org/10.1016/j.dr.2015.12.001

Scribner, S. (1975). Recall of classic syllogisms: A cross-cultural investigation of eros in logical problems. In R. J. Falmagne (Ed.), *Reasoning: Representation and process.* Hillsdale, NJ: Erlbaum.

Seed, A., Hanus, D., & Call, J. (2011). Causal knowledge in corvids, primates, and children: More than meets the eye? In T. McCormack, C. Hoerl & S. Butterfill (Eds.), *Tool use and causal cognition* (pp. 89–110). New York: Oxford University Press. https://doi.org/10.1093/acprof:oso/9780199571154.001.0001

Shanks, D. R. (1985). Forward and backward blocking in human contingency judgement. *Quarterly Journal of Experimental Psychology Section B, 37*(1b), 1–21. https://doi.org/10.1080/14640748508402082

Shanks, D. R. (1995). *The psychology of associative learning.* Cambridge, UK: Cambridge University Press.

Shanks, D. R., & Dickinson, A. (1988). Associative accounts of causality judgment. *Psychology of Learning and Motivation, 21*, 229–261. https://doi.org/10.1016/S0079-7421(08)60030-4

Shatz, M., Wellman, H. M., & Silber, S. (1983). The acquisition of mental verbs: A systematic investigation of the first reference to mental state. *Cognition, 14*(3), 301–321. https://doi.org/10.1016/0010-0277(83)90008-2

Shiffrin, R. M., & Schneider, W. (1977). Controlled and automatic human information processing: II. Perceptual learning, automatic attending and a general theory. *Psychological Review, 84*(2), 127–190. https://doi.org/10.1037/0033-295X.84.2.127

Shirilla, M., Golinkoff, R. M., Popp, J., Cheng, D., Cremin, T., Scheuer, N., . . . Puttre, H. (2019, March 23). *A quantitative investigation of children's perceptions of play and learning across five countries.* Paper presented at the Biennial Meeting of the Society for Research in Child Development, Baltimore, MD.

Shtulman, A. (2017). *Scienceblind: Why our intuitive theories about the world are so often wrong.* New York: Basic Books.

Shtulman, A., & Carey, S. (2007). Improbable or impossible? How children reason about the possibility of extraordinary events. *Child Development, 78*(3), 1015–1032. https://doi.org/10.1111/j.1467-8624.2007.01047.x

Shtulman, A., & Walker, C. M. (2020). Developing an understanding of science. *Annual Review of Developmental Psychology, 2*(1), 111–132. https://doi.org/10.1146/annurev-devpsych-060320-092346

Shultz, T. R. (1982). Rules of causal attribution. *Monographs of the Society for Research in Child Development, 47*(1). https://doi.org/10.2307/1165893

Sinatra, G. M., & Broughton, S. H. (2011). Bridging reading comprehension and conceptual change in science education: The promise of refutation text. *Reading Research Quarterly, 46*(4), 374–393. https://doi.org/10.1002/RRQ.005

Sinatra, G. M., & Chinn, C. A. (2012). Thinking and reasoning in science: Promoting epistemic conceptual change. In K. R. Harris, S. Graham, T. Urdan, A. G. Bus, S. Major & H. L. Swanson (Eds.), *APA educational psychology handbook, Volume 3: Application to learning and teaching* (pp. 257–282). Washington, DC: American Psychiatric Association. https://doi.org/10.1037/13275-011

Skinner, E. (1995). *Perceived control, motivation, and coping.* Thousand Oaks, CA: SAGE Publications.

Slaughter, V., & Lyons, M. (2003). Learning about life and death in early childhood. *Cognitive Psychology, 46*(1), 1–30. https://doi.org/10.1016/S0010-0285(02)00504-2

Smith, L. B., & Thelen, E. (2003). Development as a dynamic system. *Trends in Cognitive Sciences, 7*(8), 343–348. https://doi.org/10.1016/S1364-6613(03)00156-6

Smith, P. K., & Vollstedt, R. (1985). On defining play: An empirical study of the relationship between play and various play criteria. *Child Development, 56*(4), 1042–1050. https://doi.org/10.2307/1130114

Sobel, D. M. (2001). *Examining the coherence of young children's understanding of causality: Evidence from inference, explanation, and counterfactual reasoning.* [Doctoral dissertation, University of California at Berkeley]. ProQuest Dissertations and Theses Global.

Sobel, D. M. (2004a). Children's developing knowledge of the relationship between mental awareness and pretense. *Child Development, 75*(3), 704–729. https://doi.org /10.1111/j.1467-8624.2004.00702.x

Sobel, D. M. (2004b). Exploring the coherence of young children's explanatory abilities: Evidence from generating counterfactuals. *British Journal of Developmental Psychology, 22*(1), 37–58. https://doi.org/10.1348/026151004772901104

Sobel, D. M. (2006). How fantasy benefits young children's understanding of pretense. *Developmental Science, 9*(1), 63–75.

Sobel, D. M. (2007). Children's knowledge of the relation between intentional action and pretending. *Cognitive Development, 22*(1), 130–141. https://doi.org/10.1016/j.cogdev .2006.06.002

Sobel, D. M. (2009). Enabling conditions and children's understanding of pretense. *Cognition, 113*(2), 177–188. https://doi.org/10.1016/j.cognition.2009.08.002

Sobel, D. M. (2011). Domain-specific causal knowledge and children's reasoning about possibility. In C. Hoerl, T. McCormack & S. R. Beck (Eds.), *Understanding counterfactuals, understanding causation: Issues in philosophy and psychology* (pp. 123–146). Oxford: Oxford University Press. https://doi.org/10.1093/acprof:oso/9780199590698 .003.0007

Sobel, D. M. (2015). Can you do it? How preschoolers judge whether others have learned. *Journal of Cognition and Development, 16*(3), 492–508. https://doi.org/10 .1080/15248372.2013.815621

Sobel, D. M. (2020). Young children's sensitivity to priors in causal inference reflects their mechanistic knowledge. Unpublished manuscript on PsyArXiv. https://doi.org /10.31234/osf.io/9dcp8

Sobel, D. M., Benton, D. T., Finiasz, Z., Taylor, Y., & Weisberg, D. S. (2021). *Children's play, but not their learning from play, is influenced by their first action.* Manuscript under review.

Sobel, D. M., & Buchanan, D. W. (2009). Bridging the gap: Causality-at-a-distance in children's categorization and inferences about internal properties. *Cognitive Development, 24*(3), 274–283. https://doi.org/10.1016/j.cogdev.2009.03.003

Sobel, D. M., Erb, C. D., Tassin, T., & Weisberg, D. S. (2017). The development of diagnostic inference about uncertain causes. *Journal of Cognition and Development, 18*(5), 556–576. https://doi.org/10.1080/15248372.2017.1387117

Sobel, D. M., & Finiasz, Z. (2020). How children learn from others: An analysis of selective word learning. *Child Development, 91*(6), e1134–e1161. https://doi.org/10.1111/cdev.13415

Sobel, D. M., & Jipson, J. L. (Eds.). (2016). *Cognitive development in museum settings: Relating research and practice.* New York: Routledge.

Sobel, D. M., & Kirkham, N. Z. (2006). Blickets and babies: The development of causal reasoning in toddlers and infants. *Developmental Psychology, 42*(6), 1103–1115. https://doi.org/10.1037/0012-1649.42.6.1103

Sobel, D. M., & Kirkham, N. Z. (2007). Bayes nets and babies: Infants' developing statistical reasoning abilities and their representation of causal knowledge. *Developmental Science, 10*(3), 298–306. https://doi.org/10.1111/j.1467-7687.2007.00589.x

Sobel, D. M., & Kushnir, T. (2006). The importance of decision making in causal learning from interventions. *Memory & Cognition, 34*(2), 411–419. https://doi.org/10.3758/BF03193418

Sobel, D. M., & Kushnir, T. (2013). Knowledge matters: How children evaluate the reliability of testimony as a process of rational inference. *Psychological Review, 120*(4), 779–797. https://doi.org/10.1037/a0034191

Sobel, D. M., & Letourneau, S. M. (2015). Children's developing understanding of what and how they learn. *Journal of Experimental Child Psychology, 132*, 221–229. https://doi.org/10.1016/j.jecp.2015.01.004

Sobel, D. M., & Letourneau, S. M. (2016). Children's developing knowledge of and reflection about teaching. *Journal of Experimental Child Psychology, 143*, 111–122. https://doi.org/10.1016/j.jecp.2015.10.009

Sobel, D. M., & Letourneau, S. M. (2018). Preschoolers' understanding of how others learn through action and instruction. *Child Development, 89*(3), 961–970. https://doi.org/10.1111/cdev.12773

Sobel, D. M., & Letourneau, S. M. (2019). Children's developing descriptions and judgments of pretending. *Child Development, 90*(5), 1817–1831. https://doi.org/10.1111/cdev.13099

Sobel, D. M., Letourneau, S. M., Legare, C. H., & Callanan, M. A. (2021). Relations between parent–child interaction and children's engagement and learning at a museum exhibit about electric circuits. *Developmental Science, 24*(3), e13057. https://doi.org/10.1111/desc.13057

Sobel, D. M., Letourneau, S. M., & Meisner, R. (2016). Developing Mind Lab: A university-museum partnership to explore the process of learning. In D. M. Sobel & J. L. Jipson (Eds.), *Cognitive development in museum settings: Relating research and practice* (pp. 120–137). New York: Routledge.

Sobel, D. M., Li, J., & Corriveau, K. H. (2007). "They danced around in my head and I learned them": Children's developing conceptions of learning. *Journal of Cognition and Development, 8*(3), 345–369. https://doi.org/10.1080/15248370701446806

Sobel, D. M., & Lillard, A. S. (2001). The impact of fantasy and action on young children's understanding of pretence. *British Journal of Developmental Psychology, 19*(1), 85–98. https://doi.org/10.1348/026151001165976

Sobel, D. M., & Munro, S. E. (2009). Domain generality and specificity in children's causal inference about ambiguous data. *Developmental Psychology, 45*(2), 511–524. https://doi.org/10.1037/a0014944

Sobel, D. M., & Sommerville, J. A. (2009). Rationales in children's causal learning from others' actions. *Cognitive Development, 24*(1), 70–79. https://doi.org/10.1016/j.cogdev.2008.08.003

Sobel, D. M., & Sommerville, J. A. (2010). The importance of discovery in children's causal learning from interventions. *Frontiers in Psychology, 1*(176), 1–7. https://doi.org/10.3389/fpsyg.2010.00176

Sobel, D. M., Stricker, L. W., & Weisberg, D. S. (2021). *First person belief revision relates to reflections about learning during free exploration of a children's museum.* Manuscript under review.

Sobel, D. M., Tenenbaum, J. B., & Gopnik, A. (2004). Children's causal inferences from indirect evidence: Backwards blocking and Bayesian reasoning in preschoolers. *Cognitive Science, 28*(3), 303–333. https://doi.org/10.1016/j.cogsci.2003.11.001

Sobel, D. M., & Weisberg, D. S. (2014). Tell me a story: How children's developing domain knowledge affects their story construction. *Journal of Cognition and Development, 15*(3), 465–478. https://doi.org/10.1080/15248372.2012.736111

Sobel, D. M., Yoachim, C. M., Gopnik, A., Meltzoff, A. N., & Blumenthal, E. J. (2007). The blicket within: Preschoolers' inferences about insides and causes. *Journal of Cognition and Development, 8*(2), 159–182. https://doi.org/10.1080/15248370701202356

Sodian, B., Zaitchik, D., & Carey, S. (1991). Young children's differentiation of hypothetical beliefs from evidence. *Child Development, 62*, 753–766.

Solis, G., & Callanan, M. A. (2016). Evidence against deficit accounts: Conversations about science in Mexican heritage families living in the United States. *Mind, Culture, and Activity, 23*(3), 212–224. https://doi.org/10.1080/10749039.2016.1196493

Sommerville, J. A., & Hammond, A. J. (2007). Treating another's actions as one's own: Children's memory of and learning from joint activity. *Developmental Psychology, 43*(4), 1003–1018. https://doi.org/10.1037/0012-1649.43.4.1003

Sommerville, J. A., & Woodward, A. L. (2005). Infants' sensitivity to the causal features of means-end support sequences in action and perception. *Infancy, 8*(2), 119–145. https://doi.org/10.1207/s15327078in0802_2

Sommerville, J. A., Woodward, A. L., & Needham, A. (2005). Action experience alters 3-month-old infants' perception of others' actions. *Cognition, 96*(1), B1–B11. https://doi.org/10.1016/j.cognition.2004.07.004

Sophian, C., & Huber, A. (1984). Early developments in children's causal judgments. *Child Development, 55*(2), 512–526. https://doi.org/10.2307/1129962

Spelke, E. S., Breinlinger, K., Macomber, J., & Jacobson, K. (1992). Origins of knowledge. *Psychological Review, 99*(4), 605–632. https://doi.org/10.1037/0033-295X.99.4.605

Spelke, E. S., & Kinzler, K. D. (2007). Core knowledge. *Developmental Science, 10*(1), 89–96. https://doi.org/10.1111/j.1467-7687.2007.00569.x

Spirtes, P., Glymour, C., & Scheines, R. (1993). *Causation, prediction, and search.* New York: Springer-Verlag.

Stahl, A. E., & Feigenson, L. (2015). Observing the unexpected enhances infants' learning and exploration. *Science, 348*(6230), 91–94. https://doi.org/10.1126/science.aaa3799

Stanovich, K. E. (2004). *The robot's rebellion: Finding meaning in the age of Darwin.* Chicago: University of Chicago Press.

Stanovich, K. E. (2012). *How to think straight about psychology* (10th ed.). New York: Harper Collins.

Sternbach, R., & Okuda, M. (1991). *Star Trek: The Next Generation technical manual.* New York: Pocket Books.

Steyvers, M., Tenenbaum, J. B., Wagenmakers, E.-J., & Blum, B. (2003). Inferring causal networks from observations and interventions. *Cognitive Science, 27*(3), 453–489. https://doi.org/10.1207/s15516709cog2703_6

Stich, S., & Tarzia, J. (2015). The pretense debate. *Cognition, 143*, 1–12. https://doi.org/10.1016/j.cognition.2015.06.007

Stipek, D., & Mac Iver, D. (1989). Developmental change in children's assessment of intellectual competence. *Child Development, 60*(3), 521. https://doi.org/10.2307/1130719

Strauss, S. (2005). Teaching as a natural cognitive ability: Implications for classroom practice and teacher education. In D. B. Pillemer & S. White (Eds.), *Developmental psychology and social change* (pp. 368–388). New York: Cambridge University Press.

Strauss, S., & Ziv, M. (2012). Teaching is a natural cognitive ability for humans. *Mind, Brain, and Education, 6*(4), 186–196. https://doi.org/10.1111/j.1751-228X.2012.01156.x

Strauss, S., Ziv, M., & Stein, A. (2002). Teaching as a natural cognition and its relations to preschoolers' developing theory of mind. *Cognitive Development, 17*(3–4), 1473–1487. https://doi.org/10.1016/S0885-2014(02)00128-4

Stricker, L. W., & Sobel, D. M. (2020). Children's developing reflections on and understanding of creativity. *Cognitive Development, 55,* 100916. https://doi.org/10.1016/j.cogdev.2020.100916

Stricker, L. W., & Sobel, D. M. (2021, April 9). *Learning settings affect how children believe learning happens.* Poster presented at the Biennial Meeting of the Society for Research in Child Development, virtual conference.

Stutt, R. O. J. H., Retkute, R., Bradley, M., Gilligan, C. A., & Colvin, J. (2020). A modelling framework to assess the likely effectiveness of facemasks in combination with "lock-down" in managing the COVID-19 pandemic. *Proceedings of the Royal Society A: Mathematical, Physical and Engineering Sciences, 476*(2238), 20200376. https://doi.org/10.1098/rspa.2020.0376

Szechter, L. E., & Carey, E. J. (2009). Gravitating toward science: Parent-child interactions at a gravitational-wave observatory. *Science Education, 93*(5), 846–858. https://doi.org/10.1002/sce.20333

Taggart, J., Eisen, S., & Lillard, A. S. (2019). The current landscape of US children's television: Violent, prosocial, educational, and fantastical content. *Journal of Children and Media, 13*(3), 276–294. https://doi.org/10.1080/17482798.2019.1605916

Tang, C. M., & Bartsch, K. (2012). Young children's recognition of how and when knowledge was acquired. *Journal of Cognition and Development, 13*(3), 372–394. https://doi.org/10.1080/15248372.2011.577759

Tang, C. M., Bartsch, K., & Nunez, N. (2007). Young children's reports of when learning occurred. *Journal of Experimental Child Psychology, 97*(2), 149–164. https://doi.org/10.1016/j.jecp.2007.01.003

Taumoepeau, M., & Ruffman, T. K. (2008). Stepping stones to others' minds: Maternal talk relates to child mental state language and emotion understanding at 15, 24, and 33 months. *Child Development, 79*(2), 284–302. https://doi.org/10.1111/j.1467-8624.2007.01126.x

Taylor, M., & Carlson, S. M. (1997). The relation between individual differences in fantasy and theory of mind. *Child Development, 68*(3), 436–455. https://doi.org/10.1111/j.1467-8624.1997.tb01950.x

Taylor, M., Esbensen, B. M., & Bennett, R. T. (1994). Children's understanding of knowledge acquisition: The tendency for children to report that they have always known what they have just learned. *Child Development, 65*(6), 1581–1604. https://doi.org/10.1111/j.1467-8624.1994.tb00837.x

Taylor, M., Lussier, G. L., & Maring, B. L. (2003). The distinction between lying and pretending. *Journal of Cognition and Development, 4*(3), 299–323. https://doi.org/10.1207/S15327647JCD0403_04

Tenenbaum, J. B., & Griffiths, T. L. (2001). Generalization, similarity and Bayesian inference. *Behavioral and Brain Sciences, 24*(4), 629–640. https://doi.org/10.1017/S01 40525X01000061

Tenenbaum, J. B., Griffiths, T. L., & Kemp, C. (2006). Theory-based Bayesian models of inductive learning and reasoning. *Trends in Cognitive Sciences, 10*(7), 309–318. https://doi.org/10.1016/j.tics.2006.05.009

Tenney, E. R., Small, J. E., Kondrad, R. L., Jaswal, V. K., & Spellman, B. A. (2011). Accuracy, confidence, and calibration: How young children and adults assess credibility. *Developmental Psychology, 47*(4), 1065–1077. https://doi.org/10.1037/a0023273

Thelen, E., & Smith, L. B. (1996). *A dynamic systems approach to the development of cognition and action.* Cambridge, MA: MIT Press.

Tippett, C. D. (2010). Refutation text in science education: A review of two decades of research. *International Journal of Science and Mathematics Education, 8*(6), 951–970. https://doi.org/10.1007/s10763-010-9203-x

Tong, Y., Wang, F., & Danovitch, J. (2020). The role of epistemic and social characteristics in children's selective trust: Three meta-analyses. *Developmental Science, 23*(2). https://doi.org/10.1111/desc.12895

Tsai, C., & Liu, S. (2005). Developing a multi-dimensional instrument for assessing students' epistemological views toward science. *International Journal of Science Education, 27*(13), 1621–1638. https://doi.org/10.1080/09500690500206432

Tschirgi, J. E. (1980). Sensible reasoning: A hypothesis about hypotheses. *Child Development, 51*(1), 1–10. https://doi.org/10.2307/1129583

Tummeltshammer, K. S., Wu, R., Sobel, D. M., & Kirkham, N. Z. (2014). Infants track the reliability of potential informants. *Psychological Science, 25*(9), 1730–1738. https://doi.org/10.1177/0956797614540178

Tune, G. S. (1964). Response preferences: A review of some relevant literature. *Psychological Bulletin, 61*(4), 286–302. https://doi.org/10.1037/h0048618

Turing, A. M. (1950). Computing machinery and intelligence. *Mind, 236*(433–460).

Tversky, A., & Kahneman, D. (1971). Belief in the law of small numbers. *Psychological Bulletin, 76*(2), 105–110. https://doi.org/10.1037/h0031322

Ullman, T. D., Goodman, N. D., & Tenenbaum, J. B. (2012). Theory learning as stochastic search in the language of thought. *Cognitive Development, 27*(4), 455–480. https://doi.org/10.1016/j.cogdev.2012.07.005

Van de Vondervoort, J. W., & Friedman, O. (2014). Preschoolers can infer general rules governing fantastical events in fiction. *Developmental Psychology, 50*(5), 1594–1599. https://doi.org/10.1037/a0035717

van den Berg, L., & Gredebäck, G. (2021). The sticky mittens paradigm: A critical appraisal of current results and explanations. *Developmental Science, 24*(5), e13036. https://doi.org/10.1111/desc.13036

van der Graaf, J., Segers, E., & Verhoeven, L. (2015). Scientific reasoning abilities in kindergarten: Dynamic assessment of the control of variables strategy. *Instructional Science, 43*(3), 381–400. https://doi.org/10.1007/s11251-015-9344-y

Van Hamme, L. J., & Wasserman, E. A. (1994). Cue competition in causality judgments: The role of nonpresentation of compound stimulus elements. *Learning and Motivation, 25*(2), 127–151. https://doi.org/10.1006/lmot.1994.1008

Van Oers, B. (1990). The development of mathematical thinking in school: A comparison of the action-psychological and information-processing approaches. *International Journal of Educational Research, 14*(1), 51–66. https://doi.org/10.1016/0883 -0355(90)90016-2

van Schaik, J. E., Slim, T., Franse, R. K., & Raijmakers, M. E. J. (2020). Hands-on exploration of cubes' floating and sinking benefits children's subsequent buoyancy predictions. *Frontiers in Psychology, 11*, 1665. https://doi.org/10.3389/fpsyg.2020.01665

Vanderbilt, K. E., Heyman, G. D., & Liu, D. (2014). In the absence of conflicting testimony young children trust inaccurate informants. *Developmental Science, 17*(3), 443–451. https://doi.org/10.1111/desc.12134

Vosniadou, S. (1994). Capturing and modeling the process of conceptual change. *Learning and Instruction, 4*, 45–69.

Vosniadou, S., & Brewer, W. F. (1992). Mental models of the earth: A study of conceptual change in childhood. *Cognitive Psychology, 24*(4), 535–585. https://doi.org/10 .1016/0010-0285(92)90018-W

Vygotsky, L. S. (1978). *Mind in society: Development of higher psychological processes.* Cambridge, MA: Harvard University Press.

Walker, C. M., Goel, D., Nyhout, A., & Ganea, P. A. (2019, March 21). *Evidence for early recognition of inconclusive data in children's evaluation of evidence.* Paper presented at the Biennial Meeting of the Society for Research in Child Development, Baltimore, MD.

Walker, C. M., Gopnik, A., & Ganea, P. A. (2015). Learning to learn from stories: Children's developing sensitivity to the causal structure of fictional worlds. *Child Development, 86*(1), 310–318. https://doi.org/10.1111/cdev.12287

Walker, C. M., Lombrozo, T., Williams, J. J., Rafferty, A. N., & Gopnik, A. (2017). Explaining constrains causal learning in childhood. *Child Development, 88*(1), 229–246. https://doi.org/10.1111/cdev.12590

Walker, C. M., & Nyhout, A. (2020). Asking "why?" and "what if?" In L. P. Butler, S. Ronfard & K. H. Corriveau (Eds.), *The questioning child: Insights from psychology and education* (pp. 252–280). Cambridge, UK: Cambridge University Press. https://doi.org /10.1017/9781108553803.013

Walker, C. M., Wartenberg, T. E., & Winner, E. (2012). Engagement in philosophical dialogue facilitates children's reasoning about subjectivity. *Developmental Psychology, 49*(7), 1338–1347. https://doi.org/10.1037/a0029870

Walls, L. (2012). Third grade African American students' views of the nature of science. *Journal of Research in Science Teaching, 49*(1), 1–37.

Walsh, C., & Sloman, S. A. (2004). Revising causal beliefs. In K. Forbus, D. Gentner & T. Regier (Eds.), *Proceedings of the 26th Annual Conference of the Cognitive Science Society* (pp. 1423–1427). Chicago: Cognitive Science Society.

Walton, K. L. (1990). *Mimesis as make-believe.* Cambridge, MA: Harvard University Press.

Wason, P. C. (1960). On the failure to eliminate hypotheses in a conceptual task. *Quarterly Journal of Experimental Psychology, 12*(3), 129–140. https://doi.org/10.1080 /17470216008416717

Wason, P. C. (1966). Reasoning. In B. M. Foss (Ed.), *New horizons in psychology.* Harmondsworth, UK: Penguin.

Wason, P. C. (1968). Reasoning about a rule. *Quarterly Journal of Experimental Psychology, 20*(3), 273–281. https://doi.org/10.1080/14640746808400161

Wasserman, E. A., & Berglan, L. R. (1998). Backward blocking and recovery from overshadowing in human causal judgement: The role of within-compound associations. *Quarterly Journal of Experimental Psychology Section B, 51*(2), 121–138. https:// doi.org/10.1080/713932675

Weisberg, D. S. (2013). Distinguishing imagination from reality. In M. Taylor (Ed.), *Oxford handbook of the development of imagination* (pp. 75–93). New York: Oxford University Press. https://doi.org/10.1093/oxfordhb/9780195395761.013.0006

Weisberg, D. S. (2015). Pretend play. *Wiley Interdisciplinary Reviews: Cognitive Science, 6*(3), 249–261. https://doi.org/10.1002/wcs.1341

Weisberg, D. S. (2016). How fictional worlds are created. *Philosophy Compass, 11*(8), 462–470. https://doi.org/10.1111/phc3.12335

Weisberg, D. S. (2020). Is imagination constrained enough for science? In A. Levy & P. Godfrey-Smith (Eds.), *The scientific imagination: Philosophical and psychological perspectives* (pp. 250–261). New York: Oxford University Press.

Weisberg, D. S., Choi, E., & Sobel, D. M. (2020). Of blickets, butterflies, and baby dinosaurs: Children's diagnostic reasoning across domains. *Frontiers in Psychology, 11*, 2210. https://doi.org/10.3389/fpsyg.2020.02210

Weisberg, D. S., & Goodstein, J. (2009). What belongs in a fictional world? *Journal of Cognition and Culture, 9*(1), 69–78. https://doi.org/10.1163/156853709X414647

Weisberg, D. S., & Gopnik, A. (2013). Pretense, counterfactuals, and Bayesian causal models: Why what is not real really matters. *Cognitive Science, 37*(7), 1368–1381. https://doi.org/10.1111/cogs.12069

Weisberg, D. S., Hirsh-Pasek, K., & Golinkoff, R. M. (2013). Guided play: Where curricular goals meet a playful pedagogy. *Mind, Brain, and Education, 7*(2), 104–112. https://doi.org/10.1111/mbe.12015

Weisberg, D. S., Hirsh-Pasek, K., Golinkoff, R. M., Kittredge, A. K., & Klahr, D. (2016). Guided play: Principles and practices. *Current Directions in Psychological Science, 25*(3), 177–182. https://doi.org/10.1177/0963721416645512

Weisberg, D. S., & Hopkins, E. J. (2020). Preschoolers' extension and export of information from realistic and fantastical stories. *Infant and Child Development, 29*(4), e2812. https://doi.org/10.1002/icd.2182

Weisberg, D. S., Hopkins, E. J., & Taylor, J. C. V. (2018). People's explanatory preferences for scientific phenomena. *Cognitive Research: Principles and Implications, 3*(1), 44. https://doi.org/10.1186/s41235-018-0135-2

Weisberg, D. S., Ilgaz, H., Hirsh-Pasek, K., Golinkoff, R. M., Nicolopoulou, A. & Dickinson, D. (2015). Shovels and swords: How realistic and fantastical themes affect children's word learning. *Cognitive Development, 35*, 1–14. https://doi.org/10.1016/j.cogdev.2014.11.001

Weisberg, D. S., Keil, F. C., Goodstein, J., Rawson, E., & Gray, J. R. (2008). The seductive allure of neuroscience explanations. *Journal of Cognitive Neuroscience, 20*(3), 470–477. https://doi.org/10.1162/jocn.2008.20040

Weisberg, D. S., Landrum, A. R., Hamilton, J., & Weisberg, M. (2021). Knowledge about the nature of science increases public acceptance of science regardless of identity factors. *Public Understanding of Science, 30*(2), 120–138. https://doi.org/10.1177/0963662520977700

Weisberg, D. S., Landrum, A. R., Metz, S. E., & Weisberg, M. (2018). No missing link: Knowledge predicts acceptance of evolution in the United States. *BioScience, 68*(3), 212–222. https://doi.org/10.1093/biosci/bix161

Weisberg, D. S., & Sobel, D. M. (2012). Young children discriminate improbable from impossible events in fiction. *Cognitive Development, 27*(1), 90–98. https://doi.org/10.1016/j.cogdev.2011.08.001

Weisberg, D. S., Sobel, D. M., Goodstein, J., & Bloom, P. (2013). Young children are reality-prone when thinking about stories. *Journal of Cognition and Culture, 13*(3–4), 383–407. https://doi.org/10.1163/15685373-12342100

Wellman, H. M. (2014). *Making minds: How theory of mind develops*. New York: Oxford University Press.

Wellman, H. M., Cross, D., & Watson, J. (2001). Meta-analysis of theory-of-mind development: The truth about false belief. *Child Development, 72*, 655–684.

Wellman, H. M., & Gelman, S. A. (1992). Cognitive development: Foundational theories of core domains. *Annual Review of Psychology, 43*(1), 337–375. https://doi.org/10.1146/annurev.ps.43.020192.002005

Wellman, H. M., & Gelman, S. A. (1998). Knowledge acquisition in fundamental domains. In D. Kuhn & R. S. Siegler (Eds.), *Handbook of child psychology: Cognition, perception, and language* (pp. 523–573). New York: Wiley.

Wellman, H. M., Hickling, A. K., & Schult, C. A. (1997). Young children's psychological, physical, and biological explanations. In H. M. Wellman & K. Inagaki (Eds.), *The emergence of core domains of thought: Children's reasoning about physical, psychological, and biological phenomena (New directions for child development, No. 75.)* (pp. 7–25). San Francisco, CA: Jossey-Bass.

Wellman, H. M., & Lagattuta, K. H. (2004). Theory of mind for learning and teaching: The nature and role of explanation. *Cognitive Development, 19*(4), 479–497. https://doi.org/10.1016/j.cogdev.2004.09.003

Wellman, H. M., & Liu, D. (2004). Scaling of theory-of-mind tasks. *Child Development, 75*(2), 523–541. https://doi.org/10.1111/j.1467-8624.2004.00691.x

Wellman, H. M., & Liu, D. (2007). Causal reasoning as informed by the early development of explanations. In A. Gopnik & L. E. Schulz (Eds.), *Causal learning: Psychology, philosophy, and computation* (pp. 261–279). New York: Oxford University Press.

Wellman, H. M., & Woolley, J. D. (1990). From simple desires to ordinary beliefs: The early development of everyday psychology. *Cognition, 35*(3), 245–275. https://doi.org/10.1016/0010-0277(90)90024-E

Wente, A. O., Kimura, K., Walker, C. M., Banerjee, N., Fernández Flecha, M., MacDonald, B., Lucas, C., & Gopnik, A. (2019). Causal learning across culture and socioeconomic status. *Child Development, 90*(3), 859–875. https://doi.org/10.1111/cdev.12943

Werchan, D. M., Collins, A. G. E., Frank, M. J., & Amso, D. (2016). Role of prefrontal cortex in learning and generalizing hierarchical rules in 8-month-old infants. *Journal of Neuroscience, 36*(40), 10314–10322. https://doi.org/10.1523/JNEUROSCI.1351-16.2016

White, R. E., & Carlson, S. M. (2016). What would Batman do? Self-distancing improves executive function in young children. *Developmental Science, 19*(3), 419–426. https://doi.org/10.1111/desc.12314

White, R. E., Prager, E. O., Schaefer, C., Kross, E., Duckworth, A. L., & Carlson, S. M. (2017). The "Batman Effect": Improving perseverance in young children. *Child Development, 88*(5), 1563–1571. https://doi.org/10.1111/cdev.12695

Wiesen, S. E., Watkins, R. M., & Needham, A. W. (2016). Active motor training has long-term effects on infants' object exploration. *Frontiers in Psychology, 7,* 599. https://doi.org/10.3389/fpsyg.2016.00599

Williamson, R. A., Jaswal, V. K., & Meltzoff, A. N. (2010). Learning the rules: Observation and imitation of a sorting strategy by 36-month-old children. *Developmental Psychology, 46*(1), 57–65. https://doi.org/10.1037/a0017473

Wimmer, H., & Perner, J. (1983). Beliefs about beliefs: Representation and constraining function of wrong beliefs in young children's understanding of deception. *Cognition, 13,* 103–128.

Wittgenstein, L. (1958). *Philosophical investigations* (3rd ed.). Oxford: Blackwell.

Woith, H., Petersen, G. M., Hainzl, S., & Dahm, T. (2018). Review: Can animals predict earthquakes? *Bulletin of the Seismological Society of America, 108*(3A), 1031–1045. https://doi.org/10.1785/0120170313

Wood, D., Bruner, J. S., & Ross, G. (1976). The role of tutoring in problem solving. *Journal of Child Psychology and Psychiatry, 17*(2), 89–100. https://doi.org/10.1111/j.1469-7610.1976.tb00381.x

Wood, D., & Middleton, D. (1975). A study of assisted problem-solving. *British Journal of Psychology, 66*(2), 181–191. https://doi.org/10.1111/j.2044-8295.1975.tb01454.x

Woodward, A. L. (1998). Infants selectively encode the goal object of an actor's reach. *Cognition, 69*(1), 1–34. https://doi.org/10.1016/S0010-0277(98)00058-4

Woodward, J. (2003). *Making things happen: A theory of causal explanation.* New York: Oxford University Press.

Woolley, J. D. (1997). Thinking about fantasy: Are children fundamentally different thinkers and believers from adults? *Child Development, 68*(6), 991–1011. https://doi.org/10.1111/j.1467-8624.1997.tb01975.x

Wu, R., Gopnik, A., Richardson, D. C., & Kirkham, N. Z. (2011). Infants learn about objects from statistics and people. *Developmental Psychology, 47*(5), 1220–1229. https://doi.org/10.1037/a0024023

Wu, R., & Kirkham, N. Z. (2010). No two cues are alike: Depth of learning during infancy is dependent on what orients attention. *Journal of Experimental Child Psychology, 107*(2), 118–136. https://doi.org/10.1016/j.jecp.2010.04.014

Xu, F. (2019). Towards a rational constructivist theory of cognitive development. *Psychological Review, 126*(6), 841–864. https://doi.org/10.1037/rev0000153

Xu, F., & Denison, S. (2009). Statistical inference and sensitivity to sampling in 11-month-old infants. *Cognition, 112*(1), 97–104. https://doi.org/10.1016/j.cognition .2009.04.006

Xu, F., & Kushnir, T. (Eds.). (2012). *Rational constructivism in cognitive development* (*Advances in Child Development and Behavior*, Vol. 43). Waltham, MA: Academic Press.

Xu, F., & Tenenbaum, J. B. (2007). Word learning as Bayesian inference. *Psychological Review, 114*(2), 245–272. https://doi.org/10.1037/0033-295X.114.2.245

Yang, D. J., Bushnell, E. W., Buchanan, D. W., & Sobel, D. M. (2013). Infants' use of contextual cues in the generalization of effective actions from imitation. *Journal of Experimental Child Psychology, 116*(2), 510–531. https://doi.org/10.1016/j.jecp.2012 .09.013

Yermolayeva, Y., & Rakison, D. H. (2016). Seeing the unseen: Second-order correlation learning in 7- to 11-month-olds. *Cognition, 152*, 87–100. https://doi.org/10.1016 /j.cognition.2016.03.012

Young, A. G., Alibali, M. W., & Kalish, C. W. (2012). Disagreement and causal learning: Others' hypotheses affect children's evaluations of evidence. *Developmental Psychology, 48*(5), 1242–1253. https://doi.org/10.1037/a0027540

Younger, B. A., & Cohen, L. B. (1983). Infant perception of correlations among attributes. *Child Development, 54*(4), 858. https://doi.org/10.2307/1129890

Zehr, S. C. (2000). Public representations of scientific uncertainty about global climate change. *Public Understanding of Science, 9*(2), 85–103. https://doi.org/10.1088 /0963-6625/9/2/301

Zimmerman, C. (2000). The development of scientific reasoning skills. *Developmental Review, 20*(1), 99–149. https://doi.org/10.1006/drev.1999.0497

Zimmerman, C. (2007). The development of scientific thinking skills in elementary and middle school. *Developmental Review, 27*(2), 172–223. https://doi.org/10.1016/j .dr.2006.12.001

Ziv, M., & Frye, D. (2004). Children's understanding of teaching: The role of knowledge and belief. *Cognitive Development, 19*(4), 457–477. https://doi.org/10.1016/j .cogdev.2004.09.002

Ziv, M., Solomon, A., & Frye, D. (2008). Young children's recognition of the intentionality of teaching. *Child Development, 79*(5), 1237–1256. https://doi.org/10.1111/j .1467-8624.2008.01186.x

Index

Note: Page numbers in *italics* followed by the letters *f* or *t* refer to figures and tables, respectively.